DATA ANALYSIS TOOLS FOR DNA MICROARRAYS

CHAPMAN & HALL/CRC
Mathematical Biology and Medicine Series

Aims and scope:
This series aims to capture new developments and summarize what is known over the whole spectrum of mathematical and computational biology and medicine. It seeks to encourage the integration of mathematical, statistical and computational methods into biology by publishing a broad range of textbooks, reference works and handbooks. The titles included in the series are meant to appeal to students, researchers and professionals in the mathematical, statistical and computational sciences, fundamental biology and bioengineering, as well as interdisciplinary researchers involved in the field. The inclusion of concrete examples and applications, and programming techniques and examples, is highly encouraged.

Series Editors

Alison M. Etheridge
Department of Statistics
University of Oxford

Louis J. Gross
Department of Ecology and Evolutionary Biology
University of Tennessee

Suzanne Lenhart
Department of Mathematics
University of Tennessee

Philip K. Maini
Mathematical Institute
University of Oxford

Hershel M. Safer
Informatics Department
Zetiq Technologies, Ltd.

Eberhard O. Voit
Department of Biometry and Epidemiology
Medical University of South Carolina

Proposals for the series should be submitted to one of the series editors above or directly to:
CRC Press UK
23 Blades Court
Deodar Road
London SW15 2NU
UK

Chapman & Hall/CRC Mathematical Biology and Medicine Series

DATA ANALYSIS TOOLS FOR DNA MICROARRAYS

SORIN DRĂGHICI

CHAPMAN & HALL/CRC

A CRC Press Company
Boca Raton   London   New York   Washington, D.C.

Library of Congress Cataloging-in-Publication Data

Drăghici, Sorin.
   Data analysis for DNA microarrays / Sorin Drăghici.
      p. cm.
   Includes bibliographical references and index.
   ISBN 1-58488-315-4 (alk. paper)
      1. DNA microarrays—Methodology. 2. DNA microarrays—Statistical methods. I. Title.
II. Series.

QP624.5.D726D735 2003
572.8′636—dc21   2003046077

Visit the CRC Press Web site at www.crcpress.com

© 2003 by Chapman & Hall/CRC

No claim to original U.S. Government works
International Standard Book Number 1-58488-315-4
Library of Congress Card Number 2003046077
Printed in the United States of America   1   2   3   4   5   6   7   8   9   0
Printed on acid-free paper

To my father, Ion, and mother, Marioara-Petruta, who taught me why and how I should learn and to Althea who makes it all worthwhile.

Preface

Although the industry once suffered from a lack of qualified targets and candidate drugs, lead scientists must now decide where to start amidst the overload of biological data. In our opinion, this phenomenon has shifted the bottleneck in drug discovery from data collection to data analysis, interpretation and integration.

—Life Science Informatics, UBS Warburg Market Report, 2001.

One of the most promising tools available today to researchers in life sciences is the microarray technology. DNA chips or DNA microarrays are microscope slides (or some other solid support) containing a large number of DNA samples. In cDNA arrays, the samples are attached as a number of dots arranged in a regular pattern usually forming a rectangular array. This array is subsequently probed with complementary DNA (cDNA) obtained by reverse-transcriptase reaction. This DNA is fluorescently labeled with a dye and a subsequent illumination with an appropriate source of light will provide an image of the array of dots. Variants of the technique use radioactive substances to label the mRNA. The intensity of each dot can be related to the expression level of the particular gene corresponding to that dot. After an image processing step is completed, the result is a large number of expression values. The Affymetrix technology uses synthesized sequences spotted in small regions on the chip and compares their hybridization with the hybridization of purposely mismatched sequences. Independently of the specific technology used, the final result of a microarray experiment is a set of numbers associated with the expression levels of various genes or DNA fragments.

Typically, one DNA chip will provide hundreds or thousands of such numbers. However, the immense potential of this technology can only be realized if many such experiments are done. In order to understand the biological phenomena, expression levels need to be compared between species or between healthy and ill individuals or at different time points for the same individual or population of individuals. This approach is currently generating an immense quantity of data. Buried under this humongous pile of numbers lies invaluable biological information. The keys to understanding phenomena from fetal development to cancer may be found in these numbers. Clearly, powerful analysis techniques and algorithms are essential tools in mining these data. However, the computer scientist or statistician who has the expertise to use advanced analysis techniques usually lacks the biological knowledge necessary to understand even the simplest biological phenomena. At the same time, the scientist having the right background to formulate and test biological hypotheses may feel a little uncomfortable when it comes to analyzing the data thus generated. This is because the data analysis task often requires a good understanding of a number of different algorithms and techniques and most people usually associate such an understanding with a background in mathematics, computer science or statistics.

In spite of the huge amount of interest around the microarray technology, so far there are very few books available on this topic. Furthermore, the few texts that are available concentrate more on the wet lab techniques than on the data analysis aspects. Thus, help is available for the laboratory part in which most researchers already have a very good experience (at Ph.D. level in many cases) while little help is available on how to extract biologically interesting information out of the large amount of data generated in the lab.

Audience and prerequisites

The goal of this book is to fulfill this need by presenting the main computational techniques available in a way that is useful to both life scientists and analytical scientists. The book tries to demolish the imaginary wall that separates biology and medicine from computer science and statistics and allow the biologist to be a refined user of the available techniques, as well as be able to communicate effectively with computer scientists and statisticians designing new analysis techniques. The intended audience includes as a central figure the researcher or practitioner with a background in the life sciences that needs to use computational tools in order to analyze data. At the same time, the book is intended for the computer scientists or statisticians who would like to use their background in order to solve problems from biology and medicine. The book explains the nature of the specific challenges that such problems pose as well as various adaptations that classical algorithms need to undergo in order to provide good results in this particular field.

Finally, it is anticipated that there will be a shift from the classical compartmented education to a highly interdisciplinary approach in which people with skills across a range of disciplines cross the borders between traditionally unrelated fields such as medicine or biology and statistics or computer science. This book can be used as a textbook for a senior undergraduate or graduate course in such an interdisciplinary curriculum. The book is suitable for a data analysis and data mining course for students with a background in biology, molecular biology, chemistry, genetics, computer science, statistics, mathematics, etc.

Useful prerequisites for a biologist include elementary calculus and algebra. However, the material is designed to be useful even for readers with a shaky mathematical foundation since those elements that are crucial for the topic are fully discussed. Useful prerequisites for a computer scientist or mathematician include some elements of genetics and molecular biology. Once again, such knowledge is useful but not required since the essential aspects of the technology are covered in the book.

Aims and contents

The first and foremost aim of this book is to provide a clear and rigorous description of the algorithms without overwhelming the reader with the usual cryptic notation or with too much mathematical detail. The presentation level is appropriate for a scientist with a background in biology. Little or no mathematical training is needed in order to understand the material presented here. Those few mathematical and statistical facts that are really needed in order to understand the techniques are fully explained in the book at a level that is fully accessible to the non-mathematically minded reader. The goal here was to keep the level as accessible as possible. The mathematical apparatus was voluntarily limited to the very basics. The most complicated mathematical symbol throughout the book is the sum of n terms: $\sum_{i=1}^{n} x_i$. In order to do this, certain compromises had to be made. The definitions of many statistical concepts are not as comprehensive as they could be. In certain places, giving the user a powerful intuition and a good understanding of the concept took precedence over the exact, but more difficult to understand, formalism. This was also done for the molecular biology aspects. Certain cellular phenomena have been presented in a simplified version, leaving out many complex phenomena that we considered not to be absolutely necessary in order to understand the big picture.

A second specific aim of the book is to allow a microarray user to be in a position to make an informed choice as to what data analysis technique to use in a given situation. The existing software packages usually include a very large number of techniques which in turn use an even larger number of parameters. Thus the biologist trying to analyze DNA microarray data is confronted with an overwhelming number of possibilities. For instance, GeneSight 4.0, which is one of the data analysis packages designed especially for microarray data used in this book, offers several ways of pre-processing and normalizing the data before even applying any specific analysis techniques. In this package, there are 5 ways of doing background correction, 3 ways of treating the flagged spots, 5 choices for combining replicates, 5 choices for treating missing values, 2 for flooring the data, 2 for ratios, 2 for omitting low values and 8 different ways of normalizing the data (divide by mean, divide by percentile, subtract mean, subtract percentile, piecewise linear, Z transform, linear regression and LOWESS). In total there are $5 \times 3 \times 5 \times 5 \times 2 \times 2 \times 2 \times 8 = 24{,}000$ qualitatively different ways to do normalization in this package alone without counting the many variations that can be obtained by changing the order or the parameters of individual steps. And many other pre-processing and normalization techniques have been proposed in the literature. Such flexibility is absolutely crucial because several data sets obtained in different laboratories have different characteristics and biases and the normalization is very important if we are to compare such data sets. However, such wealth of choices can be overwhelming for the life scientist who, in most cases, is not very familiar with all intricacies of data analysis and ends up by always using the default choices. This book is designed to help such a scientist by emphasizing at a high level of abstraction the characteristics of various techniques in a biological context.

As a text designed to bridge the gap between several disciplines, the book includes chapters that would give the all the necessary information to readers with a variety of backgrounds. The book is divided into two parts. The first part is designed to offer an overview of microarrays and to create a solid foundation by presenting the elements from statistics that constitute the building blocks of any data analysis. The second part introduces the reader to the details of the techniques most commonly used in the analysis of microarray data.

Chapter 1 presents a short primer on the central dogma of molecular biology and why microarrays are useful. This chapter is aimed mostly at analytical scientists with no background in life sciences. Chapter 2 presents briefly the microarray technology. For the computer scientist or statistician, this constitutes a microarray primer. For the microarray user, this will offer a bird's eye perspective on several techniques emphasizing common as well as technology specific issues related to data analysis. This is useful since often the users of a specific technology are so engulfed in the minute details of that technology that they might not see the forest for the trees.

Chapter 3 constitutes a short primer on digital imaging and image processing. This chapter is mostly aimed at the life scientists or statisticians who are not familiar with the digital image processing.

Chapters 4 and 5 focus on some elementary statistics notions. These chapters will provide the biologist with a general perspective on issues very intimately related to microarrays. The purpose here is to give only as much information as is needed in order to be able to make an informed choice during the subsequent data analysis. The aim of the discussion here is to put things in the perspective of somebody who analyzes microarray data rather than offer a full treatment of the respective statistical notions and techniques. Chapter 5 discusses the classical hypothesis testing approach and 6 applies it to microarray data analysis.

Chapter 7 presents the family of ANalysis Of VAriance (ANOVA) methods intensively used by many researchers to analyze microarray data. Chapter 8 uses some of the ANOVA approaches in the discussion of various techniques for experiment design.

Chapter 9 discusses several issues related to the fact that microarrays interrogate a very large number of genes simultaneously and its consequences regarding data analysis.

Chapters 10 and 11 present the most widely used tools for microarray data analysis. In most case, the techniques are presented using real data. Chapter 10 includes several techniques that are used in exploratory analysis, when there is no known information about the problem and the task is to identify relevant phenomena as well as the parameters (genes) that control them. The main techniques discussed here include: box plots, histograms, scatter plots, time series, principal component analysis (PCA) and independent component analysis (ICA). The clustering techniques described in Chapter 11 include K-means, hierarchical clustering and self-organizing feature maps. Again, the purpose here is to explain the techniques in an unsophisticated yet rigorous manner. The all important issue of when to use which technique is discussed with various examples emphasizing the strengths and weaknesses of each individual technique.

Chapter 12 concentrates on data preparation issues. Although such issues are crucial for the final results of the data mining process, they are often ignored. Issues such as color swapping, color normalization, background correction, thresholding, mean normalization, etc. are discussed in detail. This chapter will be extremely useful both to the biologist who will become aware of the different numerical aspects of the various pre-processing techniques and to the computer scientist who will gain a deeper understanding of various biological aspects, motivations and meanings behind such pre-processing.

Chapter 13 presents several methods used to select differentially regulated genes in comparative experiments. Chapter 14 shows how such lists of differentially regulated genes can be translated into a better understanding of the underlying biological phenomena. Chapter 15 reverses somehow the direction considering the problem of how to select the best microarrays for a given biological hypothesis.

Most of the data analysis issues in Chapters 1–13 are treated on a theoretical level, completely independent of any commercial software package. However, the real world researcher cannot be expected to develop their own software. With this in mind, Chapter 16 was conceived as an invitation for the major players in the arena of commercial software for data analysis to showcase their products. The chapter includes a few real world data sets analyzed with software packages from BioDiscovery, Insightful, SAS and Spotfire. Clearly, these sections will have a commercial bias.

Finally, the last chapter of the book presents some conclusions as well as a brief presentation of some novel techniques expected to have a great impact on this field in the near future.

Road map

The book is laid out as follows:

- Chapter 1 – Short primer on the central dogma of molecular biology and why microarrays are useful.

- Chapter 2 – Short primer on the microarray technology.

- Chapter 3 – Short primer on basic digital imaging and image processing.

- Chapter 4 – Basic terms, computing with probabilities, common distributions, Bayes' theorem.

- Chapter 5 – Classical hypothesis testing.

- Chapter 6 – Analyzing microarray data with classical hypothesis testing.

- Chapter 7 – Analysis of variance (ANOVA).

- Chapter 8 – Experiment design.

- Chapter 9 – Corrections for multiple comparisons – very important and often neglected in microarray data analysis.

- Chapter 10 – Analysis and visualization tools.

- Chapter 11 – Cluster analysis.

- Chapter 12 – Pre-processing and normalization.

- Chapter 13 – Methods for selecting differentially regulated genes.

- Chapter 14 – Biological interpretation of data analysis results.

- Chapter 15 – Hypothesis-driven experiments using focused microarrays.

- Chapter 16 – Data analysis examples using commercial software.

- Chapter 17 – The road ahead.

The book can be used in several ways, depending on the background of the reader.

1. **Life scientists** currently using microarrays for research; excellent grasp of statistics and basic maths; very comfortable with Type I and II errors, significance level and power of a statistical test: Chapters 3, 7–17.

2. **Life scientists** currently using microarrays for research; an average understanding of statistics and basic maths: Chapters 3–17.

3. **Life scientists** considering using microarrays in their research: Chapters 2–17.

4. **Statisticians** with state of the art knowledge in their field but somewhat limited understanding of molecular biology and microarrays: Chapters 1–3, 9–17.

5. **Computer scientists** very comfortable with computer science issues (e.g. image processing techniques) but limited understanding of molecular biology and microarrays: Chapters 1–17 but can skip first half of Chapter 3.

6. **Students** of either computer science, statistics or any of the relevant life science disciplines with a view towards future use of microarrays and/or bioinformatics: Chapters 1–17.

The book is accompanied by software for image processing and data analysis kindly provided by BioDiscovery and Insightful. The two accompanying CDs include full feature trial versions of ImaGene (BioDiscovery), GeneSight (BioDiscovery), S-Plus (Insightful) and ArrayAnalyzer (Insightful). These packages, together with dChip (http://www.biostat.harvard.edu/complab/dchip/) and Onto/Tools (http://vortex.cs.wayne.edu/Projects.html) allow a full analysis of any current type of microarray covering the whole process from choosing the best microarray for a given biological hypothesis, through image processing up to and including finding the biological processes significantly impacted by a given condition.

A number of sample data sets are included with each tool. Most of the graphs and figures in the book have been obtained by using the software and data included. The others have been generated with readily available tools such as Microsoft Excel and Matlab. The reader is advised to install the software and actually use it to perform the analysis steps discussed in the book. A course with or without a laboratory component can be easily taught using this book and the software included.

Acknowledgments

The author of this book has been supported by the Biological Databases Program of the National Science Foundation under grant number NSF-0234806, the Bioinformatics Cell, MRMC US Army - grant number DAMD17-03-2-0035, National Institutes of Health - grant numbers R01-NS045207-01 and R21-EB000990-01, and Michigan Life Sciences Corridor grant number MLSC-27. Many thanks to Sylvia Spengler, Director of the Biological Databases program within the National Science Foundation for her advice, encouragement and support.

My sincere gratitude also goes to Sun Microsystem whose generous grant support provided most of the servers in my laboratory. Currently, the Onto-Express, Onto-Compare, Onto-Design and Onto-Translate databases, their servers and associated applications run entirely on Sun equipment. The rock-solid reliability of their hardware, as well as the high performance design, make their equipment ideal for high-load bioinformatics applications. Special thanks to Anna Golubeva-Faybysh and Stefan Unger who supported my ideas.

I am very grateful to several people for their help and support. This book owes a lot to Alexander Kuklin, currently Director of Product Planning at TransGenomic. Without Alexander's help this book would not have existed. First, he had the super-human patience to go through the video and audio recordings of most of my lectures in the Data Analysis course I taught in the Winter 2002 semester and transcribe them into notes that I later used. Second, Alexander wrote Chapter 3 and a first draft for part of Chapter 9. Chapter 3 is still very much the way he wrote it. Finally, after all this work without which the book would not have been possible in its present form, Alexander had the modesty to graciously decline co-authorship.

I thank Russ Wolfinger (Director of Genomics, SAS), Michael O'Connor (Director, BioPharm Solutions Insightful Corporation), Bruce Hoff (Director of Analytical Sciences, BioDiscovery Inc.) and Matt Ansett (Application Specialist, Spotfire) for their direct contribution included in Chapter 16. Also, thanks to BioDiscovery and Insightful for including some of their software with this book, thus giving the reader the opportunity to experiment at will with all techniques described in this book.

Gary Chase kindly reviewed the chapters on statistics and did so on incredibly short deadlines that allowed the book to be published on time. His comments were absolutely invaluable and made the book stronger than I initially conceived it.

I owe a lot to my colleague, friend and mentor, Michael A. Tainsky. Michael taught me everything I know about molecular biology and I continue to be amazed by his profound grasp of some difficult data analysis concepts. I have learned a lot during our collaboration over the past 4 years. I am also in debt to his collaborators including Olga Studiskaia, Jurek Wojciechowski, Mita Chaterjee, Xiaoju Wang. In particular, I am grateful to Dr. Kathryn Carolin for her review of several chapters as well as for her valuable clinical perspective that forced me to always put things in perspective. Michael's students Aviva Friedman and Lin Tan have helped me and my students with many useful comments especially related to the development of Onto-Design and Onto-Compare described in Chapter 15. Michael also introduced me to Judy Abrams who provided a lot of encouragement and has been a wonderful source of fresh and exciting ideas, many of which ended up in grant proposals.

My colleague and friend Steve Krawetz was the first one to identify the need for something like Onto-Express (Chapter 14). During my leave of absence, he continued to work with my students and participated actively in the creation of the first Onto-Express prototype. Our collaboration over the past few years produced a number of great ideas and a few papers.

Thanks also go to Jeff Loeb for our productive interaction as well as for our numerous discussions on corrections for multiple experiments. I am looking forward to our freshly funded collaboration on gene expression changes in epilepsy.

Many thanks to Otto Muzik for sticking to the belief that our collaboration is worthwhile through the first, less than productive, year and for being such a wonderful colleague and friend.

Jim Granneman, Bob Mackenzie, and Donald Rao knocked one day on my door looking for some help analyzing their microarray data. One year later, we were awarded a grant of over 3.5 million from the Michigan Life Science Corridor. This allowed me to shift some of my effort to this book. Thank you all for involving me in your work.

I am grateful to Soheil Shams, Bruce Hoff, Anton Petrov and everybody else with whom I interacted during my one year sabbatical at BioDiscovery. Bruce, Anton and I worked together at the development of the ANOVA based noise sampling method described in Chapter 13 based on some initial experiments by Xiaoman Li. Figures 11.6, 11.12 and 11.15 were initially drawn by Bruce. BioDiscovery has been the source of some state of the art software for image processing and data analysis that I am still using now, 3 years after visiting them. I am very grateful to them for promptly responding to my many questions and far too many requests for modifications in their software.

During my stay in L.A., I met some top scientists including Michael Waterman and Wing Wong. I learned a lot from our discussions and I am very grateful for that. Mike Waterman provided very useful comments and suggestions on the noise sampling method in Chapter 13 and later on Onto-Express described in Chapter 14. Wing Wong and his students at UCLA and Harvard have done a great job with their software dChip, which is my first stop every time I have to normalize Affymetrix data.

Gary Churchill provided useful comments on the work related to the ANOVA based noise sampling technique presented in Chapter 13. The discussions with him and Kathy Kerr helped us a lot. Also, their ANOVA software for microarray data analysis represent a great tool for the research community. We used some of their software to generate the images illustrating the LOWESS transform in Chapter 12.

John Quackenbush kindly provided a wonderful sample data set that we used in the PCA examples in Chapter 10. He and his colleagues at TIGR designed, implemented and made available to the community a series of very nice tools for microarray data analysis including the Multiple Experiment Viewer (MEV) used to generate some images for Chapter 10.

Thanks to Mary Ann Brown and Phillips Kuhl from Cambridge Healthtech Institute for providing a wonderful meeting place for people interested in microarray data analysis in their excellent annual conference on this topic.

I would also like to acknowledge useful discussions with Simon Twigger, Mike Thomas, Nan Jiang, Susan Bromberg, Mary Shimoyama and the other people working on the rat genome database at the Medical College of Wisconsin. Their work made possible the incorporation of rat data in the Onto-Express and Onto-Compare databases.

Special thanks go to Matthew A. Roberts, Nestle Research Center and Manuel Duval, Pfizer France for their input and valuable discussions.

Mark Hughes, David Womble, Sue Land and the other people in the Genomics Core at Wayne State have always been extremely kind. Mark provided the vision and leadership, Sue processed all our (many) arrays and David provided useful comments and advice. Their help has been very much appreciated.

Ishwar Sethi was my mentor in my previous life. In fact, I came to Wayne State several years ago in order to work with him in machine learning and neural networks. We did work together for a number of years after which he moved on. I am still grateful for his mentorship in my early research.

O. Colin-Stine from U. of Maryland, Baltimore and his colleague Violeta Rus provided excellent feedback on the Onto-Express work, feedback which led directly to the development of Onto-Design.

I have to thank David Sprecher for his friendship and patience. This book took so much of my time that I never managed to finish a research project we started together a couple of years ago.

I thank my colleague and collaborator Harvey Pass for his patience and extremely powerful insights. He is one of those very few brilliant surgeons that also manage to be very accomplished researchers. Analyzing microarray data is a pleasure when the biological questions are formulated as clearly as he does.

Jaques Reifman, Senior Scientist with the U.S. Army was my partner for a great many interesting discussions. During the summer of 2002, I visited the Bioinformatics Cell at MRMC which provided a safe heaven, far from my department and students, for a whole month. Several chapters were written during the quiet evenings and weekends in Frederick.

My Chair, Dr. William Hase my Associate Chair, Farshad Fotouhi and all my colleagues in the Computer Science Department have provided a wonderful research and teaching environment and have helped with their advice in many occasions.

Of course, this book would not have been possible without the hard work and patience of the Chapman and Hall/CRC Press editorial staff. Sunil Nair managed to convince me to work with CRC Press in the first place and went out of his way to keep me happy. Jasmin Naim did all the ground work regarding early reviews, and various software agreements. Many thanks go to Helena Redshaw and Samar Haddad for their patience with my ever changing deadlines.

Contents

List of Tables

List of Figures

Any good poet, in our age at least, must begin with the scientific view of the world; and any scientist worth listening to must be something of a poet, must possess the ability to communicate to the rest of us his sense of love and wonder at what his work discovers.

—Edward Abbey, The Journey Home

Chapter 1

Introduction

If we begin with certainties, we shall end in doubts; but if we begin with doubts, and are patient in them, we shall end in certainties.

—Francis Bacon

1.1 Bioinformatics – an emerging discipline

Life sciences are currently at the center of an informational revolution. Dramatic changes are being registered as a consequence of the development of techniques and tools that allow the collection of biological information at an unprecedented level of detail and in extremely large quantities. The human genome project is a compelling example. Initially, the plan to map the human genome was considered extremely ambitious, on the border of feasibility. The first serious effort was planned over 15 years at a cost of $3 billion. Soon after, the schedule was revised to last only 5 years. Eventually, the genome was mapped in less than 3 years, at a cost much lower than initially expected [232]. The nature and amount of information now available open directions of research that were once in the realm of science fiction. Pharmacogenomics [230], diagnostics [184, 231, 240, 281] and drug target identification [209] are just few of the many areas [115] that have the potential to use this information to change dramatically the scientific landscape in the life sciences.

During this informational revolution, the data gathering capabilities have greatly surpassed the data analysis techniques. If we were to imagine the Holy Grail of life sciences, we might envision a technology that would allow us to fully understand the data at the speed at which it is collected. Sequencing, localization of new genes, functional assignment, pathway elucidation and understanding the regulatory mechanisms of the cell and organism should be seamless. Ideally, we would like knowledge manipulation to become tomorrow the way goods manufacturing is today: high automatization producing more goods, of higher quality and in a more cost effective manner than manual production. In a sense, knowledge manipulation is now reaching its pre-industrial age. Our farms of sequencing machines and legions of robotic arrayers can now produce massive amounts of *data* but using it to manufacture highly processed pieces of *knowledge* still requires skilled masters painstakingly forging through small pieces of raw data one at a time. The ultimate goal is to automate this knowledge discovery process.

Bioinformatics is the field of research that has both the opportunity and the challenge to bring us closer to this goal. Bioinformatics is an emerging discipline situated at the interface between computer science and biological sciences such as molecular biology and genetics. Initially, the term bioinformatics was used to denote very specific tasks such as the activities related to the storage of data of biological nature in databases. As the field evolved, the term has started to encompass also algorithms and techniques used in the context of biological problems. Although there is no universally accepted definition of bioinformatics, currently the term denotes a field concerned with the application of information technology techniques, algorithms and tools to solve problems in biological sciences. The techniques currently used have their origins in a number of areas such as computer science, statistics, mathematics, etc. Essentially, **bioinformatics** is the science of refining biological information into biological knowledge using computers. Sequence analysis, protein structure prediction and dynamic modelling of complex biosystems are just few examples of problems that fall under the general umbrella of bioinformatics. However, new types of data have started to emerge. Examples include protein-protein interactions, protein-DNA interactions, enzymatic and biochemical pathways, population-scale sequence data, large-scale gene expression data and ecological and environmental data [129].

A subfield of particular interest today is genomics. The field of **genomics** encompasses investigations into the structure and function of very large number of genes undertaken in a simultaneous fashion. Structural genomics includes the genetic mapping, physical mapping and sequencing of genes for entire organisms (genomes). Comparative genomics deals with extending the information gained from the study of some organisms to other organisms. Functional genomics is concerned with the role that individual genes or subsets of genes play in the development and life of organisms.

Currently, our understanding of the role played by various genes and their interactions seems to be lagging far behind the knowledge of their sequence information. Fig. 1.1 presents some data that reflect the relationship between sequencing an organism and understanding the role of its various genes [213]. The yeast is an illustrative example. Although the 6,600 genes of its genome have been known since 1997, only approximatively 40% of them have inferred functions. The situation is similar for *E. coli, C. elegans, Drosophila* and *Arabidopsis*.[1]

Most researchers agree that the challenge of the near future is to analyze, interpret and understand all data that are being produced [39, 105, 201, 273]. In essence, the challenge faced by the biological scientists is to use the large-scale data that are being gathered to discover and understand fundamental biological phenomena. At the same

---

[1]Even genomes that are considered substantially complete, in reality may still have small gaps [3]. Until we understand better the function of various genes we cannot discount the functional relevance of the genetic material in those gaps.

Organism	Number of genes	Genes with inferred function	Genome Completion date
E. coli	4288	60%	1997
Yeast	6,600	40%	1996
C. elegans	19,000	40%	1998
Drosophila	12,000 - 14,000	25%	1999
Arabidopsis	25,000	40%	2000
Human	30,000	10-20%	2000

FIGURE 1.1: The unbalance between obtaining the data and understanding it. Although the complete genomes of several simpler organisms are available, understanding the role of various genes lags far behind.

time, the challenge faced by computer scientists is to develop new algorithms and techniques to support such discoveries [129].

The explosive growth of computational biology and bioinformatics has just started. Biotechnology and pharmaceutical companies are channelling many resources towards bioinformatics by starting informatics groups. The tendency has been noted by the academic world and there are a number of universities which have declared bioinformatics a major research priority.

1.2 The building blocks of genomic information

DNA is most commonly recognized as two paired chains of chemical bases, spiraled into what is commonly known as the double helix. DNA is a large polymer with a linear backbone of alternating sugar and phosphate residues. The sugar in DNA molecules is a 5 carbon sugar (deoxyribose), and successive sugar residues are linked by strong (covalent) phosphodiester bonds. A nitrogenous base is covalently attached to carbon atom number $1'$ (one prime) of each sugar residue. There are four different kinds of bases in DNA, and this is what makes it simple to understand its basic function and structure. The order in which the basis occur determines the information stored in the region of DNA being looked at.

The four types of bases in DNA are adenine (A), cytosine (C), guanine (G) and thymine (T), which consist of heterocyclic rings of carbon and nitrogenous atoms. The bases are divided into two classes: purines (A and G) and pyrimidines (C and T). When a base is attached to a sugar we speak of a nucleoside. If a phosphate group is attached to this nucleoside then it becomes a nucleotide. The nucleotide is the basic repeat unit of a DNA strand.

The formation of the double helix is due to the hydrogen bonding that occurs between laterally opposed bases. Two bases form a base pair. The chemical structure of the bases is such that adenine (A) specifically binds to thymine (T) and cytosine (C)

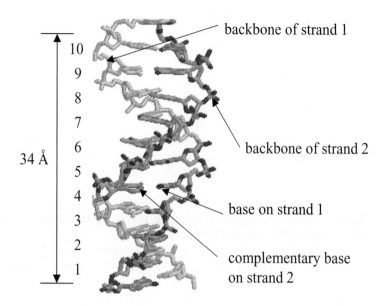

FIGURE 1.2: A short fragment (10 base pairs) of double stranded DNA. Image obtained with Protein Explorer.

specifically binds to guanine (G). These are the so called Watson-Crick rules. Since no other interactions are possible between any other combination of base pairs, it is said that A is complementary to T and C is complementary to G.[2] Two strands are called complementary if for any base on one strand, the other strand contains this base's complement. Two complementary single-stranded DNA chains that come into close proximity react to form a stable double helix (see Fig. 1.2) in a process known as hybridization or annealing. Conversely, a double-stranded DNA can be split into two complementary, single-stranded chains in a process called denaturation or melting. Hybridization and denaturation play an extremely important role both in the natural processes that happen in the living cells and in the laboratory techniques used in genomics. Due to the base complementarity, the base composition of a double-stranded DNA is not random. The amount of A equals the amount of T, and the amount of C is the same as the amount of G.

Let us look at the backbone of the DNA strand again. Phosphodiester bond link carbon atoms number $3'$ and $5'$ of successive sugar residues. This means that in the terminal sugar the $5'$ is not linked to a neighboring sugar residue. The other end is termed $3'$ end and it is characterized by the lack of a phosphodiester bond. This gives

---

[2]In fact, other interactions are possible but the A-T and C-G are the ones that occur normally in the hybridization of two strands of DNA. This is because the A-T and C-G pairings are the ones that introduce a minimal distortion to the geometrical orientation of the backbones.

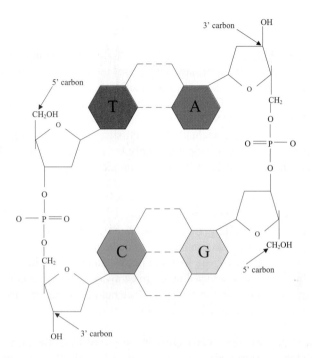

FIGURE 1.3:   Each end of a single strand of DNA is identified by the carbon atom that terminates the strand (5′ or 3′). Two strands of DNA always associate or anneal in such a way that the 5′ → 3′ direction of one DNA strand is the opposite to that of its partner.

a unique direction to any DNA strand. By convention, the DNA strand is said to run from the 5′ end to the 3′ end. The two strands of a DNA duplex are considered to be antiparallel. They always associate or anneal in such a way that the 5′ → 3′ direction of one DNA strand is the opposite to that of its partner (see Fig. 1.3).

Each base can be thought of as a specific letter of a 4-letter alphabet, combining to form words and sentences. Together, DNA works with protein and RNA (ribonucleic acid) in a manner that is similar to the way a computer works with programs and data. DNA has a code that is continuously active, having instructions and commands such as "if-then", "go to", and "stop" statements. It is involved with regulating protein levels, maintaining quality control and providing a database for making proteins. The best analogy to use when describing DNA and its function in cells is to look at a cell as a little factory. In a cell, much like in a factory, specific items are produced in specific places. There is a certain flow of materials through the cell and there are various communication and feedback mechanisms regulating the speed with which various processes happen in various parts of the cell. Such communication and feedback mechanisms also allow the cell to adapt the speed of its internal processes to the demands of the environment much like a factory can adjust its production in response to the changing needs of the market.

The two uprights of the DNA ladder are a structural backbone, supporting the rungs of the ladder. These are the informational parts of the DNA molecule. Each rung of the ladder is made up of two bases that are paired together. This is what makes the steps of the spiral staircase. The two-paired bases are called a base pair as described earlier. The length of any DNA fragment is measured in base pairs (bp), similarly to how we measure length in inches. However, since the DNA is formed with base pairs, the length of a DNA fragment can only be a discrete number of such pairs, unlike a length which can include fractions of an inch.

Each base has a discrete identity; the sequence of the identities is read by the "machinery" inside the cell. Genes, which represent large sequences of DNA, can be looked at as instructions telling the cell how much protein to make, when it should be made, and the sequence that can be used to make it. The information in the DNA is like a library. In the library you will find books, and they can be read and re-read many times, but they are never used up or given away. They are retained for further use. Similarly, the information in each gene (see below) is read, perhaps millions of times in the life of an organism, but the DNA itself is never used up.

A gene is a segment or region of DNA that encodes specific instructions, which allow a cell to produce a specific product. This product is typically a protein, such as an enzyme. There are many different types of proteins. Proteins are used to support the cell structure, break down chemicals, build new chemicals, transport items, and regulate production. Every human being has about 40,000 putative genes (at the moment of this writing there exists a controversy over the number) that produce proteins. Many of these genes are always identical from person to person, but others show variation in different people. The genes determine hair color, eye color, sex, personality, and many other traits that in combination make everyone a unique entity. Some genes are involved in our growth and development. These genes act as tiny switches that direct the specific sequence of events that are necessary to create a human being. They affect every part of our physical and biochemical systems, acting in a cascade of events turning on and off the expression, or production, of key proteins that are involved in the different steps of development.

The key term in growth and development is differentiation. Differentiation involves the act of a cell changing from one type of cell, when dividing through mitosis, into two different types of cells. The most common cells used to study differentiation are the stem cells. These are considered to be the "mother cells", and are thought to be capable of differentiating into any type of cell. It is important to understand the value of differentiation when learning about genetics and organism development. Everyone starts life as one single cell, which divides several times before any differentiation takes place. Then, at a particular instance, and in a particular cell, finally differentiation takes place, and organ cells, skin cells, muscle cells, blood cells are created.

1.3 Expression of genetic information

The flow of genetic information is from DNA to RNA to proteins. This one-way process is the expression of genetic information in all cells and has been described as the central dogma of molecular biology.

To make products from a gene, the information in the DNA is first copied, base for base, into a similar kind of information carrier, called a transcript, or RNA. The RNA copy of the gene sequence acts as a messenger, taking information from the nucleus (where the DNA is found in its chromosomal form) and transporting it into the cytoplasm of the cell (where the machinery for making gene products is found). Once in the cytoplasm, the messenger RNA is translated using a different chemical language into the product of the gene, a protein. The sequence of the protein is defined by the original sequence of the DNA bases found in the gene.

DNA is the hereditary material in all present day cells. However, in early evolution it is likely that RNA served this role. As a testimony for this, there still exist organisms, such as RNA viruses, that use RNA instead of DNA to carry the hereditary information from one generation to another. Retroviruses such as the Human Immunodefficiency Virus (HIV) are a subclass of RNA viruses, in which the RNA replicates via a DNA intermediate, using the enzyme reverse transcriptase (RT). This enzyme is a RNA dependent DNA polymerase which simply said makes DNA from RNA. This enzyme plays a very important role in microarray technology as will be discussed later.

Genes make up only a subset of the entire amount of DNA in a cell. Spacer DNA, which conveys no known protein-coding or regulatory information, makes up the bulk of the DNA in our cells. Active research in the last 20 years has identified signals (specific sequences of bases) that delimit the beginning and ending of genes. These signaling areas can be thought of as regulatory elements. They are used to control the production of protein, and work as a biofeedback system. They are usually located near the beginning of a sequence that is used to code for a protein. Through protein-DNA interactions they are used to turn on and off the production of proteins.

Let us re-examine the several processes that are fundamental for life and important to understand the need for microarray technology and its emergence. Every cell of an individual organism will contain the same DNA, carrying the same information. However, a liver cell will be obviously different from a muscle cell for example. The differentiation occurs because not all the genes are expressed in the same way in all cells. The differentiation between cells is given by different patterns of gene activations which in turn control the production of proteins.

Proteins are long, linear molecules that have a crucial role in all life processes. Proteins are chains of amino acid molecules. There are 20 amino acid molecules that can be combined to build proteins. The number of all possible sequences of amino acids is staggering. For instance, a sequence with length 10 can contain $20^{10} = 10,240$ billion different combinations of amino acids which is a very large number indeed. Although proteins are linear molecules, they are folded in complex ways. The protein folding process is a crucial step since a protein has to be folded into a very specific way for it to function properly.[3] Enzymes are specialized proteins[4] that control the internal chemistry of the cells. This is usually done by binding to specific molecules in a very specific way such as certain atoms in the molecules can form bonds. Sometimes, the molecules are distorted in order to make them react more easily. Some specialized enzymes process DNA by cutting long chains into pieces or by assembling pieces into longer chains. A gene is active, or expressed, if the cell produces the protein encoded by the gene. If a lot of protein is produced, the gene is said to be highly expressed. If no protein is produced, the gene is not expressed (unexpressed).

The process of using the information encoded into a gene to produce the coded protein involves reading the DNA sequence of the gene. The first part of this process is called **transcription** and is performed by a specialized enzyme called **RNA polymerase**. The transcription process converts the information coded into the DNA sequence of the gene into an RNA sequence. This "expression" of the gene will be determined by various internal or external factors. The objective of researchers is to detect and quantify gene expression levels under particular circumstances.

The RNA molecule is a long polynucleotide very similar to the DNA only that it has a slightly different backbone structure. The second difference between RNA and DNA is that the base uracil (U) substitutes the thymine (T) in RNA. The third difference is that RNA molecules within a cell normally exist as single molecules in contrast to the double helix structure of DNA.

The enzyme RNA polymerase attaches itself to a specific DNA nucleotide sequence situated just before the beginning of a gene. This special sequence is called a **promoter** and is such that the RNA polymerase is placed on the correct DNA strand and pointing in the right direction. The two DNA strands are separated locally such that the RNA polymerase can do its work and transcribe DNA into RNA. The RNA molecule is synthesized as a single strand, with the direction of transcription being $5' \rightarrow 3'$. The RNA polymerase starts constructing RNA using ribonucleotides moving freely in the cell. The RNA sequence constructed will be complementary to the DNA sequence read by the RNA polymerase. When the DNA sequence contains a G for instance, the polymerase will match it with a C in the newly synthesized RNA molecule; an A will be matched with a U into the chain under construction because U substitute for T in the RNA molecule. The process will continue with the RNA polymerase moving into the gene, reading its sequence and constructing a complementary RNA chain until the end of the gene is reached. The end of the gene is marked by a special sequence that signals the polymerase to stop. When this sequence is encountered, the polymerase ends the synthesis of the RNA chain and detaches itself from the DNA sequence.

---

[3]The so-called "mad cow disease" is apparently caused by a brain protein folded in an unusual way. When a wrongly folded protein comes into contact with a normal protein it induces the normal protein to fold itself abnormally. This abnormal folding prevents the protein from performing its usual role in the brain which leads to a deterioration of brain functions and, eventually, death.

[4]Most but not all enzymes are proteins. Some RNA molecules called ribozymes act like enzymes.

The RNA sequence thus constructed contains the same information as the gene. This information will be used to construct the protein coded for by the gene.

The RNA chain synthesized by the RNA polymerase is called a primary transcript and is the initial transcription product. It is interesting to note that the sequence of nucleotides in a gene may not be used in its entirety to code for the gene product. Thus, for more complex organisms, a great part of the initial RNA sequence is disposed of during a **splicing** process to yield a smaller RNA molecule called **messenger RNA** or **mRNA**. Its main role is to carry this information to some cellular structures outside the nucleus called ribosomes where proteins will be synthesized. The subsequences of the primary transcript that are eliminated in order to form the mRNA do not actually code for anything. Such sequences are called non-coding regions or **introns**. Conversely, the regions that will be used to build the gene product are called coding regions or **exons**. Thus, RNA molecules transcribed from genes containing introns are longer than the mRNA that will carry the code for the construction of the protein. The mechanism that cuts the transcribed RNA into pieces, eliminates the introns and re-assembles the exons together into mRNA is called **RNA splicing**.

Another important reaction that occurs at this stage is called **polyadenylation**. This reaction is of interest because some protocols in microarray technology utilize the final product of polyadenylation. Transcription of the RNA is known to stop after the enzymes (and some specialized small nuclear RNAs) responsible for the transcription process recognize a specific termination site. Cleavage of the RNA molecule occurs at a site with sequence AAUAAA and then about 200 adenylate (i.e. AMP) residues are sequentially added in mammalian cells by the enzyme poly(A) polymerase to form a poly(A) tail. This tail is used later in the process of reverse transcription.

After the post-transcriptional processing, mRNA transcribed from genes in nuclear DNA leaves the nucleus and moves into the cytoplasm. The mRNA containing the sequence coding for the protein attaches to subcellular structures called **ribosomes**. Here, the information contained in the mRNA is mapped from a sequence of RNA nucleotides into a sequence of amino acids forming the protein. This process is called **translation**. Each triplet of nucleotides is called a **codon** and corresponds to a certain amino acid (see Fig. 1.4). Most amino acids are coded for by more than one codon. There is a start codon which indicates where the translation should start and an end codon which indicates the end of the coding sequence.

The ribosome attaches to the messenger RNA near a specific start codon that signals the beginning of the coding sequence. The various amino acids that form the protein are brought to the ribosome by molecules of RNA that are specific to each type of amino acid. This RNA is called **transfer RNA (tRNA)**. The tRNA molecules recognize complementary specific codons on the mRNA and attach to the ribosome. The first tRNA to be used will have a sequence complementary to the sequence of the first codon of the mRNA. In turn, this first tRNA molecule will bring to the ribosome the first amino acid of the protein to be synthesized. Subsequently, a second tRNA molecule with a sequence complementary to the second codon on the mRNA will attach to the existing ribosome-mRNA-tRNA complex. The shape of the com-

		Second Position				
		T	C	A	G	
First Position	T	TTT Phe [F]	TCT Ser [S]	TAT Tyr [Y]	TGT Cys [C]	T
		TTC Phe [F]	TCC Ser [S]	TAC Tyr [Y]	TGC Cys [C]	C
		TTA Leu [L]	TCA Ser [S]	TAA *Ter* [end]	TGA *Ter* [end]	A
		TTG Leu [L]	TCG Ser [S]	TAG *Ter* [end]	TGG Trp [W]	G
	C	CTT Leu [L]	CCT Pro [P]	CAT His [H]	CGT Arg [R]	T
		CTC Leu [L]	CCC Pro [P]	CAC His [H]	CGC Arg [R]	C
		CTA Leu [L]	CCA Pro [P]	CAA Gln [Q]	CGA Arg [R]	A
		CTG Leu [L]	CCG Pro [P]	CAG Gln [Q]	CGG Arg [R]	G
	A	ATT Ile [I]	ACT Thr [T]	AAT Asn [M]	AGT Ser [S]	T
		ATC Ile [I]	ACC Thr [T]	AAC Asn [N]	AGC Ser [S]	C
		ATA Ile [I]	ACA Thr [T]	AAA Lys [K]	AGA Arg [R]	A
		ATG Met [M]	ACG Thr [T]	AAG Lys [K]	AGG Arg [R]	G
	G	GTT Val [V]	GCT Ala [A]	GAT Asp [D]	GGT Gly [G]	T
		GTC Val [V]	GCC Ala [A]	GAC Asp [D]	GGC Gly [G]	C
		GTA Val [V]	GCA Ala [A]	GAA Glu [E]	GGA Gly [G]	A
		GTG Val [V]	GCG Alo [A]	GAG Glu [E]	GGG Gly [G]	G

FIGURE 1.4:   The genetic code.

plex is such that the amino acids are brought into proximity and they bind to each other. Then, the first tRNA molecule is released, the first two amino acids linked to the second tRNA molecule are shifted on the ribosome bringing the third codon into position. The tRNA bringing the third amino acid can now attach to the third codon due to its complementary sequence and the process is repeated until the whole protein molecule is synthesized. The process stops when a special stop codon is encountered which determines the mRNA to fall off the ribosome together with the newly constructed protein.[5] After it is released, the protein may suffer a set of final changes called post-translational modifications. Such modifications might include cleavage, folding, etc. Once this is done, the protein starts performing the cellular function for which it was designed.

It must be mentioned that the process described above is greatly simplified. For instance, the complex between tRNA molecules and their corresponding amino acids is in turn controlled by another enzyme called aminoacyl-tRNA synthetase. There are at least one type of synthetase for each type of amino-acid. Since other complex steps, such as tRNA-amino acid reaction, are involved in the protein synthesis, it is clear that the amount of protein produced in the cell is also dependent on the successful completion of all these intermediate steps. However, in general there is a quantitative correspondence between the amount of mRNA produced by the enzyme reading the gene and the amount of protein produced. Therefore, the amount of

---

[5]See [86] for an excellent introduction to DNA and gene cloning for the non-specialist and [259] for a more complete treatment of the subject.

mRNA produced from various genes can be usually translated into gene expression levels.

1.4 The need for microarrays

Why bother measuring the expression of all genes? A simple answer involves the fact that the genomes of many model organisms have been sequenced and we would like to simply have the luxury of looking at the whole genome expression profile under the influence of a particular factor. Several methods have been available to look at expression levels but, alas, only of a few genes at a time. Large scale screenings of gene expression signatures were not possible the way they are routinely performed nowadays with microarrays. Therefore a need for a quick snapshot of all or a large set of genes was pressing. Another important reason for the emergence of microarrays is the necessity to understand the networks of bio-molecular interactions at a global scale. Each particular type of cell (i.e. tissue) will be characterized by a different pattern of gene expression levels, i.e. each type of cell will produce a different set of proteins in very specific quantities. A typical method in genetics was to use some method to render a gene inactive (knock it out) and then study the effects of this knockout in other genes and processes in a given organism. This approach, which was for a long time the only approach available, is terribly slow, expensive and inefficient for a large scale screening of many genes. Microarrays allow the interrogation of thousands of genes at the same time. Being able to take a snapshot of a whole gene expression pattern in a given tissue opens innumerable possibilities. One can compare various tissues with each other, or a tumor with the healthy tissue surrounding it. One can also study the effects of drugs or stressors by monitoring the gene expression levels. Gene expression can be used to understand the phenomena related to aging or fetal development. Screening tests for various conditions can be designed if those conditions are characterized by specific gene expression patterns. Drug development, diagnosis, comparative genomics, functional genomics and many other fields may benefit enormously from a tool that allows accurate and relatively inexpensive collection of gene expression information for thousands of genes at a time.[6]

---

[6]This is not to be interpreted that microarrays will substitute gene knockouts. Knocking out a gene allows the study of the more complex effects of the gene, well beyond the mRNA abundance level. Microarrays are invaluable as screening tools able to interrogate simultaneously thousands of genes. However, once interesting genes have been located, gene knockouts are still invaluable tools for a focused research.

1.5 Summary

Deciphering the genomes of several organisms, including the one of humans, led to an avalanche of data that needed to be analyzed and translated into biological meaning. This sparked the emergence of a new scientific field called bioinformatics. This term is generally used to denote computer methods, statistical and data mining techniques, and mathematical algorithms that are used to solve biological problems. The field of bioinformatics combines the efforts of experts from various disciplines who need to communicate with each other and understand the basic terms in their corresponding disciplines. This chapter was written for computer engineers, statisticians and mathematicians to help them refresh their biological background knowledge. The chapter described the basic structure of DNA, RNA and the process of gene expression. There are 4 types of DNA building blocks called nucleotide bases: A, C, G and T. These 4 bases form the genetic alphabet. The genetic information is encoded in strings of variable length formed with letters from this alphabet. The genetic information flows from DNA to RNA to proteins; this is known as the central dogma of molecular biology. The genetic information is stored in various very long strings of DNA. Various substrings of such a DNA molecule constitute functional units called genes, and contain information necessary to construct proteins. The process of constructing proteins from the information encoded into genes is called gene expression. Firstly, the information is mapped from DNA to RNA. RNA is another type of molecule used to carry genetic information. Similarly to DNA, there are 4 types of RNA building blocks and a one-to-one mapping from the 4 types of DNA bases to the 4 types of RNA bases. The process of converting the genetic information contained in a gene from the DNA alphabet to the RNA alphabet is known as transcription. The result of the transcription is an RNA molecule that has the informational content of a specific gene. The transcription process takes place in the nucleus of cells of higher organisms which hosts the DNA. The RNA molecules are subsequently exported out of the cell nucleus into the cytoplasm where the information is used to construct proteins. This process, known as translation, converts the message from the 4 letter RNA alphabet to the 20 letter alphabet of the amino-acids used to build proteins. The amounts of protein generated from each gene determine both the morphology and the function of a given cell. Small changes in expression levels can determine major changes at the organism level and induce illnesses such as cancer. Therefore, comparing the expression levels of various genes between different conditions is of extreme interest to life scientists. This need stimulated the development of high throughput techniques for monitoring gene expression such as microarrays.

Chapter 2

Microarrays

If at first you don't succeed, you are running about average.

—M.H. Alderson

2.1 Microarrays – tools for gene expression analysis

In its most general form, a DNA array is usually a substrate (nylon membrane, glass or plastic) on which one deposits single stranded DNAs (ssDNA)with various sequences. Usually, the ssDNA is printed in localized features that are arranged in a regular grid-like pattern. In this book, we will conform with the nomenclature proposed by Duggan et al. [99] and we will refer to the ssDNA printed on the solid substrate as a **probe**.

What exactly is deposited depends on the technology used and on the purpose of the array. If the purpose is to understand the way a particular set of genes function, the surface will contain a number of regions dedicated to those individual genes. However, arbitrary strands of DNA may be attached to the surface for more general queries or DNA computation. The array thus fabricated is then used to answer a specific question regarding the DNA on its surface. Usually, this interrogation is done by washing the array with a solution containing ssDNA, called a **target**, that is generated from a particular biological sample under study as described below. The idea is that the DNA in the solution that contains sequences complementary to the sequences of the DNA deposited on the surface of the array will hybridize to those complementary sequences. The key to the interpretation of the microarray experiment is in the DNA material that is used to hybridize on the array. Since the target is labelled with a fluorescent dye, a radioactive element, or another method, the hybridization spot can be detected and quantified easily.

When used in gene expression studies, the DNA target used to hybridize the array is obtained by reverse transcriptase reaction from the mRNA extracted from a tissue sample (see Fig. 2.1). This DNA is fluorescently labelled with a dye and a subsequent illumination with an appropriate source of light will provide an image of the array of features (sets of probes in GeneChips or spots in cDNA arrays). The intensity of each spot or the average difference between matches and mismatches can be related

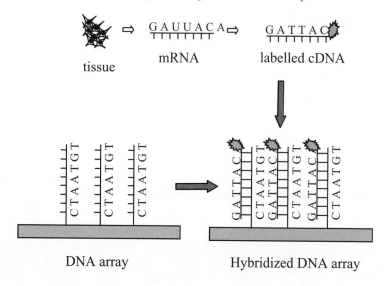

FIGURE 2.1:   A general overview of the DNA array used in gene expression studies. The mRNA extracted from tissue is transformed into complementary DNA (cDNA) which is hybridized with the DNA previously spotted on the array.

to the amount of mRNA present in the tissue and in turn, with the amount of protein produced by the gene corresponding to the given feature.

This step can also be accomplished in many different ways. For instance, the labelling can be done with a radioactive substance and the image obtained by using a photo-sensitive device. Or several targets can be labelled with different dyes and used at the same time in a competitive hybridization process in a multichannel experiment. A typical case is a two channel experiment using cy3 and cy5 as dyes but other dyes can also be used. After an image processing step is completed, the result is a large number of expression values. Typically, one DNA chip will provide expression values for hundreds or thousands of genes.

2.2   Fabrication of microarrays

Two main approaches are used for microarray fabrication: deposition of DNA frag-ments and *in situ* synthesis. The first type of fabrication involves two methods: depo-sition of PCR-amplified cDNA clones and printing of already synthesized oligonu-cleotides. *In situ* manufacturing can be divided into photolithography, ink jet printing and electrochemical synthesis.

2.2.1 cDNA microarrays

In the deposition based fabrication, the DNA is prepared away from the chip. Robots dip thin pins into the solutions containing the desired DNA material and then touch the pins onto the surface of the arrays. Small quantities of DNA are deposited on the array in the form of spots. Unlike *in situ* manufacturing where the length of the DNA sequence is limited, spotted arrays can use small sequences, whole genes or even arbitrary PCR products.

The living organism can be divided into two large categories: eukaryotes and prokaryotes. The group of **eukaryotes** includes the organisms whose cells have a nucleus. **Prokaryotes** are organisms whose cells do not have a nucleus, such as bacteria. In general, eukaryotes have a much more complex intracellular organization than prokaryotes. Gene expression in most eukaryotes is studied by starting with complementary DNA (cDNA) clones. DNA cloning involves the selective amplification of a desired fragment of DNA. This method creates an abundance of the selected DNA sequence so further genetic work can be carried out with it. If a short nucleotide sequence of a particular clone is available then oligonucleotide primers can be designed for amplification of this clone by PCR. The site of this clone can then be compared to a tag for this clone and it is called a **sequence-tagged site (STS)**.

The expressed component of eukaryote genomes represent only a fraction of the total genome. cDNA clones are created and short sequences of about 200 bp are used as tags so sequence-specific primers can be designed for a PCR assay. This implicates that an STS is available for an expressed sequence. The tag is called an **expressed sequence tag (EST)**. In other words, ESTs are single pass, partial sequences of cDNA clones. The cDNA cloned inserts are amplified by PCR from plasmid DNA, and the purified PCR products are then printed on solid support.

Another method of microarray fabrication is the attachment of synthesized oligonucleotides to the solid support. One advantage of this method is that oligonucleotide probes can be designed to detect multiple variant regions of a transcript or the so-called splice variants.

2.2.2 *In situ* synthesis

During array fabrication based on *in situ* synthesis, the probes are photochemically synthesized on the chip. There is no cloning, no spotting and no PCR carried out. The elimination of these steps, which introduce a lot of noise in the cDNA system, constitutes an advantage of this array fabrication approach.

Probe selection is performed based on sequence information alone. This means that every probe synthesized on the array is known in contrast to cDNA arrays, which deal with expressed sequence tags and in many cases the function of the sequence corresponding to a spot is unknown. Additionally, this technology can distinguish and quantitatively monitor closely related genes just because it can avoid identical sequence among gene family members.

The *in situ* synthesis approach of microarray fabrication is represented by three approaches. The first method is photolithographic (Affymetrix, Santa Clara, CA), and is similar to the technology used to build very large scale integrated (VLSI) circuits used in modern computers. This fabrication process uses photolithographic masks for each base. If a probe should have a given base, the corresponding mask will have a hole allowing the base to be deposited at that location. Subsequent masks will construct the sequences base by base. This technology allows the fabrication of very high density arrays but the length of the DNA sequence constructed is limited. This is because the probability of introducing an error at each step is very small but different from zero. In order to limit the overall probability of an error, one needs to limit the length of the sequences. To compensate for this, a gene is represented by several such short sequences. The particular sequences must be chosen carefully to avoid cross-hybridization between genes.

The second approach is the ink jet technology (Agilent, Protogene, etc.) and it employs the technology used in ink jet color printers. Four cartridges are loaded with different nucleotides (A, C, G and T). As the print head moves across the array substrate, specific nucleotides are deposited where they are needed.

Finally, the electrochemical synthesis approach (CombiMatrix, Bothel, WA) uses small electrodes embedded into the substrate to manage individual reaction sites. Solutions containing specific bases are washed over the surface and the electrodes are activated in the necessary positions in a predetermined sequence that allows the sequences to be constructed base by base.

The Affymetrix technology includes the steps outlined in Figs. 2.2, 2.3 and 2.4. Synthetic linkers modified with photochemical removable protecting groups are attached to a glass surface. Light is shed through a photolithographic mask to a specific area on the surface to produce a localized photodeprotection (Fig. 2.2). The first of a series of hydroxyl-protected deoxynucleosides is incubated on the surface. In this example it is the protected deoxynucleoside T. In the next step, the mask is directed to another region of the substrate by a new mask, and the chemical cycle is repeated (Fig. 2.3). Thus, one nucleotide after another is added until the desired chain is synthesized. Recall that the sequence of this nucleotide corresponds to a part of a gene in the organism under scientific investigation. The synthesized oligonucleotides are called probes. The material that is hybridized to the array (the reverse transcribed mRNA) is called the target or the sample.

The gene expression arrays have a match/mismatch probe strategy (Fig. 2.4). Probes that match the target sequence exactly are referred to as reference probes. For each reference probe, there is a probe containing a nucleotide change at the central base position – such a probe is called a mismatch. These two probes – reference and mismatch, are always synthesized adjacent to each other to control for spatial differences in hybridization. Additionally, the presence of several such pairs per gene (each pair corresponding to various parts – or exons – of the gene) helps to enhance the confidence in detection of the specific signal from background in case of weak signals.

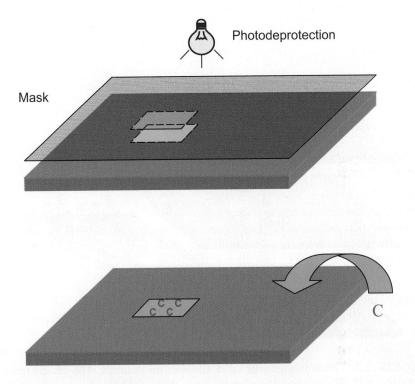

FIGURE 2.2: Photolithographic fabrications of microarrays. Synthetic linkers modified with photochemical removable protecting groups are attached to a glass surface. Light is shed through a photolithographic mask to a specific area on the surface to produce a localized photodeprotection. The first of a series of hydroxyl-protected deoxynucleosides is incubated on the surface. In this example it is the protected deoxynucleoside C. The surface of the array is protected again and the array is ready for the next mask.

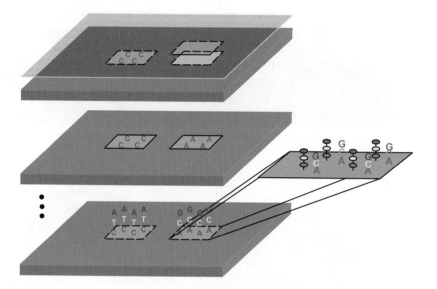

FIGURE 2.3:   Photolithographic fabrications of microarrays. The second mask is applied and light is used to deprotect the areas that are designed to receive the next nucleoside (A). The fabrication process would generally require 4 masking steps for each element of the probes. Several steps later, each area has its own sequence as designed.

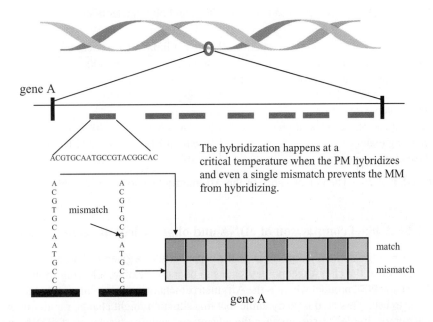

ACGTGCAATGCCGTACGGCAC

The hybridization happens at a
critical temperature when the PM hybridizes
and even a single mismatch prevents the MM
from hybridizing.

mismatch

match

mismatch

gene A

FIGURE 2.4:   The principle of the Affymetrix technology. The probes correspond to short oligonucleotide sequences thought to be representative for the given gene. Each oligonucleotide sequence is represented by two probes: one with the exact sequence of the chosen fragment of the gene (perfect match or PM) and one with a mismatch nucleotide in the middle of the fragment (mismatch or MM). For each gene, the value that is usually taken as representative for the expression level of the gene is the average difference between PM and MM. Reprinted from S. Draghici, "Statistical intelligence: effective analysis of high-density microarray data" published in Drug Discovery Today, Vol. 7, No. 11, p. S55-S63, 2002, with permission from Elsevier.

cDNA arrays	Oligonucleotide arrays
Long sequences	Short sequences due to the limitations of the synthesis technology
Spot unknown sequences	Spot known sequences
More variability in the system	More reliable data
Easier to analyze with appropriate experimental design	More difficult to analyze

TABLE 2.1:   A comparison between cDNA and oligonucleotides arrays.

	Current limit	Practical use
Density (genes/array)	40,000	20,000
Gene representation (probe pairs/gene)	4	20
Linear dynamic range	4 logs	3 logs
Fold change detection	10%	100%
Similar sequences separation	93% identical	70-80% identical
Starting material	$2ng$ total RNA	$5\mu g$ total RNA
Detection specificity	$1 : 10^6$	$1 : 10^5$

TABLE 2.2:   The performance of the Affymetrix technology [22, 199].

2.2.3   A brief comparison of cDNA and oligonucleotide technologies

It is difficult to make a judgment as to the superiority of a given technology. At this point in time, the cDNA technology seems to be more flexible, allowing spotting of almost any PCR product whereas the Affymetrix technology seems more reliable and easier to use. This field is so dynamic that this situation might change rapidly in the near future. Table 2.1 summarizes the advantages and disadvantages of cDNA and high-density oligonucleotide arrays. Table 2.2 shows the current performance of the Affymetrix oligonucleotides arrays [22, 199].

2.3   Applications of microarrays

Microarrays have been used successfully in a range of applications including sequencing [237] and single nucleotide polimorphism (SNP) detection [107, 274]. However, the mainstream application for microarrays remains the investigation of the genetic mechanisms in the living cells [104, 200, 238, 237, 246, 127, 262]. A few typical examples would include comparing healthy and malignant tissue [9, 127, 10, 38, 220], studying cell phenomena over time [80, 254] as well as study the effect of various factors such as interferons [79], cytomegalovirus infection [303] and oncogene transfection [184] on the overall pattern of expression. Perhaps even more importantly than the success in any individual application, it has been shown that microarrays can be used to generate accurate, precise and reliable gene expression data [60, 299, 237].

Microarrays can also be used for purely computational purposes such as in the field of DNA computing [172]. In these cases, the microarray can contain sequences of DNA encoding various possible solutions of the problem to be solved. Several successive steps are performed in order to solve the problem. Each such step consists of three sub-steps: a hybridization, the destruction of the single stranded DNA not hybridized and a denaturation that will prepare the chip for the next computational step. The role of the DNA used in each step is to prune the large number of potential solutions coded on the surface of the array. Specific sequences added in a specific step hybridize to the single stranded DNA attached to the surface. This marks the partial solutions by binding them in double strands. Subsequently, the chip is washed with a solution that destroys the single stranded DNA. A denaturation step will break the double stranded DNA and bring the chip to a state in which it is ready for the next computational step. When used for such a purpose, DNA microarrays lose any biological meaning and become simple tools for parallel and efficient manipulation of a large number of symbolic strings. Because of their ability to perform efficient searches in large dimensional spaces, DNA arrays and DNA computing are particularly attractive when attempting to solve computationally intractable problems [172, 173, 174].

In this book, we will concentrate on the use of microarrays in gene expression studies, focusing on specific challenges that are related to this particular application. Most techniques described here can be applied directly to protein arrays, as well.

2.4   Challenges in using microarrays in gene expression studies

Compared to other molecular biology techniques, microarrays are relatively new. As such, their users are challenged by a number of issues as follows:

1. **Noise.**

   Due to their nature, microarrays tend to be very noisy. Even if an experiment is performed twice with exactly the same materials and preparations in exactly the same conditions, it is likely that after the scanning and image processing steps, many genes will probably be characterized by different quantification values. In reality noise is introduced at each step of various procedures[1] [241]: mRNA preparation (tissues, kits and procedures vary), transcription (inherent variation in the reaction, enzymes), labeling (type and age of label), amplification, pin type (quill, ring, ink-jet), surface chemistry, humidity, target volume (fluctuates even for the same pin), slide inhomogeneities (slide produc-

---

[1]Not all steps apply to all types of arrays.

tion), target fixation, hybridization parameters (time, temperature, buffering, etc.), unspecific hybridization (labelled cDNA hybridized on areas which do not contain perfectly complementary sequences), non-specific background hybridization (e.g. bleeding with radioactive materials), artifacts (dust), scanning (gain settings, dynamic range limitations, inter-channel alignment), segmentation (feature/background separation), quantification (mean, median, percentile of the pixels in one spot), etc.

The challenge appears when comparing different tissues or different experiments. Is the variation of a particular gene due to the noise or is it a genuine difference between the different conditions tested? Furthermore, when looking at a specific gene, how much of the measured variance is due to the gene regulation and how much to noise? The noise is an unescapable phenomenon and the only weapon that the researcher seems to have against it is replication (Chap. 7 and 13).

2. **Normalization.**

The aim of the normalization is to account for systematic differences across different data sets (e.g. overall intensity) and eliminate artifacts (e.g. non-linear dye effects). The normalization is crucial if results of different experimental techniques are to be combined. While everybody agrees on the goal of normalization, the consensus seems to disappear regarding how exactly the normalization should be done. Normalization can be necessary for different reasons such as different quantities of mRNA (leading to different mean intensities), dye non-linearity and saturation toward the extremities of the range, etc. Normalization issues and procedures are discussed in detail in Chap. 12.

3. **Experimental design.**

The experimental design is a crucial but often neglected phase in microarray experiments. A designed experiment is a test or several tests in which a researcher makes purposeful changes to the input variables of a process or a system so we may observe and identify the reasons for changes in the output response. Experiment design issues are discussed in details in Chap. 8.

4. **Large number of genes.**

The fact that microarrays can interrogate thousands of genes in parallel is one of the features that led to the wide adoption of this technology. However, this characteristic is also a challenge. The classical metaphor of the needle in the haystack can easily become an accurate description of the task at hand when tens of thousands of genes are investigated. Furthermore, the sheer number of genes can change the quality of the phenomenon and the methods that need to be used. The classical example is that of the p-values in a multiple testing situation (Chap. 9).

5. **Significance.**

If microarrays are used to characterize specific conditions (e.g. [9, 127]), a crucial question is whether the expression profile differ in a significant way between the groups considered. The classical statistical techniques that were designed to answer such questions (e.g. chi-square tests) can not be applied directly because in microarray experiments the number of variables (usually thousands of genes) is much greater than the number of experiments (usually tens of experiments). Novel techniques need to be developed in order to address such questions.

6. **Biological factors.**

In spite of their many advantages, microarrays are not necessarily able to substitute completely other tools in the arsenal of the molecular biologist. For instance, knocking out genes is slow and expensive but offer an unparalleled way of studying the effects of a gene well beyond its mRNA expression levels. In the normal cell, the RNA polymerase transcribes the DNA into mRNA which carries the information to ribosomes where the protein is assembled by tRNA in the translation process. Most microarrays measure the amount of mRNA specific to particular genes and the expression level of the gene is associated directly with the amount of mRNA. However, the real expression of the gene is the amount of protein produced not the amount of mRNA. Although in most cases, the amount of mRNA reflects accurately the amount of protein, there are situations in which this may not be true. If nothing else, this is a fundamental reason for which microarrays cannot be trusted blindly. However, there are also other reasons. Even if the amount of the protein were always directly proportional to the amount of mRNA, the proteins may require a number of post-translational modifications in order to become active and fulfill their role in the cell. Any technology that works exclusively at the mRNA level, such as microarrays, will be blind with respect to these changes. In conclusion, microarrays are great tools for screening many genes and focusing hypotheses. However, conclusions obtained with microarrays must be validated with independent assays, using different techniques that investigate the phenomenon from various perspectives.

Furthermore, even if we assume that all the genes found to be differentially regulated in a microarray experiment are indeed so, a non-trivial issue is to translate this information into biological knowledge. The way various biological processes are affected and the degree to which genes interact on various regulatory pathways is much more important than which particular genes are regulated. The issue of translating lists of differentially regulated genes into biological knowledge is discussed in Chap. 14.

7. **Array quality assessment.**

It is useful if data analysis is not seen as the last step in a linear process of microarray exploration but rather as a step that completes a loop and provides the feedback necessary to fine-tune the laboratory procedures that produced the microarray. Thus, array quality assessment is an aspect that should be included among the goals of the data analysis. It would be very useful if the analysis could provide besides the expression values some quality assessment of the arrays used. Such quality measures will allow discarding the data coming from below standard arrays as well as the identification of possible causes of failure in the microarray process. A method able to provide information about the quality of the individual arrays is presented in Chap. 13.

One of the advantages of DNA arrays is that they can be used to query many genes at the same time. A single array can contain anywhere from a few hundreds to a few tens of thousands of spots. However, this advantage is also a challenge. After hybridization and image processing, each spot will generate a quantified number. Therefore, each array can generate up to a few tens of thousands of values. Furthermore, the immense potential of the DNA arrays can be realized only if many such experiments are done for various biological conditions and for many different individuals. Therefore, the amount of data generated by this type of experiments is staggering. Techniques classically developed in computer science (such as data mining and machine learning) can prove to be extremely useful for the interpretation of such data.

Another aspect of microarray experiments that relates to their use in medicine and drug discovery is their reproducibility. A DNA array experiment tends to be a complicated process as shown schematically in Fig. 2.5. Furthermore, each step in this process is usually associated to a very complicated laboratory protocol that needs to describe in great detail the substances, conditions and procedures used [236]. There are an amazing number of factors that can influence the results in a dramatic way. In order to be able to validate the results obtained using such DNA arrays (e.g. a new drug), a very large amount of data needs to be stored and made accessible in a convenient way [48]. In an effort to alleviate data sharing and storing problems, **standards** are being developed. There are current international standardization efforts undertaken jointly by various agencies in the US and Europe. The Microarray Gene Expression Data (MGED) society has on-going efforts in the following directions [48]:

- MIAME – The formulation of the minimum information about a microarray experiment required to interpret and verify the results.

- MAGE – The establishment of a data exchange format (MAGE-ML) and object model (MAGE-OM) for microarray experiments.

- Ontologies – The development of ontologies for microarray experiment description and biological material (biomaterial) annotation in particular.

- Normalization – The development of recommendations regarding experimental controls and data normalization methods.

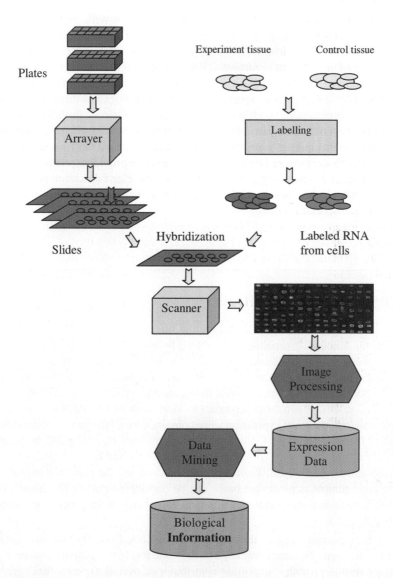

FIGURE 2.5: An overview of the cDNA array processing. The cylinders represent data that needs to be stored. The parallelepipeds represent pieces of hardware while the hexagons represent information processing steps in which the computers play an essential role.

Clearly, novel techniques in databases (e.g. multi dimensional indexing) can make life a lot easier at this stage.

According to Murphy's law, if anything can go wrong, it will. In DNA arrays, the number of things that can go wrong is very large [241, 285]. Debugging the global process involving DNA arrays requires simultaneous access to different types of data usually stored in different databases. Being able to do transversal queries across multiple databases might help locating problems. For instance, combining queries in a database containing known expression values with queries in a database containing functional information and with queries in a database containing the laboratory protocols might point out that a particular subset of experimental results are questionable. Combining further queries across the various databases might help pinpoint the problem (e.g. the clones prepared by a certain lab technician provided questionable results or the preparations involving a solution from a particular fabrication batch need to be discarded). Techniques developed for data warehouses for business applications may be used successfully here.

2.5   Sources of variability

It is interesting and important to follow the microarray process from the point of view of the amount of variability introduced by each step. Our goal is to compare gene expression in tumor tissue versus healthy tissue. The ultimate goal of such an experiment is to see the differences in gene expression between these two states. If the same experiment is performed several times, each run will provide slightly different expression values. When the experiments are done with the two RNA samples that are to be compared, the exact expression values obtained for all the genes will be different between the two samples. Unfortunately, it is difficult to distinguish between the variation introduced by the different expression levels and the variation inherent to the laboratory process itself. From this point of view, any variation that appears even if the same sample is processed twice can be considered noise. The main sources of such variations corresponding to each individual step in the process are shown in Table 2.3.

One of the important factors is the preparation of mRNA. Even if the kits used for RNA isolation are from the same company and the same batch, two preparations may yield different results. Another important contributor to overall experimental variability is the target preparation. This particular step of the microarray protocol comprises the enzyme mediated reverse transcription of mRNA and concomitant incorporation of fluorescently labelled nucleotides. A careful analysis of all components of this step has been shown to maximize the reaction and increase the signal-to-noise ratio [285].

The sources of variability can be divided into several categories as follows. The sources related to the sample preparation include the mRNA preparation, transcrip-

Factor	Comments
mRNA preparation	Tissues, kits and procedures vary
transcription	Inherent variation in the reactions, type of enzymes used
Labeling	Depends on the type of labeling and procedures as well as age of labels
Amplification (PCR protocol)	PCR is difficult to quantify
Pin geometry variations	Different surfaces and properties due to production random errors
Target volume	Fluctuates stochastically even for the same pin
Target fixation	The fraction of target cDNA that is chemically linked to the slide surface from the droplet is unknown
Hybridization parameters	Influenced by many factors such as temperature of the laboratory, time, buffering conditions and others
Slide inhomogeneities	Slide production parameters, batch to batch variations
Non-specific hybridization	cDNA hybridizes to background or to sequences that are not their exact complement
Gain setting (PMT)	Shifts the distribution of the pixel intensities
Dynamic range limitations	Variability at low end or saturation at the high end
Image alignment	Images of the same array at various wavelengths corresponding to different channels are not aligned; different pixels are considered for the same spot
Grid placement	Center of the spot is not located properly
Non-specific background	Erroneous elevation of the average intensity of the background
Spot shape	Irregular spots are hard to segment from background.
Segmentation	Bright contaminants can seem like signal (e.g. dust)
Spot quantification	Pixel mean, median, area, etc.

TABLE 2.3: Sources of fluctuations in a microarray experiment [241, 285].

tion (RT-PCR) and labelling. The slide preparation stages contribute with variability caused by pin type variation, surface chemistry, humidity, target volume, slide inhomogeneities and target fixation issues. Hybridization related variability is determined by hybridization parameters, non-specific spot hybridization and non-specific background hybridization.

One of the sources of fluctuations in microarray data is the pin geometry. As already outlined, microarrays can be obtained by printing the spots on solid surfaces such as glass slides or membranes. This printing involves deposition of small amounts of volume of DNA on a solid surface. The volumes deposited by microarray printing pins are within the range of nanoliters (10^{-9} L) or picoliters (10^{-12} L). Due to the fact that these volumes are much below the range of normal liquid dispensing systems new technologies for printing of microarrays emerged [77, 230].

The technologies for printing can be divided into contact and non-contact printing. The first category involves direct contact between the printing device and the solid support. Contact printing includes solid pins, capillary tubes, tweezers, split pins, and microspotting pins. Non-contact printing uses a dispenser borrowed from the ink-jet printing industry.

The most commonly used pins are the split pins. Such a split or a quill pin has a fine slot machined into its end. When the split pin is dipped into the well of a PCR plate, a small amount of volume (0.2 to 1.0 μl) is loaded into the slot. The pin then touches the surface of the slide, the sample solution is brought into contact with the solid surface, the attractive force of the substrate on the liquid withdraws a small amount of volume from the channel of the pin, and a spot is formed. However, the amount of transported target fluctuates stochastically even for the same pin. The geometry of the pin and its channel introduce variability into the system between the genes spotted with different pins. The surface chemistry and the humidity also play important roles in the spot formation and, in consequence, introduce variability in the process.

All the above sources of variation need to be addressed at various stages of data acquisition and analysis. Some sources of variance may not produce results observable on the scanned image of the microarray. However, several sources of variance produce results that are very clearly visible on the image of the microarray. Several examples are shown in Fig. 2.6 and include incorrect alignment, non-circular spots, spots of variable size, spots with a blank center, etc. The microarray jargon includes colorful terms that are very descriptive if somehow less technical. For instance, blooming is an extension of a particular spot in its immediate neighborhood which is very frequent on filter arrays when the labelling is done with a radioactive substance (see panel A in Fig. 2.6). If the amount of radioactive mRNA is very abundant in a particular spot, the radioactivity will affect the film in a larger area around the spot sometime covering several of the neighboring spots. Doughnuts are spots with a center having a much lower intensity than their circumference. A possible cause may be the pin touching the surface too hard, damaging it and thus preventing a good hybridization in the center of the spot. Other factors may involve the superficial tension of the liquid printed. Comets are long, tail-like marks coming out of the spots while splashes resemble the pattern created by a drop of liquid splashing on a hard

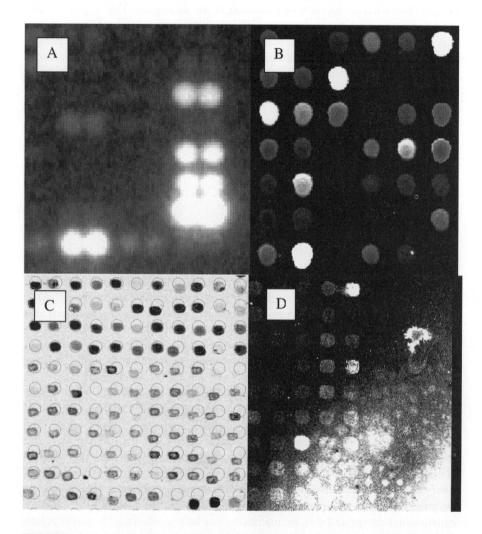

FIGURE 2.6: Examples of microarray defects. A. Radioactively labelled spots expanding over neighboring spots (blooming). B. Non-circular and weak spots as well as some weak-center spots (doughnuts). C. Imperfect alignment. D. Hybridization and contamination causing a non-specific, high background with various gradients.

surface (notice a splash on the right of panel D in Fig. 2.6). Hybridization, washing and drying can also create problems. Typically, such problems involve larger areas and many spots (lower right corner of panel D in Fig. 2.6). Dust is a problem for printing and a drastic example is illustrated in the same panel D in Fig. 2.6. This is why most arrayers are enclosed in a glass or plexiglass container to prevent dust deposition during printing. Some facilities adhere to the "clean environment policy" according to which researchers attending the arrayers and performing manipulations in the printing facility wear special uniforms and caps, the room is equipped with air filters and other technical precautions are taken.

The fact that many of these problems are so conspicuous in the image emphasizes the fact that the image processing and quantification is a crucial step in the sequence of analyzing the microarray results. These issues will be discussed in detail in Chapter 3.

2.6   Summary

Microarrays are solid substrates hosting hundreds of single stranded DNAs with a specific sequence, which are found on localized features arranged in grids. These molecules, called probes, will hybridize with single stranded DNA molecules, named targets, that have been labelled during a reverse transcription procedure. The targets reflect the amount of mRNA isolated from a sample obtained under a particular influence factor. Thus, the amount of fluorescence emitted by each spot will be proportional with the amount of mRNA produced from the gene having the corresponding DNA sequence. The microarray is scanned and the resulting image is analyzed such that the signal from each feature or probe can be quantified into some numerical values. Such values will represent the expression level of the gene in the given condition. Microarrays can be fabricated by depositing cDNAs or previously synthesized oligonucleotides; this approach is usually referred to as printed microarrays. In contrast, *in situ* manufacturing encompasses technologies that synthesize the probes directly on the solid support. Each technology has its advantages and disadvantages and serves a particular research goal. The microarray technology has a very high throughput interrogating thousands of genes at the same time. However, the process includes numerous sources of variability. Several tools such as statistical experimental design and data normalization can help to obtain high quality results from microarray experiments.

Chapter 3

Image processing

Not everything that can be counted counts, and not everything that counts can be counted.

—Albert Einstein

3.1 Introduction

The main goal of array image processing is to measure the intensity of the spots and quantify the gene expression values based on these intensities. A very important and often neglected goal is also assessing the reliability of the data, and generating warnings signalling possible problems during the array production and/or hybridization phases. This chapter is divided into two parts. Section 3.2 provides a very short description of the basic notions involved in digital imaging. Section 3.3 and following focus on image processing issues specific to microarrays.

3.2 Basic elements of digital imaging

In computer processing, the analog image must undergo an analog to digital (A/D) conversion. This procedure consists of sampling and quantification. **Sampling** is the process of taking samples or reading values at regular intervals from a continuous function (see Fig. 3.1). In image processing, the continuous function is a continuous voltage waveform that is provided by an analog sensor. The sampling process transforms the original continuous function into an array of discrete values. This is the sampled function. There is a complex theory that established how many samples need to be taken in order to capture the information present in the continuous function and what happens when fewer samples are captured. However, these issues are beyond the scope of the present text.

Analog images can be seen as 2-dimensional continuous functions. If the image is scanned along a particular direction, the variation of the intensity along that direction forms a one dimensional intensity function. For instance, if we follow a horizontal

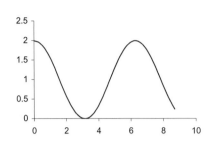 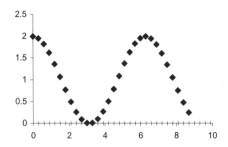

FIGURE 3.1: Sampling is the process of taking samples (or reading values) at regular intervals from a continuous function.

M	Resolution	Total number of pixels
7	128x128	$2^{14} = 16,384$
8	256x256	$2^{16} = 65,536$
9	512x512	$2^{18} = 262,144$
10	1024x1024	$2^{20} = 1,048,576$

TABLE 3.1: Total number of pixels for various resolutions.

line in an image, we can record how the intensity increases and decreases according to the informational content of the image. When the image is sampled, the continuous intensity variation along the two directions of the image is captured in a rectangular array of discrete values (see Fig. 3.2). Such rectangular array of sampled intensity values forms a **digital image**. Each of these picture elements is called a **pixel**.

Typically the image array is a rectangle with dimensions that are a power of 2: $N = 2^M$. This is done in order to simplify the processing of such images on numerical computers which use base 2. Table 3.1 shows the dimensions in pixels of a square image of size $N = 2^M$ for increasing values of M. It is important to understand that the number of pixels is completely independent of the physical dimensions of the image. The same image can be scanned at 128x128 pixels or 1024x1024. This is simply an issue of how often do we read values from any given line and how many different lines we will consider. The number of pixels used for scanning is called resolution and is usually specified as a combination of two numbers: number of lines read and number of values read for each line. For instance an image taken at 1600x1200 will contain 1600 lines and each line will have 1200 values for a total of 1600x1200=1,920,000 pixels. This is the figure used to describe the capabilities of computer monitors or digital cameras. It is essential that the image be sampled sufficiently densely or the image quality will be severely degraded.

For continuous functions that change in time, this requirement is expressed mathematically by the **sampling theorem** which relates the number of samples to the highest frequency contained in the signal. Intuitively, the sampling theorem says

columns

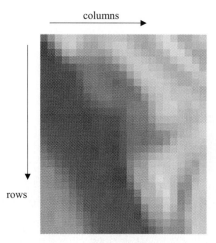

rows

FIGURE 3.2: A digital image. When the image is sampled, the continuous intensity variation along the two directions of the image is captured in a rectangular array of discrete values.

that at least two samples must be taken for each period of the highest frequency signal. If the sampling is performed at this minimum rate or better, the original signal can be reconstructed perfectly from the sampled signal. If the sampling is performed at a rate below this minimum, distortions called **aliasing effects** occur.

An analogue image can be seen as a continuous function that relates intensity to spatial coordinates. For images, performing a sufficiently dense sampling reduces to having a sufficient resolution for the digital image that will be used to capture the information in the analogue image. The importance of the resolution is illustrated in Fig. 3.3. In a low resolution image, a pixel will represent a larger area. Since a pixel can only have a unique value, when such a digital image is displayed, one can notice large rectangular areas of uniform intensity corresponding to the individual pixels. Robust quantification can be obtained only if the image is scanned with a good resolution. A rule of thumb is to have the pixel size approximatively $1/10$ of the spot diameter. If the average spot diameter is at least 10-12 pixels, each spot will have a hundred pixels or more in the signal area ($A = \pi r^2$ where A is the area and r is the radius) which is usually sufficient for a proper statistical analysis that will help the segmentation.

Another important notion in digital image processing is the color depth. This term refers to how many different values can be stores in a single pixel, i.e. how many different colors or shades of gray that pixel can take. The color depth is directly dependent on the amount of memory available for each pixel. This is another typical example of digitization. **Digitization** is the process of converting analogue data to digital data. If an analogue quantity or measurement such as the intensity

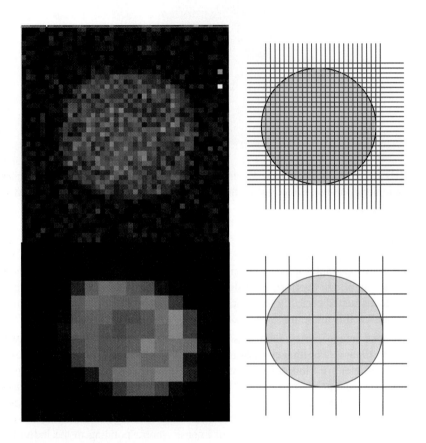

FIGURE 3.3:   The effect of the resolution. Two spots scanned at high resolution (top) and low resolution (bottom). A low resolution means that a pixel, which is a unique value, will represent a larger area of the original image. Usually the value of the pixel will be some type of average of the intensity in the area represented by the given pixel.

Color depth	Number of colors available	Colors available
1	2	$0, 1$
2	4	$0, 1, 2, 3$
8	256	$0, 1, \ldots, 255$
16	65,536	$0, 1, \ldots, 65, 535$
22	4,194,304	$0, 1, \ldots, 4, 194, 303$
24	16,777,216	$0, 1, \ldots, 16, 777, 215$

TABLE 3.2: Color depths and number of colors available.

level in a gray level image needs to be stored in a digital computer, this analogue quantity has to be converted to digital form first. The first step is to decide the amount of memory available to store one such intensity value. Let us assume that one **byte** is available for this purpose. One byte equals eight bits. Eight bits can store $2^8 = 256$ different values: $0, 1, \ldots, 254, 255$. Once this value is known, the range of the analogue value to be represented is divided into 256 equal intervals. Let us say the intensity can take values from 0 to 1000. The intervals will be $[0, 1/256], [1/256, 2/256], \ldots, [254/256, 255/256]$. This mapping between intervals of the analogue value and the fixed number of discrete values $0, 1, \ldots, 254, 255$ available to represent it is the digitization. Every time an analogue value is measured, the interval that contains it is found and the corresponding digital value is stored. For instance, if the analogue reading is in between $1/256$ and $2/256$, which is the second interval available, the value will be stored as 1 (the first value is always 0).

The **color depth** of an image is the number of bits used to store one pixel of that image. An image with a color depth of 8 will be able to use 256 different colors. The computer makes no distinction between storing colors or gray levels so 256 colors may actually be 256 different levels of gray. Table 3.2 presents a few common color depths together with the number of colors available in each case. The first row of the table corresponds to a binary image in which each pixel is represented on one bit only and can only be black or white. The last row corresponds to a color depth of 22 bits which allows for 4 million colors. A color depth of 24 bits or more is sometimes referred to as "true color" since an image using this many colors will be able to reproduce accurately the natural colors that surround us (i.e. in such a way that our eyes would not be able to distinguish the difference).

The color depth issue hides another possible danger. Fig. 3.4 shows what happens when the color depth is not sufficient. The image to the left uses 8 bits/pixel and is of reasonable quality. The middle image uses 4 bits/pixels. Artifacts can already be noted in particular on the sky behind the photographer. What was a smooth gradual transition in the original 8 bits/pixel image is now substituted by three clearly delimited areas. The original image used a range of many values to represent that particular area of the sky. When the color depth was reduced, the same interval was mapped to only three different gray levels. Thus the gradual smooth transition was replaced by three areas each displayed with one of the three available shades of gray. The examination of the third image displayed at 2 bits/pixel is even more interesting.

FIGURE 3.4:   The effects of an insufficient color depth. Left to right the images use the same resolution and a color depth of 8, 4 and 2 bits/pixel, respectively. Note what happens to the sky in the middle image (4bit/pixel) and to the tower in the background in the image to the right (2 bits/pixel). (Reprinted from Bovik, A., Handbook of Image and Video Processing, 2000, p. 500. With permission from Elsevier Science.)

Now, only 4 different shades of gray are available. A superficial look may lead to the conclusion that this image is actually better than the 4 bits/pixels image in the middle or even the original at 8 bits/pixels. This is due to the fact that the contrast was increased by bringing the coat of the photographer to a darker level (actually black) while bringing the sky to a lighter level. However, a more thorough examination shows that whole areas of the image have lost their information content (see for instance the tower behind the photographer which disappeared completely in this image). This is again due to the fact that different shades of gray (the ones used for the tower and the sky immediately behind it) are now mapped to the same value, making those pixels indistinguishable. Thus, pixels which were part of different objects now have exactly the same color and the objects simply disappear.

It is important to note that the color depth and resolution are orthogonal. In other words, what is lost in terms of resolution cannot be recovered by using a higher color depth or vice versa. In practical terms, this means that we need to make sure that *both* color depth and resolution are suitable for our purposes when using digital images. For microarray applications the usual color depth is 16 bits/pixels which allows for 65,535 shades of gray. As mentioned elsewhere, the scanning resolution should be such that the diameter of a spot is 10 pixels or more.

3.3   Microarray image processing

A typical two-channel or two-color microarray experiment involves two samples such as a tumor sample and a healthy tissue sample. RNA is isolated from both samples. Reverse transcription is carried out in order to obtain cDNA and the products are labelled with fluorescent dyes. One sample (for instance the tumor) is labelled with the red fluorescent dye and the other (the healthy tissue) with the green dye. The labelled cDNAs are hybridized to the probes on the glass slides and the slides are scanned to produce digital images.

For each array, the scanning is done in two phases. Firstly, the array is illuminated with a laser light that excites the fluorescent dye corresponding to one channel, for instance the red channel corresponding to the tumor sample. An image is captured for this wavelength. In this image, the intensity of each spot is theoretically proportional to the amount of mRNA from the tumor with the sequence matching the given spot. Subsequently, the array is illuminated with a laser light having a frequency that excites the fluorescent dye used on the green channel corresponding to the healthy tissue. Another image is captured. The intensity of each spot in this second image will be proportional to the amount of matching mRNA present in the healthy tissue. Both images are black and white and usually stored as high resolution Tag Image File Format (.tiff) files. For visualization purposes, most of the software available create a composite image by overlapping the two images corresponding to the individual channels. In order to allow a visual assessment of the relationship between the quantities of mRNA corresponding to a given gene in the two channels, the software usually uses a different artificial color for each of the two channels. Typically, the colors used are red and green to make the logical connection with the wavelength of the labelling dye. If these colors are used, overlapping the images will produce a composite image in which spots will have colors from green through yellow to red as in Fig. 3.5. Let us consider a certain gene that is expressed abundantly in the tumor tissue and scarcely in the healthy (e.g. spot 2 in Fig. 3.5). The spot corresponding to this gene will yield an intense spot on the red channel due to the abundant mRNA labelled with red coming from the tumor sample (upper right in Fig. 3.5). The same spot will be dark on the green channel since there is little mRNA from this gene in the healthy tissue (upper left in Fig. 3.5). Superposing the two images will produce a red spot (lower panel in Fig. 3.5). A gene expressed in the healthy tissue and not expressed in the tumor will produce a green spot (e.g. spot 3 in Fig. 3.5); a gene expressed in both tissues will provide equal amounts of red and green and the spot will appear as yellow (spot 4), and a gene not expressed in either tissue will provide a black spot (spot 1).

The main steps of data handling in a microarray process are described in Fig. 3.6. The microarray process is initiated as two threads: the production of the microarray itself and the collection and treatment of the sample(s) which will provide the mRNA that will be queried. The two threads are joined in the hybridization step in which the labelled mRNA is hybridized on the array. Once this is done, the hybridized array is scanned to produce high resolution tiff files. These files need to be processed by some specialized software that quantifies the intensity values of the spots and their local background on each channel. Ideally, this step would provide expression data as a large matrix traditionally visualized with genes as rows and conditions as columns. A final process of data analysis/data mining should extract the biologically relevant knowledge out of these quantitative data.

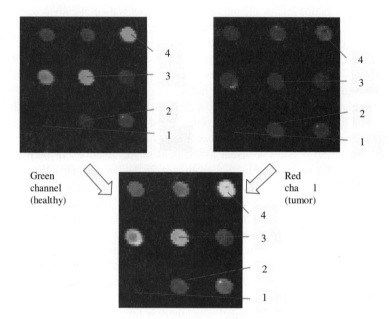

FIGURE 3.5:   A synthetic image is obtained by overlapping the two channels. A gene that is expressed abundantly in the tumor tissue and scarcely in the healthy will appear as a red spot (e.g. spot 2), a gene expressed in the healthy tissue and not expressed in the tumor will appear as green spot (e.g. spot 3), a gene expressed in both tissues will provide equal amounts of red and green and the spot will appear as yellow (spot 4) and a gene not expressed in either tissue will provide a black spot (spot 1).

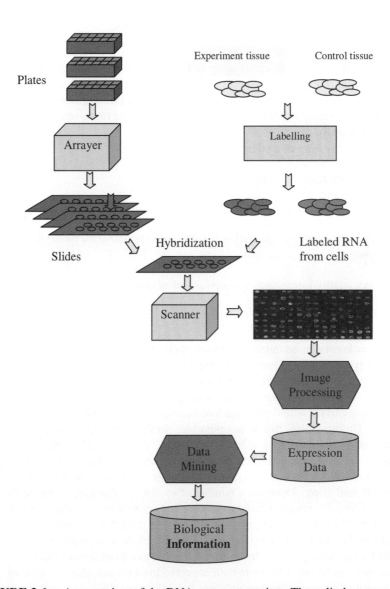

FIGURE 3.6: An overview of the DNA array processing. The cylinders represent data that needs to be stored. The parallelepipeds represent pieces of hardware while the hexagons represent information processing steps in which the computers play an essential role.

3.4 Image processing of cDNA microarrays

cDNA microarray images consist of spots arranged in a regular grid-like pattern. Often, it is useful to think of the array as being organized in **sub-grids**. Such sub-grids are usually separated by small spaces from neighboring sub-grids and form a **meta-array**. Each sub-grid is created by one pin of the printing-head. A spot can be localized on the array by specifying its location in terms of **meta-row**, **meta-column**, **row** and **column**).

The image processing of a microarray can be divided into:

1. Array localization – spot finding.

2. Image segmentation – separating the pixels into signal, background and other.

3. Quantification[1] – computation of values representative for the signal and background levels of each spot.

4. Spot quality assessment – computations of quality measures.

3.4.1 Spot finding

Since it is known in advance how many spots there are, the pattern according to which they were printed, as well as their size, a simple computer program could apparently accomplish the image processing task by superimposing an array of circles with the defined dimensions and spacing on the given image. The pixels falling inside these circles would be considered signal and those outside would be background. Unfortunately, this is not possible. In the real world, the exact location of each grid may vary from slide to slide even if the grid itself were perfect. Furthermore, the relative position of the sub-grids may again vary even if the sub-grids themselves were perfect. Finally, within a single sub-grid, individual spots can be severely misaligned. There are a number of sources contributing to the problem of an imperfect grid, mainly related to mechanical constraints in the spotting process, hybridization inconsistencies, and the necessity to print dense arrays in order to increase the throughput of the approach. Because of these reasons, the first step of the image processing has to deal with finding the exact position of the spots which sometimes can be rather far from their expected location according to the grid.

The spot finding operation aims to locate the signal spots in images and estimate the size of each spot. There are three different levels of sophistication in the algorithms for spot finding, corresponding to the degree of human intervention in the process. These are described below in the order given by the most to the least amount of manual intervention.

---

[1]The term of *quantitation* is sometimes used in the microarrray jargon to describe the same process.

Manual spot finding. This method is essentially a computer-aided image processing approach. The computer itself does not have any abilities to "see" the spots. It merely provides tools to allow the users to tell the software where each of the signal spots are in the image. Historically, this was the first method used only in the very early days of microarray technology. This method is prohibitively time-consuming and labor intensive for images that have thousands of spots. Users had to spend a day or so to adjust the circles over the spots such that an acceptable level of accuracy was achieved. Furthermore, considerable inaccuracies may be introduced at this time due to human errors, particularly with arrays having irregular spacing between the spots and large variation in spot sizes.

Semi-automatic spot finding. The semi-automatic method requires some level of user interaction. This approach typically uses algorithms for automatically adjusting the location of the grid lines, or individual grid points after the user has specified the approximate location of the grid. What the user needs to do is to tell the program where the outline of the grid is in the image. For example, the user may need to put down a grid by specifying its dimensions (in number of rows and columns) and by clicking on the corner spots. Subsequently, the spot finding algorithm adjusts the location of the grid lines, or grid points, to locate the arrayed spots in the image. User interface tools are usually provided by the software in order to allow for manual adjustment of the grid points if the automatic spot finding method fails to correctly identify each spot. This approach offers great timesaving over the manual spot finding method since the user needs only to identify a few points in the image and make minor adjustments to a few spot locations if required. Such capabilities are offered by most software vendors.

Automatic spot finding. The ultimate goal of array image processing is to build an automatic system, which utilizes advanced computer vision algorithms, to find the spots reliably without the need for any human intervention. This method would greatly reduce the human effort, minimize the potential for human error, and offer a great deal of consistency in the quality of data. Such a processing system would require the user to specify the expected configuration of the array (e.g., number of rows and columns of spots), and would automatically search the image for the grid position. Having found the approximate grid position, which specify the centers of each spot, the neighborhood can be examined to detect the exact location of the spot. Knowledge about the image characteristics should be incorporated to account for variability in microarray images. The spot location, size, and shape should be adjusted to accommodate for noise, contamination and uneven printing.

3.4.2 Image segmentation

Image segmentation is the process of partitioning an image into a set of non overlapping regions whose union is the entire image. The purpose of segmentation is to decompose the image into parts that are meaningful with respect to a particular application, in this case spots separated from background. Once the spots have been found, an image segmentation step is necessary in order to decide which pixels form the spot and should be considered for the calculation of the signal, which pixels form the background and which pixels are just noise or artifacts and should be eliminated. In the following discussion of various approaches used for segmentation, we will assume that the image contains high intensity pixels (white) on a low intensity background (black) as in Fig. 3.7 A, B and D. This is usually the case and all the computation is done on such data. However, images are often displayed in negative (i.e. with low intensity pixels on high intensity background – black pixels on white background as in Fig. 3.7 C) or in false color where the intensity of the spots is displayed as a color (e.g. green or red). False colors allow composite displays where two or more images each corresponding to a different mRNA and each encoded with a different false color are overlapped. In such composite images, the relative intensity between the various channels is displayed as a range of colors obtained from the superposition of various amounts of the individual colors, as in Fig. 3.5.

Pure spatial-based signal segmentation. The simplest method is to place a circle over a spot and consider that all the pixels in this circle belong to the signal. In fact, it is safer to use two circles, one slightly smaller than the other (see Fig. 3.8). The pixels within the inner circle are used to calculate the signal value while the pixels outside the outer circle are used to calculate the background. The pixels between the two circles correspond to the transition area between the spot and its background and are discarded in order to improve the quality of the data. All the pixels outside the circle within the boundary of a square determined by the software are considered as background. This approach is very simple but poses several big problems as illustrated in the figure. Firstly, several spots are smaller than the inner circle they are in. Counting the white pixels as part of the spots just because they fall in the circle will artifactually lower the signal values for those spots. Secondly, there are dust particles, contaminants and sometimes fragments of spots which are outside the outer circle and will be counted as background. Again, this would artificially increase the background and distort the real relationship between the spot and its local background. For these reasons, this method is considered not to be very reliable.

Intensity based segmentation. Methods in this category use exclusively intensity information to segment out signal pixels from the background. They assume that the signal pixels are brighter on average than the background pixels. As an example, suppose that the target region around the spot taken from the image consists of 40x40 pixels. The spot is about 20 pixels in diameter. Thus, from the total of 1600 (40x40) pixels in the region, about 314 ($\pi \cdot 10^2$) pixels, or 20%, are signal pixels and they are expected to have their intensity values higher than that of the background pixels. To identify these signal pixels, all the pixels from the target region are ordered in a one dimensional array from the lowest intensity pixel to the highest one, $\{p_1, p_2, p_3, ..., p_{2500}\}$, in which p_i is the intensity value of the pixel of the i-th lowest intensity among all the pixels. If there is no contamination in the target region, the top 20% pixels in the intensity rank may be classified as the signal pixels. The advantage of this method is its simplicity and speed; it is good for obtaining results using computers of moderate computing power. The method works well when the spots are of high intensities as compared to the background. However, the method has disadvantages when dealing with spots of low intensities, or noisy images. In particular, if the array contains contaminants such as dust or other artifacts, the method

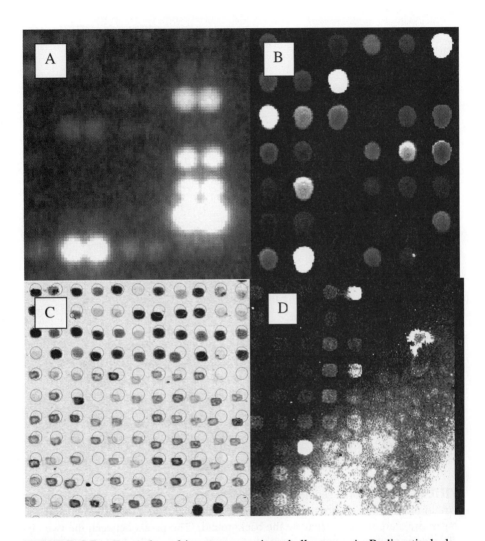

FIGURE 3.7: Examples of image processing challenges. A. Radioactively labelled spots expanding over neighboring spots (blooming). B. Non-circular and weak spots as well as some weak-center spots (doughnuts). C. Imperfect alignment. D. Hybridization and contamination causing a non-specific, high background with various gradients.

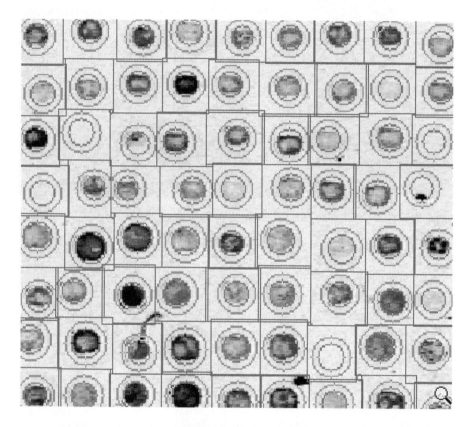

FIGURE 3.8: Spatial segmentation. A circle is placed over each spot. The pixels within the inner circle are used to calculate the signal value, the pixels outside the outer circle are used to calculate the background. The pixels between the two circles correspond to the transition area between the spot and its background and are discarded in order to improve the quality of the data.

will perform poorly since any pixel falling in the top 20% of the intensity range will be classified as signal even if it is situated way out of any spot.

Mann-Whitney segmentation. This approach combines the use of spatial information with some intensity based analysis. Based on the result of the spot finding operation, a circle is placed in the target region to include the region in which the spot is expected to be found. Since the pixels outside of the circle are assumed to be the background, the statistical properties of these background pixels can be used to determine which pixels inside the circle are signal pixels. A Mann-Whitney test is used to obtain a threshold intensity level that will separate the signal pixels from other pixels (e.g background) even if they are inside the expected area of the spot. Pixels inside of the circle having a higher intensity than the threshold intensity are treated as signal. This method works very well when the spot location is found correctly and there is no contamination in the image. However, when contamination pixels exist inside of the circle, they can be treated as signal pixels. This is due to the fact that some contamination can have an intensity higher than the background. Furthermore, if there are contamination pixels outside of the circle, or the spot location is incorrect such that some of the signal pixels are outside of the circle, the distribution of the background pixels can be incorrectly calculated resulting in a threshold level higher than appropriate. Consequently, signal pixels with intensity lower than the threshold will be misclassified as background. This method also has limitations when dealing with weak signals and noisy images. When the intensity distribution functions of the signal and background are largely overlapping with each other, classifying pixels based on an intensity threshold is prone to classification errors, resulting in measurement biases. This is similar to what has been discussed in the pure intensity based segmentation method.

Combined intensity-spatial segmentation (or the trimmed measurements approach). This approach combines both spatial and intensity information in segmenting the signal pixels from the background in a manner similar to the Mann-Whitney approach. The logic of this method proceeds as follow. Once the spot is localized and a target circle is placed in the target region, most of the pixels inside of the circle will be signal pixels and most of the pixels outside of the circle will be background. However, due to the shape irregularity, some signal pixels may leak out of the circle and some background pixels may get into the circle. Background pixels within the circle may be considered as outliers in the intensity distribution of the signal pixels. Similarly, signal pixels that fall outside the circle will also appear as outliers with respect to the intensity distribution of the background pixels. Contamination pixels anywhere will appear as outliers in the intensity domain for either signal and background. These outliers would severely change the measurement of the mean and total signal intensity if they are not eliminated. To remove the effect of outliers on these measurements, one may simply "trim-off" a fixed percentage of pixels from the intensity distribution of the pixels for both signal and background regions.

The Mann-Whitney approach described above performs a statistical analysis on the pixels outside the presumed spot area and then uses the threshold calculated there to segment the pixels inside the target area. The trimmed measurements approach performs a statistical analysis of both distribution (outside as well as inside the presumed spot) and eliminates the outliers from each such distribution without making the leap of faith that the characteristics of the distribution outside will also reflect the properties of the distribution inside. Eliminating approx. 5-10% of each distributions allows this approach to cope very elegantly with artifacts such as doughnuts, dust particles or other impurities even if they lay both inside and outside the spot.

Although this method performs extremely well in general, a potential drawback is related to its statistical approach if the spots are very small (3-4 pixels in diameter) since in this case the distribution will have relatively few pixels.

Two circles can be used in order to further improve the accuracy of this method. The inside of the inner circle will be the area of the prospective spot while the outside of the outer circle will be the area considered as background. The small region between the two circles is considered a buffer zone between the spot and its background. In this area, the intensities may vary randomly due to an imperfect spot shape and are considered unreliable. The accuracy of the analysis is improved if the pixels in this area are discarded from the analysis.

A comparison between spatial segmentation and trimmed measurement segmentation is shown in Fig. 3.9. The figure shows the spatial segmentation on top and the trimmed measurement segmentation at the bottom. The first row of each image represents the cy3 image (Sample A) while the second row represents the cy5 image (Sample B). This spot has a stronger signal on the cy3 channel and a weaker signal on the cy5 channel. However, the cy3 channel also has a large artifact in the background to the left of the spot. This artifact also covers the spot partially. This is a good example of a spot on which many methods will perform poorly. Each row shows from left to right: i) the raw image, ii) the segmented image, iii) the overlap between the raw image and the segmented image and iv) the histograms corresponding to the background and signal. Signal pixels are red, background pixels are green and ignored pixels are black. The pixels that are black in the segmented image are eliminated from further analysis. The spatial segmentation will only consider the spatial information and will consider all pixels marked red in the top two images as signal. These pixels also include a lot of pixels belonging to the artifact as can be seen from the overlap of the segmented image and the original image (image 3 in the top row). The same phenomenon happens for the background. Note that the spatial segmentation will consider all green pixels in the top two images as background. These pixels include many high intensity pixels due to the artifact and the average of the background will be inappropriately increased. The trimmed measurement segmentation does an excellent job of eliminating the artifact from both background and signal. Note in row 3, image 2 (second from the left), the discontinuity in the green area and note in the overlapped image (same row 3, image 3) that this discontinuity corresponds exactly to the artifact. Similarly, the red circle showing the segmentation of the signal pixels has a black area corresponding to the same artifact overlapping the spot itself. In effect, this method has carved a good approximation of the shape of the artifact and has prevented its pixels from affecting the computation of the hybridization signal.

The two methods discussed above use minimal amount of spatial information, i.e. the target circle obtained from spot localization is not used to improve the detection

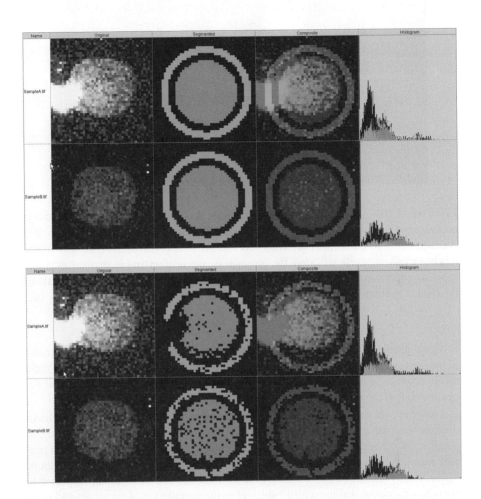

FIGURE 3.9: Spatial segmentation (top) compared to trimmed measurement segmentation (bottom). First row of each image represents the cy3 while the second row represents the cy5 image. Each row shows from left to right: i) the raw image, ii) the segmented image, iii) the overlap between the raw image and the segmented image and iv) the histograms corresponding to the background and signal. Signal pixels are red, background pixels are green, ignored pixels are black. The pixels that are black in the segmented image are eliminated from further analysis. Note that the spatial segmentation includes the artifact pixels both in the computation of the background and in the computation of the signal. The trimmed measurement segmentation does an excellent job of eliminating the artifact from both background and signal. Image obtained using ImaGene, BioDiscovery Inc.

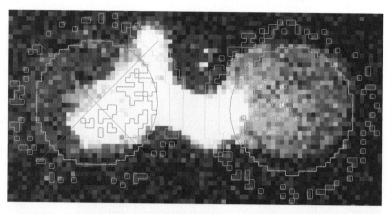

FIGURE 3.10: Artifact removal and bad spot detection in fully automatic image processing. The spot to the right is the same as the one in Fig. 3.9. The pixels corresponding to the artifact are removed and the spot is salvaged. The spot to the left has too few pixels left after the removal of the artifact and is marked as a bad spot. Its quantification value will be associated to a flag providing information about the type of problem encountered. Image obtained using ImaGene, BioDiscovery Inc.

of signal pixels. Their design priority is to make the measurements of the intensity of the spots with minimal computation. These methods are useful in semi-automatic image processing because the speed is paramount (the user is waiting in front of the computer) and the user can visually inspect the quality of data.

In a fully automated image processing system, the accuracy of the signal pixel classification becomes a central concern. Not only that the correct segmentation of signal pixels must offer accurate measurement of the signal intensity, but it must also provide multiple quality measurements based on the geometric properties of the spots. These quality measures can be used to draw the attention of a human inspector to spots having questionable quality values after the completion of an automated analysis. Fig. 3.10 shows the result of the trimmed measurement segmentation on two adjacent spots partially covered by an artifact. The spot to the right is the same as the one in Fig. 3.9. The pixels corresponding to the artifact are removed, a reliable measurement can be extracted from it and the spot is salvaged. The spot to the left has too few pixels left after the removal of the artifact and is marked as a bad spot. Its quantification value will be associated to a flag providing information about the type of problem encountered.

3.4.3 Quantification

The final goal of the image processing is to compute a unique value that hopefully is directly proportional with the quantity of mRNA present in the solution that hybridized the chip. One such value needs to be obtained for each gene on the chip.

The purpose of the **spot quantification** is to combine pixel intensity values into a unique quantitative measure that can be used to represent the expression level of a gene deposited in a given spot. Such a unique value could be obtained in several ways. Typically, spots are quantified by taking the mean, median or mode of the intensities of all signal pixels. Note that a simple sum of the pixel intensities would be dependent on the size of the spot through the number of pixels in a spot. Therefore, values obtained from microarrays printed at different spot densities could not be compared directly.

The key information that needs to be recorded from microarrays is the expression strength of each target. In gene expression studies, one is typically interested in the difference in expression levels between the test and reference mRNA populations. On two channel microarrays, each channel of the control (reference) and experiment (test) is labelled with a different fluorochrome and the chip is scanned twice, once for each channel wavelength. The difference in expression levels between the two conditions under study now translates to differences in the function of intensities on the two images. Under idealized conditions, the total fluorescent intensity from a spot is proportional to the expression strength. These idealized conditions are:

- The preparation of the target cDNA (through reverse transcription of the extracted mRNA) solution is done appropriately, such that the cDNA concentration in the solution is proportional to that in the tissue.

- The hybridization experiment is done appropriately, such that the amount of cDNA binding on the spots is proportional to the target cDNA concentration in the solution.

- The amount of cDNA deposited on each spot during the chip fabrication is constant.

- There is no contamination on the spots.

- The signal pixels are correctly identified by image analysis.

In the following discussion, we assume that the first two conditions are satisfied. Whether these two conditions are truly satisfied should be controlled through the design of the experiments. For the measurements obtained based on image analysis algorithms, we are mainly concerned about the remaining three conditions. In most cases, the last three conditions are all violated. The amount of DNA deposited during the spotting procedure may vary from time to time and spot to spot. Higher amounts may result in larger spot sizes so the size of the spots cannot be considered constant, not even for spots on the same array. When a spot is contaminated, the signal intensity covered by the contaminated region is not measurable. The image processing may not correctly identify all signal pixels; thus, the quantification methods should not assume an absolute accuracy in the segmentation stage.

The values commonly computed for individual spots are: total signal intensity, mean signal intensity, median signal intensity, mode signal intensity, volume, intensity ratio, and the correlation ratio across two channels. The underlying principle for judging which one is the best method is based on how well each of these measurements correlates to the amount of the DNA target present at each spot location.

The **total signal intensity** is the sum of the intensity values of all pixels in the signal region. As it has been indicated above, this total intensity is sensitive to the variation of the amount of DNA deposited on the spot, the existence of contamination and the anomalies in the image processing operation. Because these problems occur frequently, the total signal intensity is not an accurate measurement and is rarely used.

The **mean signal intensity** is the average intensity of the signal pixels. This method has certain advantages over the total. Very often the spot size correlate to the DNA concentration in the wells during the spotting processing. Using the mean will reduce the error caused by the variation of the amount of DNA deposited on the spot by eliminating the differences introduced by the size of the spot. With advanced image processing allowing for accurate segmentation of contamination pixels from the signal pixels, the mean should be a very good measurement method.

The **median of the signal intensity** is the intensity value that splits the distribution of the signal pixels in halves. The number of pixels above the median intensity is the same as the number of pixels below. Thus, this value is a landmark in the intensity distribution profile. An advantage of choosing this landmark as the measurement is its resistance to outliers. As it has been discussed in the previous section, contamination and problems in the image processing operation introduce outliers in the sample of identified signal pixels. The mean measurement is very vulnerable to these outliers. A unique erroneous value much higher or much lower than the others can dramatically affect the mean. However, such a unique outlier will not affect the median.[2]

When the distribution profile is unimodal, the median intensity value is very stable and it is close to the mean. In fact, if the distribution is symmetric (in both high and lower intensity sides) the median is equal to mean. Thus, if the image processing techniques used are not sophisticated enough to ensure the correct identification of signal, background and contamination pixels, the median might be better choice than the mean. An alternative to the median measurement is to use a trimmed mean. The trimmed mean estimation is done after certain percentage of pixels have been trimmed from the tails of the intensity distribution.

The **mode of the signal intensity** is the "most-often-found" intensity value and can be measured as the intensity level corresponding to the peak of the intensity histogram. The mode is another landmark in the intensity distribution enjoying the same robustness against outliers offered by the median. The trade-off is that the mode becomes a biased estimate when the distribution is multi-modal, i.e. when the intensity histogram has more than one peak. This is because the mode will be equal to one of peaks in the distribution, more specifically to the highest. When the distribution is uni-modal and symmetric, mean, median, and mode measurements are equal. Often the difference between mode and median values can be used as an

---

[2]See also the detailed discussion of the mean, median and mode in Chap. 4.

indicator of the degree to which a distribution is skewed (elongated on one side) or multi-modal (have several peaks).

The **volume of signal intensity** is the sum of the signal intensity above the background intensity. It may be computed as (mean of signal – mean of background) × area of the signal. This method adopts the argument that the measured signal intensity has an additive component due to the non-specific binding and this component is the same as that from the background. This argument may not be valid if the non-specific binding in the background is different from that in the spot. In this case, a better way is to use blank spots for measuring the strength of non-specific binding inside of spots. It has been shown that the intensity on the spots may be smaller than it is on the background, indicating that the nature of non-specific binding is different between what is on the background and inside of the spots. Furthermore, by incorporating the area of the signal, the volume becomes sensitive to the spot size. Thus, if an intelligent segmentation algorithm detects a defect on a spot and removes it, the remaining signal area will be artificially smaller. In consequence, the volume will be artificially decreased even if the expression level of the gene is high.

If the hybridization experiments are done in two channels, then the **intensity ratio between the channels** might be a quantified value of interest. This value will be insensitive to variations in the exact amount of DNA spotted since the ratio between the two channels is being measured. This ratio can be obtained from the mean, median or mode of the intensity measurement for each channel.

Another way of computing the intensity ratio is to perform **correlation analysis** across the corresponding pixels in two channels of the same slide. This methods computes the ratio between the pixels in two channels by fitting a straight line through a scatter plot of intensities of individual pixels. This line must pass through the origin and the slope of it is the intensity ratio between the two channels. This is also known as **regression ratio**. This method may be effective when the signal intensity is much higher than the background intensity. Furthermore, the assumption is that the array was scanned in such a way that the pixels in the two channels can be mapped exactly to each other. This may not always be possible depending on the type of scanner used. The motivation behind using this method is to bypass the signal pixel identification process. However, for spots of moderate to low intensities, the background pixels may severely bias the ratio estimation of the signal towards the ratio of the background intensity. Then the advantage of applying this method becomes unclear and the procedure suffers the same complications encountered in the signal pixel identification methods discussed above. Thus its theoretical advantage over intensity ratio method may not be present. One remedy to this problem is to identify the signal pixels first before performing correlation analysis. Alternatively, one could identify the pixels that deviate within a specified amount from the mean of the intensity population.

3.4.4   Spot quality assessment

In a fully automated image processing system, the accuracy of the quantification becomes a central concern. Not only that the correct segmentation of signal pixels must offer accurate measurement of the signal intensity, but it must also permit multiple quality measurements based on the geometric properties of the spots. These quality measures can be used to draw the attention of a human inspector to spots having questionable quality values. The following quality measures are of interest in microarray image analysis.

Spot signal area to spot area ratio. Spot area is the spot signal area plus the area occupied by ignored regions caused by contamination or other factors, which are directly connected to the signal region ("touching" the signal area as in Fig. 3.10). This measure provides information about the size of the ignored area located nearby the signal. The smaller the ratio the larger this ignored area is and therefore the lower the quality of the spot. The signal area to spot area ratio is a measure of local contamination. A researcher may generate a scatter plot of data from two sub-grids (pins) and evaluate the quality of the spot printing. This is a very convenient way to identify defective pins that might fail to be revealed by a visual inspection.

Shape regularity. This measure considers all the ignored and background pixels that fall within the circle proposed by the spot-finding method. The shape regularity index can be calculated as the ratio of number of those pixels to the area of the circle. This technique measures how deformed the actual signal region is with respect to the expected circular shape. Clearly, round, circular spots are to be trusted more than badly deformed ones so the lower the shape regularity ratio, the better the spot is.

Spot Area to perimeter ratio. This ratio will be maximum for a perfectly circular spot and will diminish as the perimeter of the spot becomes more irregular. This measure is somewhat overlapping with the shape regularity above in the sense that both will allow the detection of the spots with a highly irregular shape.

Displacement. This quality measure is computed using the distance from the expected center of the spot to its actual location: 1 - (offset to grid). The expected position in the grid is computed by the grid placement algorithm. This measure can be normalized to values from 0 to 1, by dividing it by half of the snip width (grid distance). A spot closer to its expected position will be more trustworthy than a spot far away from it.

Spot uniformity. This measure can be computed as: 1-(signal variance to mean ratio). This measure uses the ratio of the signal standard deviation and the signal mean. The ratio is subtracted from 1 such that a perfect spot (zero variance) will yield a uniformity indicator of 1. A large variation of the signal intensity will produce lower values for this quality measure and will indicate a less trustworthy spot. Dividing by the mean is necessary because spots with higher mean signal intensity also have stronger signal variation.

All of the above measures have been or can be normalized to vary from 0 to 1. Furthermore, they can also be adjusted such that a value of 1 corresponds to an ideal spot and a lower value shows spots degradation. Once this is done, a weighted sum of these measures can provide a global, unifying quality measure ranging from 0 (bad spot) to 1 (good spot). This approach also allows users to vary the relative weights of the various individual quality indicators to reflect their own preferences.

Automatic flagging of spots that are not distinguishable from the background (empty spots) or have questionable quality (poor spots) is a necessity in high throughput gene expression analysis. Flagged spots can be excluded in the data mining procedure as part of the data preparation step. In spite of the great importance of assessing the quality of the spots and carrying this information forward throughout the data analysis pipeline, relatively few software packages provide such measurements. It is hoped that, as the users become more aware of the issues involved, they will require software companies to provide such quality measures in a larger variety of products.

As a final observation, note that typically a gene is printed in several spots on the same array. Such spots are called replicates and their use is absolutely necessary for quality control purposes. Quantification values obtained for several individual spots corresponding to a given gene still need to be combined in a unique value representative for the gene. Such a representative value can in turn be computed as a mean, median or mode of the replicate values. Alternatively, individual replicate values can be analyzed as a part of a more complex model such as ANalysis Of Variance (ANOVA) (see Chap. 7). More details about the image processing of cDNA arrays are available in the literature [169].

3.5   Image processing of Affymetrix arrays

Due to the fact that the Affymetrix technology is proprietary, virtually all image processing of the Affymetrix arrays is done using the Affymetrix software. The issues are slightly different due to several important differences between technologies.

A first important difference between cDNA and oligonucleotide arrays (oligo arrays) is the fact that cDNA arrays can use long DNA sequences while oligonucleotide arrays can ensure the required precision only for short sequences. In order to compensate for this, oligo arrays represent a gene using several such short sequences. A first challenge is to combine these values to obtain a meaningful value that is proportional to the level of expression of the gene.

A second important difference is that in oligo arrays, there is no background. The entire surface of the chip is covered by probes and a background value cannot be used as an indication of the level of intensity in the lack of hybridization. However, as described in Chap. 2, the Affymetrix arrays represent a gene using a set of match/mismatch probes. In general, these arrays contain 20 different probes for each gene or EST. Each probe consists of 25 nucleotides (thus called a 25-mer). Let the reference probe be called a perfect match and denoted as PM, and the partner probe containing a single different nucleotide be called a mismatch (MM) (see Fig.3.11). RNAs are considered present on the target mixture if the signals of the PM probes are significant above the background after the signal intensities from the MM probes have been subtracted.

Thus, a first analysis of the image can provide two types of information for each gene represented by a set of PM/MM probes. The number of probes for which the PM value is considerably higher than the MM value can be used to extract a qualita-

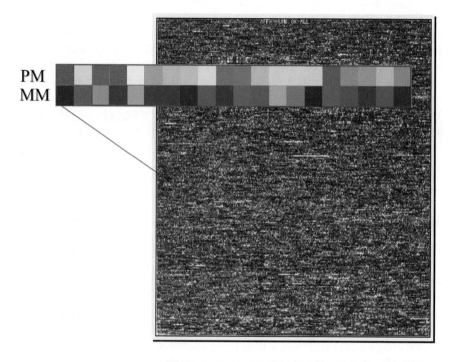

FIGURE 3.11:   The image of an Affymetrix microarray. A gene is represented by a set of 20 probes. Each probe consists of 25 nucleotides. The top row contains the perfect match (PM) probes while the bottom row contains the mismatch (MM) probes. The MM probes are different from the PM probes by a single nucleotide. If the mRNA corresponding to a gene was present during the hybridization, the PM probes have a higher intensity than the MM probes. The average difference between the PM and MM probes is considered proportional to the expression level of the gene.

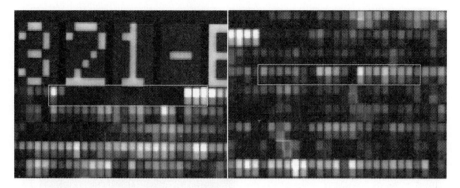

FIGURE 3.12: Two areas on an Affymetrix array corresponding to two different genes. The gene represented to the left is reported absent while the one to the right is reported as present. However, the average difference calculated for the gene on the left is higher than the average difference calculated for the gene on the right. Image obtained with dChip [194].

tive information about the gene. The Affymetrix software captures this information in a set of "calls": if many PM values are higher than their MM values, the gene is considered "present" (P); if only a few (or no) PM values are higher than their MM values, the gene is called "absent" (A). If the numbers are approximately equal the gene is called "marginal" (M). The exact formulae for determining the calls are provided by Affymetrix and can change in time.

The ultimate goal of the microarray technology is to be quantitative, i.e. to provide a numerical value directly proportional to the expression of the gene. A commonly used value provided by the Affymetrix software is the average difference between PM and MM:

$$AvgDiff = \frac{\sum_i^N (PM_i - MM_i)}{N} \tag{3.1}$$

where PM_i is the PM value of the i-th probe, the MM_i is the corresponding MM value and N is the number of probes. In general, this value is high for expressed genes and low for genes that are not expressed.

The fact that there are two measures of which one is qualitative (the call) and the other one is quantitative (the average difference) can lead to situations apparently contradictory. Let us consider the example given in Fig. 3.12. We have two cases. For the gene to the left, calculating the pixel intensities from the array features yields an average difference of 1,270. The call of the software is "Absent" for this RNA. The gene on the right has an average difference of 1,250 but this time the software indicates a "present" RNA. This is an apparent counter-example to the general rule that expressed genes have higher average differences than not expressed genes. How can we explain this? What is happening here?

Let us consider two genes, A and B. Gene B is not expressed, but has a very short

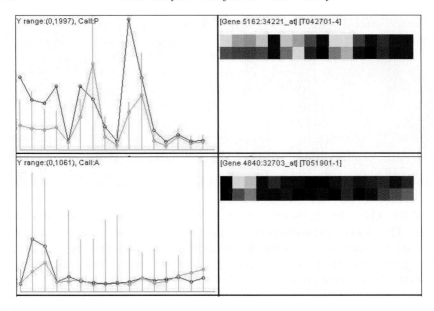

FIGURE 3.13:   An apparent contradiction between the calls of two genes and their average differences. The figure shows the hybridization of the probes corresponding to each gene to the right and a graphical representation of the intensities corresponding to each probe to the left. The gene on top is reported as present, the gene on the bottom is reported as absent. However, the average difference of the gene on top is lower (1250) than the average difference of the gene at the bottom (1270). Image obtained with dChip [194].

sequence, which is identical to a subsequence of gene A which is highly expressed. In this case, the probes of B identical to the substrings of A will give a strong signal artificially increasing the average difference for B. At the same time, if a different gene C is expressed at a moderate level, all its probes might have moderate intensities. The average difference calculated for C might be comparable or even lower than the average difference calculated for B. Such an example is illustrated in Fig. 3.13.

3.6   Summary

Digital images are rectangular arrays of intensity values characterized by several resolution and color depth. Each intensity value corresponds to a point in the image and is called a pixel. The resolution is the number of pixels in the image and is usually expressed as a product between the number of rows and the number of columns (e.g. 1024x768). The color depth is the number of bits used to store the intensity value of a single pixel. The digital image of a microarray has to satisfy minimum requirements of resolution and color depth. The resolution used for cDNA arrays should be such that the diameter of a spot measured in pixels is at least 10 pixels (but more is preferable). The usual color depth used for microarrays is 16 bits which allows each pixel to represent 65,536 different intensity levels. The image processing of microarray digital images aims at obtaining numerical values proportional to the level of mRNA present in the tested sample(s) for each of the genes interrogated. For cDNA arrays, this is done in several steps: array localization, image segmentation, quantification and spot quality assessment. The array localization is the process that localizes the spots (spot finding). The spot finding can be done manually (obsolete), semi-automatically or automatically. The image segmentation is the process that decides which pixels belong to the spots and which pixels belong to the background. The segmentation can be done based on spatial information, intensity information, using a Mann-Whitney analysis of the distribution of the pixels or using a combined intensity-spatial approach. The quantification combines the values of various pixels in order to obtain a unique numerical value characterizing the expression level of the gene. Quantification must be done selectively such that the pixels corresponding to various possible contaminations or defects are not taken into consideration in the computation of the value representative for the expression level of the gene. This representative value can be calculated as the total, mean, median or mode of the signal intensity, the volume of the signal intensity of the intensity ratio between the channels. The spot quality can be assessed using the ratio between the signal area and the total spot area, the shape regularity, ratio between spot area to perimeter, displacement from the expected position in the grid and spot uniformity. The image processing of Affymetrix arrays produces several types of information including calls and average differences. Calls are meant to provide qualitative information about genes. A gene can be present, marginally present or absent. The average difference is calculated as the average of the differences between the perfect match (PM) and mismatch (MM) of all probes corresponding to a given gene. The average difference is commonly used as the quantitative measure of the expression level of the gene.

Chapter 4

Elements of statistics

"Organic chemist!" said Tilley expressively. "Probably knows no statistics whatever."

—Nigel Balchin, The Small Back Room

Definition of a Statistician: A man who believes figures don't lie, but admits that under analysis some of them won't stand up either.

—Evan Esar

The most important questions of life are, for the most part, really only problems of probability.

—Pierre Simon Laplace

4.1 Introduction

The basics of microarray data analysis historically relied on the ratio of two signals from a spot. One signal came from a red dye and the other from a green dye. Initially, it was postulated that the relevant information from two-dye microarray experiments is captured in the ratio of these two signals from each spot. Additionally, tools such as clustering techniques were applied to data analysis without too much attention to classical statistical analysis. Gary Churchill and his postdoctoral researcher Kathleen Kerr (Jackson Laboratories) were arguably the first to observe that the needs and challenges of the data coming from microarray experiments are about as old as statistics itself.

The dictionary definition of statistics is: "a branch of mathematics dealing with the collection, analysis, interpretation, and presentation of masses of numerical data" [210]. This definition places statistics within the realm of mathematics thus emphasizing its rigor and precision. However, statistics has also been defined as *"the art and science* of collecting, analyzing, presenting, and interpreting data" [53]. This is perhaps a better definition since it also emphasizes the fact that statistical analysis is far from being a mechanical process, and that creativity and imagination are often required in order to extract most information from the data.

Nowadays, statistics has become an integral part of microarray data analysis and it is of paramount importance for anybody dealing with such data to understand the basic statistical terms and techniques. The objective of this chapter is to review some fundamental elements of statistics as related to microarray experiments. Our primary goal here is not to construct a rigorous mathematical edifice but to give the reader a few working definitions and useful intuitions. This chapter is intended as a statistical primer for the life scientist. As such we preferred the use of simpler mathematical tools even if we had to sacrifice at times the generality of some definitions and proofs.

4.2 Some basic terms

Define your terms, you will permit me again to say, or we shall never understand one another.

—Voltaire

The term **population** denotes the ensemble of entities considered or the set of all measurements of interest to the sample collector. If one were to study the opinion of the American people regarding a new presidential election in the USA then one would need to study the opinion of the whole population of the country about their votes for their future president.[1] In the same way, if one wanted to see what genes are over-expressed in people suffering from obesity, one would need to consider the whole population of clinically overweight people registered in the USA.

In the above example it is obvious that, in most cases, one cannot study directly whole populations. Therefore, a subset of the population, called **a sample**, is considered instead. A sample is any subset of measurements selected from the population. The objective of statistics is to make an inference about a population based on information contained in the sample. However, choosing this sample is a science by itself since **bias** can affect the results. A classic example is the sampling that occurred in the US in 1936. A magazine contacted its readers and subscribers who numbered 10 million people and asked them whom they were going to vote for. Based on the 2,300,000 replies, a prediction was made that the Republican candidate Landon would be elected. However, it turned out that the Democratic candidate Franklin D. Roosevelt won the elections by a very large margin. The error occurred because the over 2 million who responded did not represent correctly the American population at that time. In some situations, the bias might be very apparent. For instance, let us consider a potential bill banning possession of firearms. If a sample chosen for an opinion poll were to be taken exclusively from the members of the National Rifle Association, a majority rejecting such a bill would not be surprising. However, should the sample be taken from the members of a support group for firearms victims, the exactly opposed conclusion might be expected. Of course, both such polls would

---

[1] This assumes that everybody votes.

provide little information about the results of a potential referendum involving the entire US population.

A variable whose values are affected by chance is called a **random variable**. To convey an idea about the general appearance of objects or phenomena people need numerical descriptive measures. Numerical descriptive measures of random variables for a population are called **parameters**. For example, the mean weight of people in the US is a parameter of the US population. A numerical descriptive measure of a sample is called a **statistic**. In most cases, this numerical value is obtained by applying a mathematical function (e.g. the average) to the values in the sample. For example, the mean weight of people in a classroom is a statistic calculated for the sample of the US population given by their presence in the classroom. These people represent a subset of the whole population living in the US. The very same numerical descriptor of a group of measurements can be a parameter or a statistic depending on the given context. For instance, the mean weight of the people in the US is a parameter of the US population, but only a statistic in the larger context of the world population.

The data can be of different types as follows:

1. Quantitative

   - Continuous: these data take continuous values in an interval. For example, spot intensity levels, gene expression levels.

   - Discrete: these data take discrete values such as number of spots, number of genes, number of patients, etc.

2. Ordinal (ranked). In this type of data there is ordering but no numerical value. For example the top ten rock hits on the music charts, top ten companies performing on the stock market, etc.

3. Categorical data: genotype, phenotype, healthy/disease groups and others.

Subtle differences exist between discrete data such as the number of times a student has attempted a given exam before passing (taking the discrete values 1,2,3,4,...) and ordinal data such as stocks ordered by their performance over a given period. At a superficial view, stocks can be ordered with the best performer as 1, second best as 2, etc. Students can also be ordered in the order given by their number of attempts. The crucial difference between the stocks (ordinal data in this context) and students (discrete data) is that discrete data provide more information than categorical ones. Thus, stock ordering tells us that the second best is worse than the best but it does not give us any information about *how much worse it is*. The difference may be very small or may be huge. We just cannot tell. However, knowing the number of times various students attempted an exam before passing will allow us to get an idea about the difference between consecutive students. If the best student passed at the first attempt and the second best student passed only after 6 attempts, we will know that the second best student is actually much worse than the best one. In essence, discrete, quantitative data can be ranked but ranked data cannot become quantitative.

Measurements are affected by **accuracy** and **precision**. Accuracy refers to how close the measurement value is to its true value. Precision refers to how close repeated measurements are to each other. A biased but sensitive instrument will give precise but inaccurate values. An insensitive instrument might provide an accurate value on a given occasion, but the value would be imprecise (a repeated measurement would probably provide a different value). There is a trade off related to the precision. Too much precision is cumbersome. What is a "good" precision? In general, the measurements should be such that an error in the last digit would introduce an error of less than 5%, preferably less than 1%. In microarray experiments, there are also other issues. As shown in Chapter 2, microarrays involve many steps and a rather complicated data processing pipeline (Fig. 2.5). The focus has to be on improving the accuracy and precision of the overall results as opposed to improving the accuracy of a single step. Many times, a small increase in accuracy in all steps involved can be translated in an improvement of the overall performance while a large increase in accuracy or precision in a single specific step may or may not improve the global performance. As a specific example, increasing the performances of the scanning without improving the printing, mRNA quality, hybridization procedure, etc., might actually degrade the overall performance by adding more noise into the subsequent data analysis.

When comparing samples, it is important to evaluate their **central tendency** and **variability**. Central tendency measures provide information regarding the behavior of the group of measurements (sample) as a whole, while variability measures provide information regarding the degree of variation between members of the same sample.

4.3 Elementary statistics

4.3.1 Measures of central tendency: mean, mode and median

At this time, we will consider that the number of values n is finite. This is done in order to keep the mathematical apparatus simple and make the material as accessible as possible. All terms below also have more general definitions that are applicable to infinite sets, as well.

4.3.1.1 Mean

The most widely used measure of central tendency is the arithmetic **mean** or average. This is defined as the sum of all values divided by their number:

$$\frac{\sum_{i=1}^{n} x_i}{n} \tag{4.1}$$

In statistics, in spite of the fact that there is only one formula for the arithmetic mean, it is customary to differentiate between the mean of a population which is the mean of all values in the population considered:

$$\mu = \frac{\sum_{i=1}^{N} X_i}{N} \tag{4.2}$$

and the mean of a sample, which is the mean of the subset of the population members included in the sample under consideration:

$$\overline{X} = \frac{\sum_{i=1}^{n} X_i}{n} \tag{4.3}$$

The sample mean is usually reported to one more decimal place than the data. Also, the sample mean is measured in the same measurement units as the data.

There are a few interesting phenomena that are relevant to the mean as a measure of central tendency. Let us consider the following set of measurements for a given population: 55.20, 18.06, 28.16, 44.14, 61.61, 4.88, 180.29, 399.11, 97.47, 56.89, 271.95, 365.29, 807.80, 9.98, 82.73.

The population mean can be computed as:

$$\mu = \frac{\sum_{i=1}^{15} X_i}{15} = \frac{55.20 + 18.06 + 28.16 + \cdots + 82.73}{15} = 165.570 \tag{4.4}$$

Let us now consider two samples from this population. The first sample is constructed by picking the first 4 measurements in the population: 55.20, 18.06, 28.16, 44.14 and has the mean 36.640. The second sample is obtained by picking measurement 8 through 12: 399.11, 97.47, 56.89, 271.95, 365.29, 807.80. This sample has the mean 333.085. An immediate observation is that the mean of any particular sample $\overline{X}$ can be very different from the true population mean μ. This can happen because of chance or because the sampling process is biased. The larger the sample, the closer its mean will be to the population mean. At the limit, a sample including all members of the population will provide exactly the population mean. This phenomenon is related to the very important notion of expected value. The **expected value** of a random variable X, denoted $E[X]$, is the average of X in a very long run of experiments. This notion can be used to define a number of other statistical terms, as it will be discussed in the following sections.

If a sample is constructed by picking a value and then eliminating that value from the population in such a way that it cannot be picked again, it is said that the sampling is done **without replacement**. If a value used in a sample is not removed from the population such that the same value can potentially be picked again, it is said that the sampling is done **with replacement**. In the example above, the sampling was probably done without replacement because in such a small population it is likely that sampling with replacement would have picked a value more than once. However, there is no way to be sure. Conversely, if a sample includes a value more than once and we know a priori that the population has only distinct values, we can conclude that the sampling was done with replacement.

4.3.1.2 Mode

The **mode** is the value that occurs most often in a data set. For instance, let us consider the following data set: 962, 1005, 1033, 768, 980, 965, 1030, 1005, 975, 989, 955, 783, 1005, 987, 975, 970, 1042, 1005, 998, 999. The mode of this sample is 1005 since this value occurs 4 times. The second most frequent value would be 975 since this value occurs twice.

According to the definition above, any data set can only have one mode since there will be only one value that occurs most often. However, many times certain distributions are described as being "bimodal." This formulation describes a situation in which a data set has two peaks. Sometimes, this is due to the fact that the sample includes values from two distinct populations, each characterized by its own mode. Also, there are data sets that have no mode, for instance if all values occur just once.

An easy way of finding the mode is to construct a histogram. A histogram is a graph in which the different measurements (or intervals of such measurements) are represented on the horizontal axis and the number of occurrences is represented on the vertical axis.

4.3.1.3 Median and percentile

Given a sample, the **median** is the value situated in the middle of the ordered list of measurements. Let us consider the example of the following sample:

$$96, 78, 90, 62, 73, 89, 92, 84, 76, 86$$

In order to obtain the median, we first order the sample as follows:

$$62, 73, 76, 78, \mathbf{84}, \mathbf{86}, 89, 90, 92, 95$$

Once the sample is ordered, the median is the value in the middle of this ordered sequence. In this case, the number of values is even (10) so the median is calculated as the midpoint between the two central values 84 and 86:

$$median = \frac{84 + 86}{2} = 85 \tag{4.5}$$

The median can also be described as the value that is lower than 50% of the data and higher than the other 50% of it. A related descriptive statistic is the percentile. The **p-th percentile** is the value that has p% of the measurements below it and 100-p% above it. Therefore, the median can also be described as an estimate of the 50-th percentile.

4.3.1.4 Characteristics of the mean, mode and median

It is useful to compare the properties of the mean, mode and median. The mean is the arithmetic mean of the values. The mean has a series of useful properties from a statistical point of view and because of this, it is the most used measure of central tendency. However, if even a single value changes, the mean of the sample will change because the exact values of the measurements are taken into account in the computation of the mean. In contrast, the median relies more on the ordering of the measurements and thus it is less susceptible to the influence of noise or outliers. The mode also takes into consideration some properties of the data set as a whole as opposed to the individual values so it should also be more reliable. However, the mode is not always available. Let us consider the sample in the example above:

$$96, 78, 90, 62, 73, 89, 92, 84, 76, 86$$

The mean of this sample is 82.6, the median is 85 and the mode does not really exist since all values occur just once. Now, let us assume that, due to a measurement error, the lowest value of the sample was recorded as 30 instead of 62. The "noisy" sample is:

$$96, 78, 90, 30, 73, 89, 92, 84, 76, 86$$

which can be ordered as follows:

$$30, 73, 76, 78, \mathbf{84}, \mathbf{86}, 89, 90, 92, 96$$

The measures of the central tendency for this "noisy" data set are as follows: $mean = 79.4$ and $median = 85$. Note that the mean was changed by approximately 4% while the median remained the same. The fact that the median was not affected at all is somehow due to fortune. This happened because the noisy measurement substituted one of the lowest values in the set with a value which was even lower. Thus, the ordering of the data was not changed and the median was preserved. In many cases, the median will be perturbed by noisy or erroneous measurements but never as much as the mean. In general, the median is considered the most reliable measure of central tendency. For larger samples which have a mode, the mode tends to behave similarly to the median inasmuch that it changes relatively little if only a few measurements are affected by noise. These properties of the mean, median and mode are particularly important for microarray data which tend to be characterized by a large amount of noise.

A symmetric distribution will have the mean equal to the median. If the distribution is unimodal (i.e. there is only one mode), the mode will also coincide with the mean and the median. When a distribution is skewed (i.e. has a longer tail on one side), the mode will continue to coincide with the peak of the distribution, but the mean and median will start to shift towards the longer tail.

Another issue worth discussing is the informational content of the measurements of central tendencies. A well known joke tells the story of two soldiers on the battlefield, waiting to ambush enemy tanks. A tank appears and one of the soldiers aims his artillery gun and shoots at the tank. Unfortunately he misses the target and his missile impacts half a mile to the left of the tank. His comrade scorns him and says: "Now let me show you how this is done." He aims and misses too, but this time the missile hits half a mile to the right. The officer looks at them and says, "Well done! On average, the tank is destroyed!" Another classical joke, relying on the same idea, tells the story of the statistician who drowned while trying to cross a river with an average depth of 6 inches. The moral of both stories is that the data are not completely described by measures of central tendency. A very important piece of information that is not captured at all by such measures is the degree of variation, or dispersion, of the samples.

4.3.2   Measures of variability

I abhor averages. I like the individual case. A man may have six meals one day and none the next, making an average of three meals per day, but that is not a good way to live.

— Louis D. Brandeis, quoted in Alpheus T. Mason's Brandeis: A Free Man's Life

The arithmetic mean is a single number, which reflects a central tendency of a set of numbers. For example, the mean of the intensities of several replicates of a gene (microarray spots corresponding to the same gene whether on a single array or from several arrays) will give us an idea about the strength of signal or expression of this gene. However, the mean is incomplete as a descriptive measure, because it does not disclose anything about the scatter or dispersion of the values in the set of numbers from which it is derived. In some cases these values will be clustered closely to the arithmetic mean, whereas in others, they will be widely scattered.

4.3.2.1   Range

The simplest measurement of variability is the **range**:

$$X_{max} - X_{min} \tag{4.6}$$

The range is simply the interval between the smallest and the largest measurement in a group. A wider range will indicate a larger variability than a narrower range. This indicator can be quite informative. For instance, the statistician above might not have attempted to cross the river if she knew the range of the depth measurements was, for instance, 1 inch to 15 feet. However, it is still possible to have very different data sets that have exactly the same measures of central tendency *and* the same range. Therefore, more measures of variability are needed.

4.3.2.2   Variance

Measurements in a sample differ from their mean. For instance, microarray spot replicates differ from their computed mean. The differences between each individual value and the mean is called the **deviate**:

$$X_i - \overline{X} \tag{4.7}$$

where X_i is an individual measurement and $\overline{X}$ is the mean of all X_i measurements. The deviates will be positive for those values above the mean and negative for those values below it. Many different measures of variability can be constructed by using the deviates. Let us consider, for example, that a set of microarray spots have their mean intensities as follows:

$$435.02, \ 678.14, \ 235.35, \ 956.12, \ldots, 1127.82, \ 456.43$$

The mean of these values is 515.13 and their deviates are as follows:

$$435.02 - 515.13 = -80.11$$
$$678.14 - 515.13 = 163.01$$
$$235.35 - 515.13 = -279.78$$
$$956.12 - 515.13 = 440.99$$
$$\vdots$$

We would like to have a measure for the deviations able to tell us about their magnitude. A first idea would be to calculate the mean of these deviations. However, a moment of thought reveals that such a mean would be zero since the arithmetic mean is exactly what the name suggests: a mean with respect to addition. Thus, the sum of all positive deviations will equal the sum of all negative deviations and the mean deviation will be zero. This can also be shown by trying to calculate such a mean of deviates:

$$\frac{\sum_{i=1}^{n}(X_i - \overline{X})}{n} = \frac{(X_1 - \overline{X}) + (X_2 - \overline{X}) + \cdots + (X_n - \overline{X})}{n} =$$

$$= \frac{(X_1 + X_2 + \cdots + X_n) - (\overline{X} + \overline{X} + \cdots + \overline{X})}{n} = \frac{\sum_{i=1}^{n} X_i - n\overline{X}}{n} =$$

$$= \frac{\sum_{i=1}^{n} X_i}{n} - \frac{n\overline{X}}{n} = \overline{X} - \overline{X} = 0 \qquad (4.8)$$

This difficulty can be overcome in two ways: by taking the average of the absolute values of the deviations or by using their squared values. It turns out that the most useful measure has the following expression:

$$\sigma^2 = \frac{\sum_{i=1}^{N}(X_i - \mu)^2}{N} \qquad (4.9)$$

This is called **population variance** and is a measure that characterizes very well the amount of variability of a population. The variance is usually reported with two more decimal places than the data and has as measurement units the square of the measurement units of the data. This is actually the average square distance from the mean. Note that this is the *population* variance. The **variance of a sample** is calculated as:

$$s^2 = \frac{\sum_{i=1}^{n}(X_i - \overline{X})^2}{n - 1} \qquad (4.10)$$

Note that now each individual measurement is corrected by subtracting the *sample mean* $\overline{X}$ instead of population mean μ, the variable used to refer to this measurement is s^2 instead of σ^2, the sum is over the n measurements in the sample instead of the N measurements of the population and the denominator is $n - 1$. This latter difference has to do with the bias of the sample but this is beyond the scope of this overview. The quantity:

$$\sum_{i=1}^{n}(X_i - \overline{X})^2 \tag{4.11}$$

is called the **corrected sum of squared** (CSS) because each observation is adjusted for its distance from the mean.

A widely used measure of variability is the **standard deviation**, which involves the square root of the variance. The standard deviation of a set of measurements is defined to be the positive square root of the variance. Thus, the quantity:

$$s = \sqrt{\frac{\sum_{i=1}^{n}(X_i - \overline{X})^2}{n - 1}} \tag{4.12}$$

is the **sample standard deviation** and:

$$\sigma = \sqrt{\frac{\sum_{i=1}^{N}(X_i - \mu)^2}{N}} \tag{4.13}$$

is the corresponding **population standard deviation**.

In most practical cases, the real population parameters such as population mean, μ, or variance σ^2 are unknown. The goal of many experiments is precisely to estimate such population parameters from the characteristics of one or several samples drawn from the given population.

4.3.3 Some interesting data manipulations

In certain situations, all measurements of a sample are affected in the same way by a constant quantity. As an example, microarrays are quantified by scanning them and then processing the intensity values corresponding to individual spots. However, the device used to scan these arrays has some parameters that can be chosen by the user. One of these settings is the gain of the photomultiplier tube (PMT). If this value is increased between two scans of the same array, all spots intensities will be higher. Such a transformation is sometimes called **data coding**. Similar transformations are often used to change systems of measurements such as converting from miles to kilometers or from degrees Fahrenheit to degrees Celsius. It is interesting to see how descriptive statistics such as mean and variance behave when such data manipulations are applied.

Let us consider first the case in which all measurements are changed by **adding** a constant value. We would like to express the mean of the changed values $\overline{X}_c$ as a function of the mean of the original values $\overline{X}$:

$$\overline{X}_c = \frac{\sum_{i=1}^{n}(X_i + c)}{n} = \frac{\sum_{i=1}^{n}X_i + \sum_{i=1}^{n}c}{n} = \frac{\sum_{i=1}^{n}X_i}{n} + \frac{n \cdot c}{n} = \overline{X} + c \tag{4.14}$$

Therefore, if we know that all values have been changed by addition and we have the old mean, we can calculate the new mean just by adding the offset to the old mean. The computation of the new variance can be performed in a similar way. However, we can now use the fact that $\overline{X}_c = \overline{X} + c$:

$$s_c^2 = \frac{\sum_{i=1}^n \left((X_i + c) - (\overline{X} + c)\right)^2}{n-1} = \frac{\sum_{i=1}^n \left(X_i + c - \overline{X} - c\right)^2}{n-1} =$$

$$= \frac{\sum_{i=1}^n (X_i - \overline{X})^2}{n-1} = s^2 \qquad (4.15)$$

Therefore, the variance of a sample coded by addition remains unchanged.

The results are a bit different when the data are changed by multiplying it with a constant value:

$$\overline{X}_c = \frac{\sum_{i=1}^n c \cdot X_i}{n} = \frac{c \cdot \sum_{i=1}^n X_i}{n} = c\overline{X} \qquad (4.16)$$

The corresponding computation for the variance yields:

$$s_c^2 = \frac{\sum_{i=1}^n \left(cX_i - c\overline{X}\right)^2}{n-1} = \frac{c^2 \cdot \sum_{i=1}^n \left(X_i - \overline{X}\right)^2}{n-1} = c^2 \cdot s^2 \qquad (4.17)$$

In consequence, the standard deviation will be affected by the same multiplicative factor as the data:

$$s_c = c \cdot s \qquad (4.18)$$

4.3.4 Covariance and correlation

The covariance is a measure that characterized the degree to which two different variables are linked in a linear way. Let us assume that an experiment measures the height of people in a population sample x. In this experiment, another interesting variable might be the height of the knee. Let us consider this as another variable y. Such a variable is called a covariate or concomitant variable. Ultimately, we want to characterize the relationship between the variables x and y.

Let X and Y be two random variables with means μ_X and μ_Y, respectively. The quantity:

$$Cov(X, Y) = E\left[(X - \mu_X)(Y - \mu_Y)\right] = E[XY] - E[X]E[Y] \qquad (4.19)$$

is the covariance between X and Y where $E[X]$ is the expectation of variable X.

In practice, if the two variables X and Y are sampled n times, their covariance can be calculated as:

$$Cov_{xy} = \frac{\sum_{i=1}^n (x_i - \overline{x})(y_i - \overline{y})}{n-1} \qquad (4.20)$$

which is the sample **covariance**. This expression is very similar to the expression of the variance 4.10. Indeed, the covariance of a variable with itself is equal to its variance. Unlike the variance which uses the square of the deviates, the covariance multiplies the deviates of two different variables. The product of these deviates can be positive or negative. In consequence, the covariance can be positive or negative. A positive covariance indicates that the two variables vary in the same way: they are above and below their respective means at the same time which suggests that they might increase and decrease at the same time. A negative covariance shows that one variable might increase when the other one decreases.

This measure is very informative about the relationship between the two variables but has the disadvantage that is unbounded. Thus it can take any real value depending on the measurements considered. Furthermore, if we were to calculate the covariance between the size of apples and their weight on the one hand and between the size of oranges and the thickness of their skin on the other hand, the two covariance values obtained could not be compared directly. The diameter of the apples is a length and would be measured in meters whereas the weight would be a mass and would be measured in kilograms. Therefore, the first covariance would be measured in meters x kilograms. The second covariance would be measured in meters (orange diameter) x meters (skin thickness) = square meters. Comparing directly the numerical values obtained for the two covariances would be meaningless and worthy of being characterized as a comparison between "apples and oranges." In consequence, it would be impossible to assess whether the linear relationship between the size of the apples and their weight is stronger or weaker than the linear relationship between the size of the oranges and the thickness of their skin.

In order to be able to compare directly the degree of linear relationship between heterogenous pairs of variables, one would need a measure that is bounded in absolute value, for instance between 0 and 1. Such a measure can be obtained by dividing the covariance by the standard deviation of the variables involved.

Let X and Y be two random variables with means μ_X and μ_Y and variances σ_X^2 and σ_Y^2, respectively. The quantity:

$$\rho_{xy} = \frac{Cov(X,Y)}{\sqrt{(VarX)(VarY)}} \tag{4.21}$$

is called correlation coefficient (Pearson correlation coefficient, to be exact).

The Pearson correlation coefficient takes care of both measurement unit and range. The correlation coefficient does not have a measurement unit and takes absolute values between 0 and 1. In order to see this, let us compute the correlation coefficient as:

$$\rho_{xy} = \frac{Cov(XY)}{s_x \cdot s_y} = \frac{\frac{\sum_{i=1}^n (X_i - \overline{X})(Y_i - \overline{Y})}{n-1}}{\sqrt{\frac{\sum_{i=1}^n (X_i - \overline{X})^2}{n-1}} \cdot \sqrt{\frac{\sum_{i=1}^n (Y_i - \overline{Y})^2}{n-1}}} =$$

$$= \frac{\sum_{i=1}^n (X_i - \overline{X})(Y_i - \overline{Y})}{\sqrt{\sum_{i=1}^n (X_i - \overline{X})^2} \cdot \sqrt{\sum_{i=1}^n (Y_i - \overline{Y})^2}} \tag{4.22}$$

In this expression it is clear that both the numerator and the denominator have the same measurements units and therefore their ratio is without dimension. Furthermore, it is clear that this expression will take a maximum value of 1 if the two

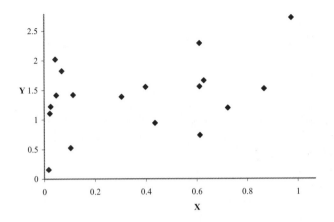

FIGURE 4.1: An example of two variables that have a low positive correlation $(r_{xy} = 0.16)$.

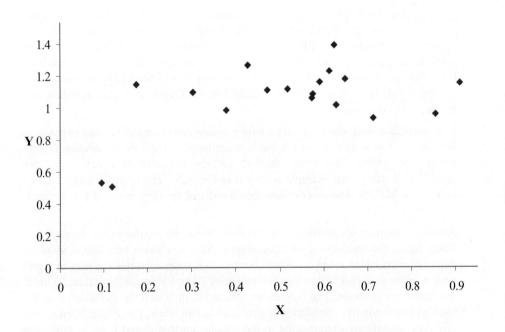

FIGURE 4.2: An example of two variables that have a medium positive correlation $(r_{xy} = 0.70)$.

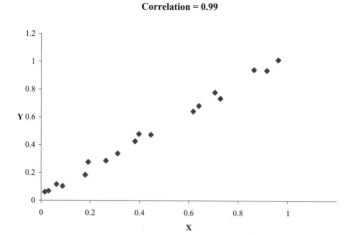

FIGURE 4.3:   An example of two variables that have a high positive correlation $(r_{xy} = 0.99)$.

variables are "perfectly correlated" and a minimum value of -1 if the two variables are exactly opposite. Values towards both extremes show the two variables have a strong linear dependence. Values near zero show the variables do not have a strong linear dependence. Fig. 4.1 shows an example of two variables with a correlation of 0.16. Fig. 4.2 shows two variables with a correlation of 0.70 and Fig. 4.3 shows two variables with a correlation of 0.99. Finally, Fig. 4.4 shows two variables with a high but negative correlation.

The correlation coefficient is a very useful measure and is very often used in practice. However, a few words of caution are in order here. Firstly, the correlation is very sensitive to outliers. Two or three outliers can bring the correlation coefficient down quite considerably. An example is shown in Fig. 4.5. The original data set had a correlation of 0.99. The two outliers shown reduced the correlation of the X and Y variables to 0.51.

Another important observation is that *correlation is not equivalent to dependence*. Many times, the results of a correlation study have a negative conclusion, showing that certain variables x and y are not correlated. A superficial reading may interpret such a result as a conclusion that the variables X and Y are independent. This is not true. For instance, Fig. 4.6 shows a data set in which the variables x and y have an extremely low correlation (correlation coefficient $r_{xy} = 0.03$). Such results might be mistakenly extrapolated to the conclusion that x and y are independent variables. However, Fig. 4.7 shows that the data plotted in Fig. 4.6 are in fact 166 samples from the function $y(x) = sin(x^2) \cdot e^{-x} + cos(x^2)$. The two variables are functionally dependent: the value of x determines exactly the value of y through the given function. The low value of the *linear* correlation coefficient is explained by

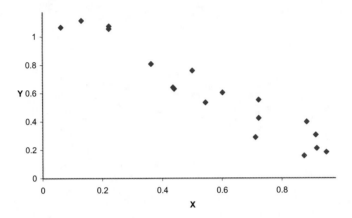

FIGURE 4.4: An example of two variables that have a high negative correlation $(r_{xy} = -0.94)$.

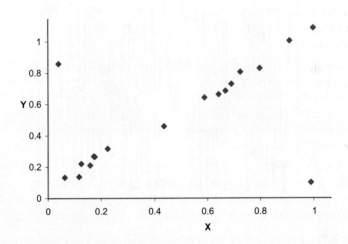

FIGURE 4.5: The effect of outliers on the correlation. Only two outliers reduce the correlation of this data set from 0.99 to 0.51.

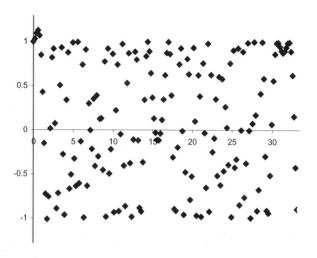

FIGURE 4.6: Two variables with a very low correlation (correlation coefficient $r_{xy} = 0.03$). Such results might be mistakenly extrapolated to the conclusion that x and y are independent variables. Fig. 4.7 shows that the two variables are highly dependent: all values of y are in fact precisely determined by the values of x.

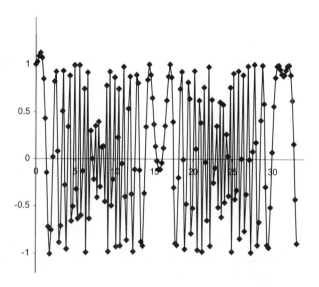

FIGURE 4.7: Limitations of the correlation analysis. The figure shows that the data plotted in Fig. 4.6 are in fact 166 samples from the function $y(x) = sin(x^2) \cdot e^{-x} + cos(x^2)$. The two variables are functionally dependent: the value of x determines exactly the value of y through the given function. The low value of the *linear* correlation coefficient is explained by the high degree of non-linearity of the function.

the high degree of non-linearity of the function.

In order to understand this phenomenon, it is important to keep in mind that the linear correlation coefficient defined above only tries to answer the question whether the variables X and Y are *linearly* dependent. In other words, the question is whether there is a dependency of the form $Y = aX + b^2$. The answer to this question will be negative if either there is no dependency between X and Y or if the dependency between X and Y exist but is approximated only very poorly by a linear function. In conclusion, if two variables are independent, they will probably have a low linear correlation coefficient. However, if two variables have a low correlation coefficient they may or may not be independent. In other words, a high correlation coefficient is, in most cases, a sufficient but not necessary condition for variable dependency.

Given a set of variables $x_1, x_2, \ldots, x_k$, one could calculate the covariances and correlations between all possible pairs of such variables x_i and x_j. These values can be arranged in a matrix in which each row and column corresponds to a variable. Thus, the element σ_{ij} situated at the intersection of row i and column j will be the covariance of variables x_i and x_j while the elements situated on the diagonal will be the variances of their respective variables:

$$\Sigma = \begin{bmatrix} \sigma_1 & \sigma_{12} & \sigma_{13} & \cdots & \sigma_{1k} \\ \sigma_{21} & \sigma_2 & \sigma_{23} & & \sigma_{2k} \\ \vdots & & & & \\ \sigma_{k1} & \sigma_{k1} & \sigma_{k3} & & \sigma_k \end{bmatrix} \tag{4.23}$$

The correlation matrix is formed by taking the ij^{th} element from Σ and dividing it by $\sqrt{\sigma_i^2 \sigma_j^2}$:

$$\rho_{ij} = \sigma_{ij} / \sqrt{\sigma_i^2 \sigma_j^2} \tag{4.24}$$

A relation such as covariance is symmetric if $\sigma_{ij} = \sigma_{ji}$. Since correlations (and covariances) are symmetrical (check in Eq. 4.20 and Eq. 4.22 that nothing changes if X and Y are swapped), the correlation and covariance matrices will be symmetrical, i.e. the part above the first diagonal will be equal to the part below it.

4.4 Probabilities

The **probability** of an event is the numerical measure of a likelihood (or degree of predictability) that the event will occur. From this point of view, an **experiment** is an activity with an observable outcome. A **random experiment** is an experiment

---

[2]Note that if Y can be expressed as $aX + b$ then it is also possible to express X as $X = cY + d$ where c and d are appropriately determined.

whose outcome cannot be predicted with certainty. Examples of experiments can include:

- Flip a coin and record how many times heads or tails are up.

- Perform several microarray experiments and record if a certain gene is up or down regulated.

- Perform several microarray experiments with the same array and see if a particular spot has high or low intensities.

The set of all possible outcomes of an experiment is called the **sample space**. An **event** is a subset of the sample space. We say that an event occurred every time the outcome of the experiment is included in the subset of the sample space that defines the event. For instance, the sample space of rolling a die is the set of all possible outcomes: $\{1, 2, 3, 4, 5, 6\}$. An event can be "obtaining a number less than 4". In this case, the event is the subset $\{1, 2, 3\}$. When we roll a die, we compare the outcome with the subset of the sample space corresponding to the event. For instance, if we rolled a "2," we check to see whether 2 is in the set $\{1, 2, 3\}$. In this case the answer is affirmative so we say that the event "obtaining a number less than 4" occurred. Note that an event has been defined as a set. If this is the case, mutually exclusive events will be events whose subsets of the sample space do not have any common elements. For instance the event of obtaining an odd number after rolling a die ($\{1, 3, 5\}$) is mutually exclusive with the event of obtaining an even number ($\{2, 4, 6\}$). However, the events "number less than or equal to 4" and "greater than or equal to 4" are not mutually exclusive since if 4 is rolled, both events occur at the same time.

Probabilities play an important role in understanding the data mining and the statistical approaches used in analyzing microarray data. Therefore, we will briefly introduce some basic terms, which will be useful for our further discussions.

There are at least two different probabilities: empirical and classical. The **empirical probability** is the relative frequency of the occurrence of a given event. Let us assume that we are interested in an event A. The probability of A can be estimated by running a trial. If there are n cases in the sample and n_A cases in which event A occurred, then the empirical probability of A is its relative frequency of occurrence:

$$P(A) = \frac{n_A}{n} \qquad (4.25)$$

Let us consider the following example. We are trying to estimate probability of occurrence of an albino pigeon in a given area. This probability can be estimated by catching a number of pigeons in the area. Most of the pigeons will be grey but a few will also be white (albino). If 100 birds are caught of which 25 birds are white, the empirical probability of an albino pigeon in the area is 0.25 or 25%.[3]

---

[3]This is just a pedagogical example. Real world estimates of such quantities are calculated using more sophisticated techniques.

The **classical probability** of an event A is defined as:

$$P(A) = \frac{n(A)}{n(S)} \tag{4.26}$$

where S is the entire sample space. The definition simply says that the classical probability can be computed as the number of elements in event A divided by the number of elements in the whole sample space. In the example above, the classical probability will be the actual number of albino pigeons divided by the total number of pigeons in the area. In this example, as in many other cases, the exact probability cannot be computed directly and is either estimated or computed by some other means. If the entire sample space is completely known, the exact probability can be calculated directly and no trials are necessary.

The classical probability relies on certain axioms. These axioms are:

1. For any event, $0 \le P(A) \le 1$

2. $P(\Phi) = 0$ and $P(S) = 1$

3. If $\{A_1, A_2, \ldots, A_n\}$ are mutually exclusive events, then

$$P(A_1 \cup A_2 \cup \ldots \cup A_n) = P(A_1) + P(A_2) + \ldots + P(A_n)$$

4. $P(A') = 1 - P(A)$

where A' is the complement of A.

These axioms might look intimidating but, in reality, they merely formalize our intuitions. Thus, the first axiom says that a probability is a real number between 0 and 1. The second axiom says that the probability of an event that is defined as the empty set (Φ) is zero. This is easy to understand. No matter what we do in a trial, we will observe an outcome. That value will not be included in the empty set because the empty set does not have any elements so the event corresponding to the empty set will never occur. In a similar way, the probability associated with the entire sample space S is 1 because no matter what the outcome of the trial is, it will be an element from the sample space. An event with a zero probability is an impossible event whereas an event with a probability of 1 is certain. The third axiom formalizes the idea that the probability of a union of disjoint events is the sum of the probabilities of the individual events. As an example, if A is the event of "rolling an odd number" and B is the event of "rolling an even number", then the probability of "A or B" is $P(A) + P(B) = 1/2 + 1/2 = 1$. In general, if the sample space is small and countable (e.g. 10 possible outcomes), it is very intuitive that if an event A has 2 favorable outcomes, an event B has, let us say, 3 favorable outcomes and we are interested in either A or B, then it is clear that the number of favorable outcomes is now $2 + 3 = 5$. Since the classical probability is defined as the ratio of the favorable outcomes to the total number of outcomes in the sample space, the probability of A or B is the sum of the two individual probabilities corresponding to A and B. Finally, the fourth axiom states that the probability of the complement of an event A, denoted by A', can be computed as 1 minus the probability of the event A. This is because the complement A′ of an event A is defined as the opposite event. If the sample space is S and A is a subset, the subset corresponding to A′ will be the complement of A with respect to S, i.e. the set of those elements in S that do not belong to A.

The connection between the two types of probabilities, empirical and classical, is made by the **law of large numbers** which states that as the number of trials of the experiment increases, the observed empirical probability will get closer and closer to the theoretical probability.

4.4.1 Computing with probabilities

4.4.1.1   Addition rule

Let us consider the following example.

EXAMPLE 4.1

We decide to declare that a microarray spot can be trusted (i.e. we have high confidence in it) if the spot is perfectly circular (the shape is a circle), perfectly uniform (the standard deviation of the signal intensities is less than a chosen threshold), or both. Let us assume that there are 10,000 spots printed on the array; 6,500 are circular, 7,000 spots are uniform and 5,000 spots are both circular and uniform. What is the probability that a randomly chosen spot is trustworthy according to our chosen standard of high confidence?

SOLUTION   The probability of a spot being circular is 6,500/10,000 = 0.65. The probability of a spot being uniform is 7,000/10,000 = 0.70. Note that these are classical probabilities, not estimates, because in this case the spots on the array constitute the entire sample space. We need to calculate the probability of the spot being either circular or uniform. However, it is clearly stated in the problem that some spots are both circular and uniform. If we just add the probabilities of circular spots and uniform spots, the result would not be correct since spots having both properties will be counted twice, as shown in Fig. 4.8. The correct result can be obtained by adding the number of circular and uniform spots but subtracting the number of spots that are both circular and uniform. Thus, there will be 6,500 + 7,000 - 5,000 acceptable spots. In consequence, the probability that a spot is trustworthy on this array is:

$$P(trustworthy) = 0.65 + 0.70 - 0.50 = 0.85$$

☐

In general, the **addition rule** states that if A_1 and A_2 are two events, the probability of either A_1 or A_2 equals the probability of A_1, plus the probability of A_2 minus the probability of both happening at the same time:

$$P(A_1 \cup A_2) = P(A_1) + P(A_2) - P(A_1 \cap A_2) \tag{4.27}$$

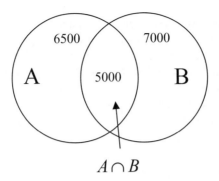

$$A \cap B$$

FIGURE 4.8: A Venn diagram. A is the set of 6,500 circular spots, B is the set of 7,000 uniform spots and $A \cap B$ is the set of 5,000 spots that are both circular and uniform. Just adding the probabilities of A and B would count the spots in $A \cap B$ twice.

4.4.1.2 Conditional probabilities

Sometimes, two phenomena influence each other. Let us consider two events caused by these two phenomena, respectively. When one such event occurs, the probability of the second event is changed. The notion of conditional probability is needed in order to describe such dependencies. Let us consider two events, i.e. A and B with non-zero probabilities. The **conditional probability** of event B given that event A has already happened is:

$$P(B|A) = \frac{P(A \cap B)}{P(A)} \tag{4.28}$$

Intuitively, the conditional probability of B given A, $P(B|A)$ stands for the probability of B and A occurring together as a percentage of the probability of A. The notion of conditional probability is useful in various situations as illustrated in the following example.

EXAMPLE 4.2
Let us consider the following experiment. During a reverse transcription reaction of a mRNA fragment to a cDNA, a dye is incorporated into the newly synthesized fragment, called a target. Dyes hybridize differentially depending on their chemical properties. Let the dye Y be incorporated with probability 0.9 into a given target. The probability that we read a signal from a spot is 0.8. What is the probability that a labelled target hybridizes on a spot?

SOLUTION Let labelling be event A and hybridization event B. In order to detect a signal, it is necessary for the spot to be hybridized with a labelled target. Therefore, if we observed a signal, both events A and B occurred for that particular spot. In other words, 0.8 is the probability of A and B occurring together $P(A \cap B)$. The labelling occurs first and does not depend on anything else. Its probability is $P(A) = 0.9$. The question refers to a labelled target hybridizing on the array. In other words, the requested quantity is the probability of the hybridization occurring given that the labelling occurred already. This is $P(B|A)$. From Eq. 4.28:

$$P(B|A) = \frac{P(A \cap B)}{P(A)} = \frac{0.8}{0.9} = 0.88 \qquad (4.29)$$

$\square$

Two events that do not influence each other are called **independent events**. Suppose that the probability of event B is not dependent on the occurrence of event A or, simply, that B and A are independent events. Two events A and B are independent if and only if (iff):

$$P(B|A) = P(B) \qquad (4.30)$$

This is read as "the probability of B occurring given that A has already occurred is equal to the probability of B alone." In other words, B "does not care" whether A has happened or not. The "if and only if" formulation has the precise meaning of logical equivalence. This means that the definition can be used both ways. If we know the probabilities involved, $P(B|A)$ and $P(B)$ and they do equal each other, then we can conclude that the two events A and B are independent. Conversely, if we know a priori that A and B are independent and we know only one of $P(B|A)$ and $P(B)$, we can calculate the unknown one based on (4.30) above.

Let us use the definition of independent events (4.30) together with the definition of conditional probabilities (4.28):

$$P(B|A) = P(B)$$

$$P(B|A) = \frac{P(A \cap B)}{P(A)}$$

Essentially, these are two different expressions for the same quantity $P(B|A)$. Therefore, the two expressions must be equal in the given circumstances:

$$P(B) = \frac{P(A \cap B)}{P(A)} \qquad (4.31)$$

From this, we can extract $P(A \cap B)$ as:

$$P(A \cap B) = P(A) \cdot P(B) \qquad (4.32)$$

This is important. If two events A and B are independent, the probability of them happening together is the simple product of their probabilities. This result can be used in various ways as shown by the following two examples.

EXAMPLE 4.3

Assume that the probability that a certain mRNA fragment incorporates the dye is 0.9, the probability that the fragment hybridizes on a given spot is 0.95 and the probability that we observe it using a microarray experiment is 0.8. Are hybridization and labelling independent?

SOLUTION   If they were independent, the probability of them happening at the same time would be:

$$P(A \cap B) = 0.9 \cdot 0.95 = 0.855$$

Since this probability was given as 0.8, the two protocol steps are not independent. ▯

EXAMPLE 4.4

Assume two unrelated genes are expressed 85% and 75% of the time in the condition under study. What is the probability that both genes are expressed at the same time in a given sample?

SOLUTION   Since the genes are unrelated, the two events are considered to be independent. The probability of both genes being expressed is then:

$$P(A \cap B) = 0.85 \cdot 0.75 = 0.6375$$

▯

4.4.1.3   General multiplication rule

Let us revisit briefly the definition of the conditional probability in (4.28). From this expression, we can extract the probability of the intersection of two events as:

$$P(A \cap B) = P(A) \cdot P(B|A) \qquad (4.33)$$

This can be interpreted as follows: the probability of A and B occurring together is equal to the probability of A occurring times the probability of B occurring given that A has already occurred. This is **the general multiplication rule**.

EXAMPLE 4.5

Assume that the probability that a certain mRNA fragment incorporates the dye is 0.9 and the probability that the labelled fragment hybridizes on a given spot is 0.888. What is the probability to obtain a spot with a non-zero signal?

SOLUTION   The probability of incorporating the dye is the probability of event A (labelling). The probability that the labelled fragment hybridizes is the prob. that the hybridization takes place given that the labelling is successful, i.e. the conditional probability of event B (hybridization) given A (labelling). In order to have a non-zero signal we need both A (labelling) and B (hybridization). The probability of this is:

$$P(A \cap B) = P(B|A) \cdot P(A) = 0.888 \cdot 0.9 = 0.7999$$

☐

4.5 Bayes' theorem

In 1763, the Royal Society published an article entitled "An Essay towards Solving a Problem in the Doctrine of Chances" by the Reverend Thomas Bayes [23, 24, 25]. This article and the approach presented in it has had the most profound effect on the science of the next 250 years. Today, perhaps more than ever, statisticians divide themselves into Bayesians and non-Bayesians according to how much they use Bayes' approach.

Bayes' theorem can be explained in very technical terms. The truth is that, like many great ideas, Bayes' theorem is really simple. Let us consider two events, A and B. Let us assume that both events have been observed. This situation could have come into existence in two ways: i) either event A happened first followed by event B or ii) event B happened first followed by event A. Let us follow the first alternative. The probability of event A happening on its own is $P(A)$. The probability of B happening after A is, by definition, the conditional probability $P(B|A)$ or the probability of B given A. Since we observed both A and B, they both must have happened so the probability of this event can be obtained by multiplying the two probabilities above:

$$P(A) \cdot P(B|A) \tag{4.34}$$

Let us now follow the second possible chain of events: event B happened first, followed by event A. The probability of event B happening on its own is $P(B)$. The probability of A happening after B is again by definition, $P(A|B)$. The probability of observing both A and B in this case is:

$$P(B) \cdot P(A|B) \tag{4.35}$$

From a probabilistic standpoint, it does not matter whether the chain of events was A followed by B or B followed by A. Therefore, the two probabilities must be equal:

$$P(A) \cdot P(B|A) = P(B) \cdot P(A|B) \tag{4.36}$$

From this, we can extract one of the terms, for instance $P(B|A)$:

$$P(B|A) = \frac{P(B) \cdot P(A|B)}{P(A)} \tag{4.37}$$

This is Bayes' theorem. Now that we gained an understanding of it, we can read it the way it is usually read: the probability of an event B occurring given that event A has occurred is equal to the probability of event A occurring given that event B has occurred, multiplied by the probability of event B occurring and divided by the probability of event A occurring.

The event B is usually associated to a cause whereas A is usually an observed event. Upon the observance of A, one goes back and assesses the probability of the causal hypothesis B. The terms $P(A)$ and $P(B)$ are called the a priori[4] probabilities of A and B, respectively. The term $P(B|A)$ is sometime called the a posteriori probability of B. The same equation may appear more intuitive if written using the notation H for hypothesis, and O for observed event:

$$P(H|O) = \frac{P(H) \cdot P(O|H)}{P(O)} \tag{4.38}$$

Furthermore, the probability of the observed event $P(O)$ can be written as:

$$P(O) = P(H) \cdot P(O|H) + P(nonH) \cdot P(O|nonH) \tag{4.39}$$

which is to say that the event O may happen if the hypothesis H is true (first term) or if the hypothesis H is not true (second term). Using this for the denominator of Eq. 4.38, the Bayes' theorem can now be written as:

$$P(H|O) = \frac{P(H) \cdot P(O|H)}{P(H) \cdot P(O|H) + P(nonH) \cdot P(O|nonH)} \tag{4.40}$$

The following example will illustrate a typical use of Bayes' theorem.

EXAMPLE 4.6
Let us assume a disease such as cervical cancer, with an incidence of about 1 case in 5000 women, is to be screened for using a testing procedure that provides positive results for 90% of the women who have the disease (true positives) but also for 0.5% of the women who do not have the disease (false positives). Let O be the event of a positive test outcome and H the event corresponding of having cervical cancer. What is the probability that a woman who tested positive actually has the disease?

SOLUTION We would like to use Bayes' theorem as expressed in Eq. 4.40. The term $P(H)$ is the probability of a woman having cancer which is

$$P(H) = \frac{1}{5000} = 0.0002 \tag{4.41}$$

The term $P(O|H)$ is the probability of a positive test given cancer, i.e.

$$P(O|H) = 0.9 \tag{4.42}$$

---
[4]The terms "a priori" and "a posteriori" actually mean "before the event" and "after the event," respectively.

The denominator of Eq. 4.40 is the probability of observing a positive test result $P(O)$ which can be expressed as:

$$P(O) = P(H) \cdot P(O|H) + P(nonH) \cdot P(O|nonH) \qquad (4.43)$$

The first two terms in this expression can be calculated using Eq. 4.41 and 4.42. The term $P(nonH)$ is the probability of a woman not having cancer. This is:

$$1 - P(H) = 0.9998 \qquad (4.44)$$

Finally, the term $P(O|nonH)$ is the probability of a positive outcome given that there is no cancer. This is the probability of having a false positive and is:

$$P(O|nonH) = 0.005 \qquad (4.45)$$

From Eq. 4.40, we can now calculate the probability for a woman who tested positive actually to have the disease as:

$$P(H|O) = \frac{0.0002 \cdot 0.9}{0.0002 \cdot 0.9 + 0.9998 \cdot 0.005} = \frac{0.00018}{0.00018 + 0.0049999} = 0.034$$
$$(4.46)$$

The conclusion is that fewer than 4% of the women with a positive test result actually have the disease. An examination of Eq. 4.46 reveals that the problem is not the test itself. A test yielding 90% true positives and 0.5% false positives is not necessarily a bad test. The problem is caused by the very low probability of having the disease $P(H) = 0.0002$. It is said that such a disease has a low prevalence (see also Chap. 13).
$\square$

Interestingly, in problems such as the one in the example above, most intuitive assessments would estimate the probability of having cancer in the case of a positive test result to a value very close to the true positive rate of the test (90% in this example) [123]. This tendency of our intuition to underestimate, sometimes even ignore, certain population characteristics such as the prior probability of an event is known as **base-rate fallacy**. Gigerenzer et al. argue that the people's poor performance in estimating probabilities such as in the case of base-rate fallacy is related to the "natural sampling" approach. The argument is that as humans evolved, the natural format of information was frequency. Thus, the human brain would be very accurate in acquiring and processing information provided as frequencies and much less accurate when the information is provided as probabilities [123].

4.6 Probability distributions

This section will introduce the notions of random variables and probability distributions. A few classical distributions including the binomial, normal and standard normal distributions will be discussed. These distributions are particularly important due to the large variety of situations for which they are suitable models. The discussion will focus on the general statistical reasoning as opposed to the particular characteristics of any given distribution.

A **random variable** is a variable whose numerical value depends on the outcome of the experiment, which cannot be predicted with certainty before the experiment is run. Examples include the outcome of rolling a die or the amount of noise on a given microarray spot. Random variables can take **discrete** values such as the die roll or the number of spots contaminated by noise on a given microarray, or **continuous** values such as the amount of noise on a given microarray spot. The behavior of a random variable is described by a function called a density function or probability distribution.

4.6.1 Discrete random variables

The **probability density function (pdf)** of a discrete variable X, usually denoted by f, is given by the probability of the variable X taking the value x:

$$f(x) = P(X = x) \tag{4.47}$$

where x is any real number.

The pdf of a discrete variable is also known as **discrete mass function**. In this analogy, $f(x)$ is the mass of the point x.

The pdf of a discrete variable is defined for all real numbers but is zero for most of them since X is discrete and can take only integer values. Also the pdf is a probability and therefore $f(x) \geq 0$. Finally, summing f over all possible values has to be 1. This corresponds to the probability of the universal event containing all possible values of X. No matter which particular value X will take, that value will be from this set and therefore this event is certain and has probability 1. In other words:

$$\sum_{all\ x} f(x) = 1$$

.

EXAMPLE 4.7
A fair (unbiased) 6 face die is rolled. What is the discrete pdf (mass function) of the random variable representing the outcome?

SOLUTION   This variable is uniformly distributed since the die is unbiased. The pdf is zero everywhere with the exception of the following values:

value	1	2	3	4	5	6
probability	1/6	1/6	1/6	1/6	1/6	1/6

▯

The **mean** or **expected value** of a variable can be calculated from its pdf as follows:

$$\mu = E(x) = \sum_x x \cdot f(x) \tag{4.48}$$

Furthermore, if a function H depends on the discrete random variable X, the expected (mean) value of H can also be calculated using the pdf of X as follows:

$$E(H(x)) = \sum_x H(x) \cdot f(x) \tag{4.49}$$

The typical example is the computation of the expected value to be gained by buying raffle tickets. Let us assume there is only one prize, the big jackpot, and this prize is worth \$1,000,000. Let us also assume that there are 10,000,000 tickets sold. The probability of having the winning ticket is 1/10,000,000 and the expected value is \$1,000,000/10,000,000 = \$0.10. Clearly, such a raffle would not help saving money for one's retirement if the ticket costs more than 10 cents.

EXAMPLE 4.8
What is the expected value of the sum of two unbiased dice?

SOLUTION   The possible outcomes are:

$$1 + 1, 1 + 2, 2 + 1, 2 + 2, 1 + 3, 3 + 1, \ldots$$

The probabilities are shown in the table below. The pdf is shown in Fig. 4.9.

sum	2	3	4	5	6	7	8	9	10	11	12
prob.	1/36	2/36	3/36	4/36	5/36	6/36	5/36	4/36	3/36	2/36	1/36

The expected value can be calculated as:

$$\mu = 2 \cdot \frac{1}{36} + 3 \cdot \frac{2}{36} + 4 \cdot \frac{3}{36} + \cdots + 11 \cdot \frac{2}{36} + 12 \cdot \frac{1}{36} = \frac{252}{36} = 7 \tag{4.50}$$

Note that the value 7 corresponds to the peak in Fig. 4.9

$\square$

The **cumulative distribution function** (or **cdf**) of a discrete random variable X is:

$$F(x) = P(X \le x) \tag{4.51}$$

for all real values x.

For a specific real value x_0 and a discrete random variable X, $P[X \le x_0] = F(x_0)$ can be found by summing the density $f(x)$ over all possible value of X less than or equal to x_0. This can be written as:

$$F(x_0) = \sum_{x \le x_0} f(x) \tag{4.52}$$

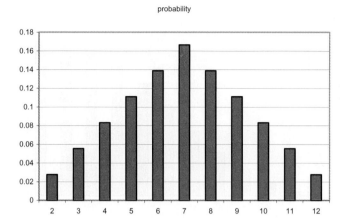

probability

FIGURE 4.9: The pdf of a discrete random variable representing the sum of two fair dice. The horizontal axis represents the possible values of the variable; the vertical axis represents the probability.

EXAMPLE 4.9

What is the probability of rolling a value less or equal than 4 using a pair of unbiased dice?

SOLUTION   From the table in example 4.8, the values of the pdf are as follows:

sum	2	3	4	5	6	7	8	9
prob.	1/36	2/36	3/36	4/36	5/36	6/36	5/36	...
cum. prob.	0.027	0.083	0.166	0.277	0.416	0.583	0.722	...

The cdf is shown in Fig. 4.10.

☐

4.6.2   Binomial distribution

The **binomial distribution** is the most important discrete variable distribution in life sciences. This makes it the preferred example distribution in the context of this book. However, the main issues to be learned from this section are related to the general concept of a distribution and the use of a distribution to answer particular statistical questions.

Any distribution will be appropriate for use in a particular statistical framework. Usually, this framework is defined by a set of assumptions. When addressing a specific statistical question, identifying the correct characteristics of the problem is essential in order to choose the appropriate distribution. Furthermore, any distribution

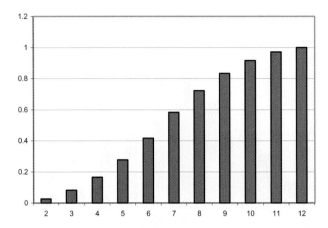

FIGURE 4.10:   The cdf of a discrete random variable representing the sum of two fair dice.

will be characterized by a probability density function (pdf) and a cumulative probability density function (cdf). It is important to recall that given a real value x, **the pdf is the function that gives the probability that a variable following the given distribution (e.g. binomial) takes the value** x. Given the same real value x, **the cdf is the function that gives the probability that the variable following the given distribution takes a value less than or equal to** x. According to the definition, the cdf can be obtained by calculating the sum of all pdf values from $-\infty$ to the value x. In other words, the value of the cdf in any particular point x is equal to the area under the curve of the pdf to the left of the given point x. This general relationship between the pdf and cdf of a given distribution is illustrated in Fig. 4.11. A few other general properties can be observed in the same figure. Most pdfs will tend to zero towards the extremities of their domain (x axis). This is because in most distributions, really extreme values have a low probability of occurring. A notable example of a distribution for which this is not true is the uniform distribution (see Example 4.7) in which the pdf is constant for the whole range of the variable.

Most cdfs will start at zero and will increase to one. This is because when we are considering an extremely large value (a large value of a in Fig. 4.11) it is very likely to get a value less than or equal to it.

The binomial distribution takes is name from "bi" which means two and "nomen" which means name. Like any other distribution, the binomial distribution describes a phenomenon with certain characteristics. The characteristics of the phenomenon are also assumptions that one makes when one applies the binomial distribution to any specific problem. The results obtained by using the binomial distribution will be accurate only if the phenomenon studied satisfies these assumptions.

The binomial distribution describes a model with the following four assumptions:

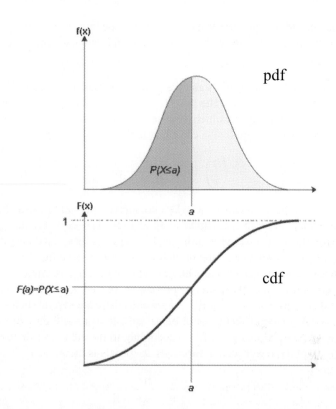

FIGURE 4.11: The relationship between the pdf and the cdf of a given distribution. The value of the cdf is at any point a, equal to the area under the curve of the pdf to the left of point a. This is because the pdf provides the probability that the variable takes precisely the value a whereas the cdf gives the probability that the variable takes a value less than or equal to a.

1. A **fixed** number of trials are carried out.

2. The trials are **independent**, i.e. the outcome of one trial does not influence in any way the outcome of successive trials.

3. The outcome of each trial can be **only one of two mutually exclusive categories**. Usually, these outcomes are labelled "success" and "failure."

4. The **probability of success is fixed**. If we assume this probability to be p, then the probability of failure will be $1 - p$ since the two categories are mutually exclusive.

Let us assume we carry out n trials and we have X successes. Clearly, X has to be less than or equal to n: $X \leq n$. The pdf $f(x)$ is the probability of x successes in n trials. There are several ways in which x successes can occur in n trials. The number of such combinations is given by the combinatorial formula:

$$\binom{n}{x} = \frac{n!}{x! \cdot (n - x)!} \tag{4.53}$$

The notation $\binom{n}{x}$ is read "n choose x." The notation $n!$ is read "n factorial" and is equal to the product of all integers up to n: $n! = 1 \cdot 2 \cdot 3 \cdots (n - 1) \cdot n$.

We would like to calculate the probability of having x successes and $n - x$ failures. Since the events are independent, the probabilities multiply (see Eq. 4.32). We need to have x successes when each success has probability p and $n - x$ failures when each failure has probability $1-p$. We obtain the probability of having such a combination by multiplying these values: $p^x \cdot (1 - p)^{n-x}$. However, there are several such alternative combinations. Any one of them would be acceptable to us and they cannot occur together, so the probabilities will add up according to the addition rule in Eq. 4.27. The probability of having exactly x successes in n trials is therefore:

$$f(x) = \binom{n}{x} \cdot p^x \cdot (1 - p)^{n-x} = \frac{n!}{x! \cdot (n - x)!} \cdot p^x \cdot (1 - p)^{n-x} \tag{4.54}$$

This is **the probability distribution function (probability mass function) of the binomial distribution**. The exact shape of the distribution depends on the probability of success in an individual trial p. Fig. 4.12 illustrates the probability of having x successes in 15 trials if the probability of success in each individual trial is $p = 0.5$. Fig. 4.13 shows the cdf of the same distribution. Fig. 4.14 and Fig 4.15 show the pdf and cdf corresponding to having x successes in 15 trials if the probability of success in each individual trial is $p = 0.2$. Note that the mode (the peak) of the pdf corresponds to the expected value. In the first case, the mode is $15 \cdot 0.5 = 7.5$ whereas in the second case the maximum probability is at $15 \cdot 0.2 = 3$.

In general, if X is a random variable that follows a binomial distribution with parameters n (number of trials) and p (probability of success), X will have the **expected value**:

$$E[X] = np \tag{4.55}$$

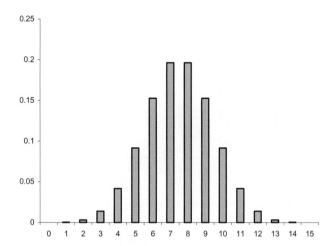

FIGURE 4.12: The pdf of a binomial distribution with the probability of success $p = 0.5$. The horizontal axis represents the possible values of the variable; the vertical axis represents the probability.

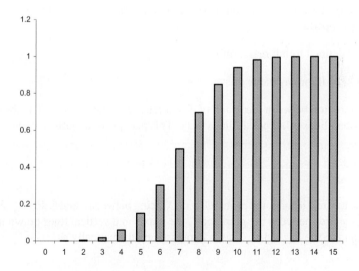

FIGURE 4.13: The cdf of a binomial distribution with the probability of success $p = 0.5$.

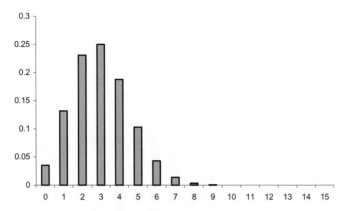

FIGURE 4.14:   The pdf of a binomial distribution with the probability of success $p = 0.2$.

and the **variance**:

$$Var(X) = np(1 - p) \tag{4.56}$$

4.6.3   Continuous random variables

Continuous random variables are random variables that can take any values in a given interval. The pdf $f(x)$ for a continuous random variable X is a function with the following properties:

1. f is defined for all real numbers.

2. $f(x) \geq 0$ for any value of x.

3. The region between the graph of f and the x axis has an area of 1 (the sum of probabilities of all possibilities is 1). This is expressed mathematically as:

$$\int_{all\ x} f(x) = 1 \tag{4.57}$$

4. For any real a and b, the probability of X being between a and b, $P(a \leq X \leq b)$, is the area between the graph of f, the x axis and vertical lines drawn through a and b (see Fig. 4.16). This is expressed mathematically as:

$$P(a \leq X \leq b) = \int_{x=a}^{b} f(x)\,dx \tag{4.58}$$

Note that if we tried to apply the definition of the discrete pdf to a continuous variable (see Eq. 4.47), the result would be zero:

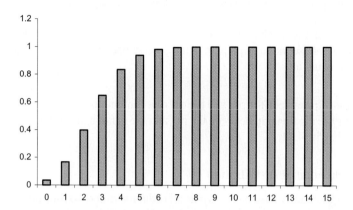

FIGURE 4.15:   The cdf of a binomial distribution with the probability of success $p = 0.2$.

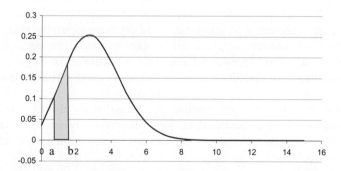

FIGURE 4.16:   For any real a and b, the probability of X being between a and b, $P(a \le X \le b)$, is the area between the graph of f, the x axis and vertical lines drawn through a and b.

$$f(x) = P(X = x) = 0 \tag{4.59}$$

since the probability of a continuous variable taking any given precise numerical value is zero.

The cdf for a continuous random variable X is (same as for a discrete random variable) a function F whose value at point x is the probability the variable X takes a value less than or equal to x:

$$F(x) = P(X \leq x) \tag{4.60}$$

The relationship between the pdf and the cdf of a continuous random variable remains the same as the relationship between the pdf and the cdf of a discrete random variable illustrated in Fig. 4.11.

4.6.4   The normal distribution

The most important continuous distribution is the **normal distribution**. The normal distribution is important because many natural phenomena follow it. In particular, most biological variables, such as the variation of anatomical and physiological parameters between different individuals of the same species, follow this distribution. It is also reasonable to believe that differences in the gene expression levels between various individuals in the same conditions also follow a normal distribution. Another common assumption is that the noise in a microarray experiment is normally distributed although this may not always be true.

The normal distribution is also called a Gaussian distribution and has a characteristic bell shape shown in Fig. 4.18. Its corresponding cdf is shown in Fig. 4.19. The normal distribution has two parameters: the mean and the standard deviation. The mean is also called the location of the curve since changing this value will move the whole curve to a different location on the x axis. The standard deviation determines the exact shape of the curve. A smaller standard deviation means the values are less dispersed. In consequence, the graph will be narrower and taller, because the total area under the curve has to remain the same (see the definition of the pdf). A larger standard deviation will determine a shorter but wider curve. This is illustrated in Fig. 4.17.

There is also an analytical expression that give the pdf of the normal distribution as a function of its mean and standard deviation. The probability density function for a normal random variable with mean μ and standard deviation σ has the form:

$$f(x) = \frac{1}{\sigma\sqrt{2\pi}} \cdot e^{-\frac{(x-\mu)^2}{2\sigma^2}} \tag{4.61}$$

The normal distribution has several important characteristics as follows:

1. The distribution is symmetrical around the mean.

2. Approximatively 68% of the values are within 2 standard deviations from the mean (i.e. less than 1 standard deviation either to the left or to the right of the mean: $x - \mu < \pm\sigma$).

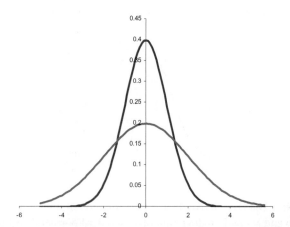

FIGURE 4.17: The probability density function of two normal distributions. The taller one has a mean of zero and a standard deviation of 1. The shorter one has a mean of zero and a standard deviation of 2.

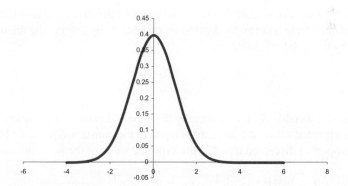

FIGURE 4.18: The probability density function of a normal distribution with a mean of zero and a standard deviation of 1 (standard normal distribution).

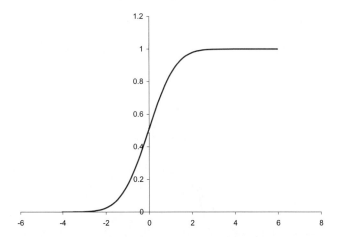

FIGURE 4.19: The cumulative distribution function of a normal distribution with a mean of zero and a standard deviation of 1 (standard normal distribution).

3. Approximatively 95% of the values are within 4 standard deviations from the mean ($x - \mu < \pm 2\sigma$).

4. Approximatively 99% of the values are within 6 standard deviations from the mean ($x - \mu < \pm 3\sigma$).

5. The inflexion points of the curve (where the curve changes from bending up to bending down) occur at $\mu \pm \sigma$.

As an exercise, let us calculate the mean and standard deviation of a new variable, Z, that is obtained from a normally distributed variable X by subtracting the mean and dividing by the standard deviation:

$$Z = \frac{X - \mu}{\sigma} \tag{4.62}$$

The mean of a variable X, μ_X, is actually the expected value of that variable, $E[X]$. Also, the expected value of a constant is equal to the constant $E(c) = c$. Using this and the properties discussed in 4.4.1, the expected value of the new variable Z is:

$$\mu_Z = E\left[\frac{X - \mu_X}{\sigma_X}\right] = \frac{1}{\sigma_X} E\left[X - \mu_X\right] = \frac{1}{\sigma_X} \left(E\left[X\right] - E\left[\mu_X\right]\right) =$$

$$= \frac{1}{\sigma_X}(\mu_X - \mu_X) = 0 \tag{4.63}$$

Therefore, the mean of the new variable Z will be zero. Note that this will happen independently of the mean of the initial variable X.

The standard deviation of the new variable will be:

$$\sigma_Z^2 = E\left[(Z - \mu_Z)^2\right] = E[Z] = E\left[\left(\frac{X - \mu_X}{\sigma_X}\right)^2\right] =$$

$$= \frac{1}{\sigma_X^2} E\left[(X - \mu_X)^2\right] = \frac{1}{\sigma_X}\sigma_X = 1 \qquad (4.64)$$

Therefore, any normal variable can be mapped into another variable distributed with mean zero and standard deviation of one. This distribution is called **the standard normal distribution** and its probability density function can be expressed as:

$$f(x) = \frac{1}{\sqrt{2\pi}} \cdot e^{-\frac{x^2}{2}} \qquad (4.65)$$

This distribution is shown in Fig. 4.18.

4.6.5 Using a distribution

Distributions can be used directly to answer a number of simple questions. However, in order to use a given distribution to answer such questions, one has to make sure that the distribution is appropriate by checking the validity of its assumptions. Typical questions that can be asked directly are:

1. What is the probability that the variable has a value higher than a certain threshold t?

2. What is the probability that the variable has values between two thresholds t_1 and t_2?

3. What is the threshold that corresponds to a certain probability?

Tables summarizing various distributions or computer programs able to calculate pdfs and cdfs are used in order to answer such questions. Tables are usually organized by the values of the random variable. For instance, a table for the normal distribution will have rows for different values of the variable Z (e.g. 0.0, 0.1, 0.2, etc.). The columns will correspond to the third digit of the same variable Z (e.g. the value situated at the intersection of row 1.1 and column 0.05 corresponds to the probability of $Z \leq 1.15$). Most general purpose spreadsheets and pocket calculators with statistical capabilities will also be able to provide the necessary values. In Excel, such values can be obtained quickly using the function "normdist(x, mean, standard deviation, TRUE)." If Excel is available, it is not necessary to standardize the variable since the mean and standard deviation can be specified explicitly. The last parameter chooses between pdf ("FALSE") and cdf ("TRUE").

EXAMPLE 4.10
The distribution of mean intensities of cDNA spots corresponding to genes that are expressed can be assumed to be normal [193]. Let us assume that the mean of this

distribution is 1000 and the standard deviation is 150. What is the probability that an expressed gene has a spot with a mean intensity of less than 850?

SOLUTION   Firstly, let us note that the problem statement contains two different means: the mean intensity of a spot and the mean of the spot distribution. The mean intensity of the spot is not particularly relevant for the question considered. It merely informs us that given a particular spot, the various pixel intensities of a spot have been combined in a single number by taking their mean. The mean of the pixel intensities is the value chosen to represent the expression level of a gene in this particular example. The value that we need to focus on is the mean of the spot intensities distribution. If we are to use a table, we need to transform this problem involving a particular normal distribution ($\mu = 1000, \sigma = 150$) into an equivalent problem using the standard normal distribution ($\mu = 0, \sigma = 1$). In order to do this, we apply the transformation in Eq. 4.63 and Eq. 4.64 by subtracting the mean and dividing by the standard deviation:

$$Z = \frac{X - \mu}{\sigma} = \frac{850 - 1000}{150} = -1 \qquad (4.66)$$

The Z transformation above maps problems such as this from any normal distribution to the one *standard* normal distribution. The normal distribution had its horizontal axis graded in units corresponding to the problem, in our case, pixel intensities. If we were to represent the Z variable, the distribution (see Fig. 4.20) would have the axis graded in standard deviations. Equation 4.66 emphasized that our problem corresponds to determining the probability that a variable following a normal distribution has a value less than 1 standard deviation below its mean:

$$P(X < 850) = P(Z < -1.0) \qquad (4.67)$$

This corresponds to the shaded area in Fig. 4.20. From a table or from a pocket calculator, we can obtain the exact value of this probability: 0.1587. In other words, approximatively 15% of the spots corresponding to expressed genes will have a mean intensity below 850 or below one standard deviation from their mean.   ▯

EXAMPLE 4.11
The distribution of mean intensities of cDNA spots corresponding to genes that are not expressed can be assumed to be normal [193]. Let us assume that the mean of this distribution is 400 and the standard deviation is 150. What is the probability that an unexpressed gene has a spot with a mean intensity of more than 700?

SOLUTION   We transform the variable X into a standard normal variable by subtracting the mean and dividing by the standard deviation:

$$Z = \frac{X - \mu}{\sigma} = \frac{700 - 400}{150} = 2 \qquad (4.68)$$

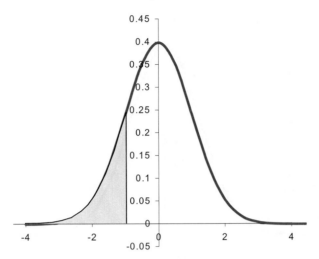

FIGURE 4.20: The shaded area under the graph of the pdf and to the left of the -1 value corresponds to the probability of the X variable having a value less than one standard deviation below its mean.

From a table of the standard normal distribution or from a calculator, we obtain the value 0.9772. A mechanical application of the reasoning used in the example above might conclude that the probability of unexpressed genes having spots with an average intensity higher than 700 is 97.72%. Clearly, this cannot be true since it is impossible that 97% of normally distributed values be higher than their own mean of 400.

Fig. 4.21 illustrates the problem: we are interested in the probability of our variable being higher than the given threshold (in the shaded area on the figure); however, the cdf provides by definition the probability of the variable Z being less than or equal to the threshold $z = 2$ which corresponds to the unshaded area in the figure. However, the figure also suggests the solution to this problem. Since we know that the area under the curve is equal to 1 (from the definition of a pdf), we can obtain the value of interest by subtracting the value provided by the cdf (unshaded) from 1:

$$P(Z > 2) = 1 - P(Z \le 2) = 1 - 0.9772 = 0.0228 \qquad (4.69)$$

In conclusion, the probability that an unexpressed gene has a high intensity spot (higher than their mean plus 2 standard deviations) is about 2.3%. ▯

The two examples above showed that it is possible for expressed genes to have lower than usual intensity spots as well as for unexpressed genes to have spots with an intensity higher than usual. Let us now consider an array containing both expressed and unexpressed genes and let us also assume that the parameters of each distribution are not changed by the presence of the other. In other words, we assume the two

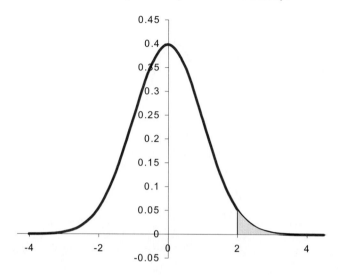

FIGURE 4.21: The shaded area under the graph of the pdf corresponds to the probability that we are interested in ($P(Z > 2)$). However, the cdf of the function provided a value corresponding to the unshaded ares to the left of 2 ($F(x) = P(Z \leq 2)$). The value of interest can be calculated as $1 - cdf = 1 - P(Z \leq 2)$.

distributions are independent. This situation is illustrated in Fig. 4.22. As before, the unexpressed genes are normally distributed with a mean of 400 and a standard deviation of 150. The expressed genes are normally distributed with a mean of 1000 and a standard deviation of 150. Note that there are expressed genes with values less than 700 and unexpressed genes with a value less than 700. This shows that in most cases, distinguishing between expressed and unexpressed genes is not as simple as it might seem. The following chapters will discuss in detail how this can be done. However, a conclusion that should already be apparent is that we will have to make some compromises and be prepared to accept the fact that we might make some mistakes. This, in itself, is not very troublesome as long as we know exactly how likely we are to make a mistake when we draw a given conclusion.

4.7   Central limit theorem

Let us revisit the example used to illustrate the concept of mode. We considered the following set of measurements for a given population: 55.20, 18.06, 28.16, 44.14, 61.61, 4.88, 180.29, 399.11, 97.47, 56.89, 271.95, 365.29, 807.80, 9.98, 82.73. We calculated the population mean as 165.570, and we considered two samples from this

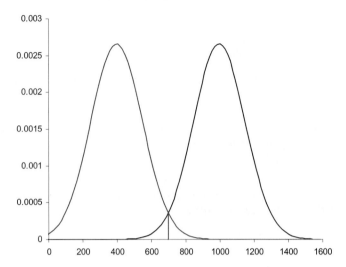

FIGURE 4.22: An array having both expressed and unexpressed genes. The unexpressed genes are normally distributed with a mean of 400 and a standard deviation of 150. The expressed genes are normally distributed with a mean of 1000 and a standard deviation of 150. Note that there are expressed genes with values less than 700 and unexpressed genes with a value less than 700.

population. We observed that two different samples can have means very different from each other and also very different from the true population mean. An interesting question is: what would happen if we considered, not only two samples, but *all possible samples of the same size*? In other words, what happens if we consider all possible ways of picking n measurements out of the population?

The answer to this question is one of the most fascinating facts in statistics. If the answer to this question were different from what it is, statistics as a science would probably not exist as we know it. It turns out that if we calculate the mean of each sample, those mean values tend to be distributed as a normal distribution, *independently on the original distribution*. Furthermore, the mean of this new distribution of the means *is exactly the mean of the original population* and the variance of the new distribution *is reduced by a factor equal to the sample size n*.

There are two important observations to be made. Firstly, the distribution of the original values X *can have any shape* (e.g. the skewed binomial distribution corresponding to a success rate of 0.2 in Fig. 4.14). Secondly, the result is true for *the means $\overline{X}$ of the subsets picked from the original X distribution*. For each subset picked, the mean $\overline{X}_i$ is calculated; this value is one sample in the new distribution.

This result is known as **the central limit theorem**:

THEOREM 4.1

*When sampling from a population with mean μ_X and variance σ_X^2, the distribution of the sample mean (or the **sampling distribution**) will have the following properties:*

1. *The distribution of $\overline{X}$ will be approximately normal. The larger the sample is, the more will the sampling distribution resemble the normal distribution.*

2. *The mean $\mu_{\overline{X}}$ of the distribution of $\overline{X}$ will be equal to μ_X, the mean of the population from which the samples were drawn.*

3. *The variance $\sigma_{\overline{X}}^2$ of the distribution of $\overline{X}$ will be equal to $\frac{\sigma_X^2}{n}$, the variance of the original population of X's divided by the sample size. The quantity $\sigma_{\overline{X}}$ is called **the standard error of the mean**.*

4.8   Are replicates useful?

There is a joke about an American tourist visiting a remote town in England. Strolling through the village, the tourist gets to the main town square where he notices two large towers, each of them having a large clock. He also notices that the two clocks are not showing the same time. Puzzled, he stops a local person and asks the time. "Can't you see the clocks?" the English man replies. "Yes, but they show different times!" says the confused tourist. "Which one is the correct one?" "Neither of them," replies the local. "If we could build a clock that could show us the exact time, we would only have that one."

Tongue in cheek, we can now use the powerful results of the central limit theorem to investigate this phenomenon through an example.

EXAMPLE 4.12

The distribution of mean intensities of cDNA spots corresponding to genes that are not expressed can be assumed to be normal [193]. Let us assume that the mean of this distribution is 400 and the standard deviation is 150. We are using an array that has each gene spotted in quadruplicates. We calculate the expression value of a gene as the mean of its replicates.

1. *What is the probability that a spot corresponding to an unexpressed gene has a value higher than 700?*

2. *What is the probability that an unexpressed gene has an expression value of 700 or higher?*

SOLUTION   In order to calculate the probability for one individual spot, we transform the variable X into a standard normal variable by subtracting the mean and dividing by the standard deviation as we did in example 4.11:

$$Z = \frac{X - \mu}{\sigma} = \frac{700 - 400}{150} = 2 \tag{4.70}$$

The probability of this happening is:

$$P(Z > 2) = 1 - P(Z \leq 2) = 1 - 0.9772 = 0.0228 \tag{4.71}$$

This value of 0.0228 or 2.28% may appear to be relatively small and, in consequence, not a cause for concern. However, an array typically contains thousands of spots. Two percent means 20 in 1,000 and 200 in 10,000. This *is* a good cause for concern. In a typical array, a researcher might look at 10,000 genes with the hope of finding about 50 or 100 genes that are differentially regulated in the condition under study. Two hundred unexpressed genes that have spots with high intensities due to various random causes will mix the 50 or 100 real positives with an additional 200 false positives. The researcher will have no way to distinguish between spots that have high intensity due to a genuinely higher level of expression and spots that have high intensity due to other random factors.

In order to answer the second question, we need to apply the central limit theorem since the second question is referring to the expression level of the gene which is obtained as a mean of 4 measurements. The standard error of the mean is:

$$\sigma_{\overline{X}} = \frac{\sigma_X}{\sqrt{n}} = \frac{150}{\sqrt{4}} = 75 \tag{4.72}$$

This means that the variance in the expression levels will be four times smaller than the variance of the spot intensities. With the standard error, we can now calculate the Z variable, mapping the problem to a standard normal distribution:

$$Z = \frac{\overline{X} - \mu_{\overline{X}}}{\sigma_{\overline{X}}} = \frac{\overline{X} - \mu_{\overline{X}}}{\frac{\sigma_X}{\sqrt{n}}} = \frac{700 - 400}{\frac{150}{\sqrt{4}}} = \frac{700 - 400}{150} \cdot 2 = 4 \tag{4.73}$$

This is equivalent to asking the probability of a normally distributed variable to take a value 4 standard deviations away from its mean. The probability of this happening is:

$$P(Z > 4) = 1 - F(4) = 1 - 0.999968 = 0.000032 \tag{4.74}$$

This probability is only 0.003%. An array containing 10,000 spots will only contain 2,500 genes (there is no such thing as a free lunch!!) but this array will give $0.000032 \cdot 2500 = 0.079$ false positives, i.e. *no false positives*.

In the given hypotheses, the choices are as follows. On the one hand, an array with 10,000 genes, each spotted once, interrogates 10,000 genes at the same time but probably many of the genes selected as expressed will not be so. On the other hand, an array spotting each gene in quadruplicates interrogates only 2,500 genes at a time but if the mean of the replicate spots of a gene indicate a high expression value, the genes will probably be truly highly expressed.

⬜

4.9   Summary

This chapter introduced a few necessary statistical terms such as: population, sample, parameters, random variables, etc. Several measures of central tendency (mean, median, mode, percentile) and variability (range, variance, standard deviation) were presented. The discussion emphasized various interesting properties of these quantities, the relationship between them and their utility in the context of microarray data. The chapter also introduced the notions of covariance and correlation, discussed their common usage as well as common mistakes associated to their usage.

The next section of the chapter introduced the notion of probabilities, conditional probabilities and operations with such. Bayes' theorem and a few related issues as well as a typical example of Bayesian reasoning were briefly reviewed. The chapter also discussed the notions of probability distributions, as well as a few specific distributions including the binomial and normal distributions.

Finally, the chapter discussed the central limit theorem and used it to explain through some examples why repeated measurements are more reliable than individual measurements.

4.10   Solved problems

1. We decide to declare a microarray spot as having high confidence if it is either perfectly circular or perfectly uniform (the standard deviation of the signal intensities is less than a chosen threshold) or both. There are 10000 spots of which 8500 are circular and 7000 are uniform. 5000 spots are both circular and uniform. What is the probability that a randomly chosen spot is good (has high confidence)?

   SOLUTION   P(circular) = 8500/10000 = 0.85

   P(uniform) = 7000/10000 = 0.70

   P(circular and uniform) = 5000/10000 = 0.50

   P(high confidence) = P(circular) + P(uniform) − P(circular and uniform)

   P(high confidence) = 0.85 + 0.70 − 0.5 = 1.05

   However, a probability cannot be larger than 1. There must be something wrong. Let us calculate the number of spots that are circular and not uniform: 8500 − 5000 = 3500.

   We can also calculate the number of spots that are uniform but not circular: 7000 − 5000 = 2000.

We can now sum up the circular but not uniform spots (3500) with the uniform but not circular spots (2000) and with the spots that are both circular and uniform (5000): $3500 + 2000 + 5000 = 10500$. This number is greater than the number of spots on the array which clearly indicates that at least one of the given measurements was incorrect.

This example emphasized the fact that in microarray applications, it is a good idea to put the results of the data analysis in the context of the problem and use them to verify the consistency of the data gathered. □

4.11 Exercises

1. The distribution of mean intensities of cDNA spots corresponding to genes that are not expressed can be assumed to be normal [193]. Let us assume that the mean of this distribution is 400 and the standard deviation is 150. We are using an array that has each gene spotted in triplicates. We calculate the expression value of a gene as the mean of its replicates.

   (a) What is the probability that a spot corresponding to an unexpressed gene has a value higher than 700?

   (b) What is the probability that an unexpressed gene has an expression value of 700 or higher?

2. The distribution of mean intensities of cDNA spots corresponding to genes that are expressed can be assumed to be normal [193]. Let us assume that the mean of this distribution is 1000 and the standard deviation is 100. We are using an array that has each gene spotted in triplicates. We calculate the expression value of a gene as the mean of its replicates.

   (a) What is the probability that a spot corresponding to an expressed gene has a value lower than 700?

   (b) What is the probability that an expressed gene has an expression value of 700 or lower?

3. The distribution of mean intensities of cDNA spots corresponding to genes that are expressed can be assumed to be normal [193]. Consider an array on which there are both expressed and unexpressed genes. Let us assume that the mean of the distribution corresponding to the expressed genes is 1000 and the standard deviation is 100; the mean of the distribution corresponding to the unexpressed genes is 400 and the standard deviation is 150 (usually there is higher variability in the low intensity spots). We are using an array that has each gene spotted in triplicates. We calculate the expression value of a gene as the mean of its replicates.

(a) What is the probability that a spot corresponding to an expressed gene has a value lower than 700?

(b) What is the probability that an expressed gene has an expression value of 700 or lower?

(c) Find the threshold T in the intensity range corresponding to the intersection point of the two distributions (the point for which the probability of having an expressed gene is equal to the probability of having an unexpressed gene).

Hint: this point is marked by a vertical bar in Fig. 4.22.

(d) What is the probability that an expressed gene has the mean of its replicates lower than T? (This would be a false negative.)

(e) What is the probability that an unexpressed gene has the mean of its replicates higher than T? (This would be a false positive.)

(f) What is the probability that an individual spot corresponding to an expressed gene has an intensity lower than T? (This would be a false negative on an array not using replicates.)

(g) What is the probability that an individual spot corresponding to an unexpressed gene has an intensity higher than T? (This would be a false positive on an array not using replicates.)

Chapter 5

Statistical hypothesis testing

The manipulation of statistical formulas is no substitute for knowing what one is doing.

—Hubert Blalock, Jr., Social Statistics, Chapter 19

5.1 Introduction

The goal of this chapter is to illustrate the process of testing a statistical hypothesis. Unfortunately, just being familiar with a number of statistical terms and a few distributions is not sufficient in order to draw valid conclusions from the data gathered. A clear formulation of the hypotheses to be tested as well as a clear understanding of the basic *mathematical* phenomena involved are absolutely necessary in order to be able to extract facts from data.

5.2 The framework

In order to illustrate the basic ideas involved in hypothesis testing, we will consider the following example.

EXAMPLE 5.1
The expression level of a gene in a given condition is measured several times. A mean $\overline{X}$ of these measurements is calculated. From many previous experiments, it is known that the mean expression level of the given gene in normal conditions is μ_X. We formulate the following hypotheses:

1. The gene is up-regulated in the condition under study: $\overline{X} > \mu_X$

2. The gene is down-regulated in the condition under study: $\overline{X} < \mu_X$

3. The gene is unchanged in the condition under study: $\overline{X} = \mu_X$

4. *Something has gone awry during the lab experiments and the gene measurements are completely off; the mean of the measurements may be higher or lower than the normal:* $\overline{X} \neq \mu_X$

This is already a departure from the lay thinking. Given two values $\overline{X}$ and μ_X, one could think that there are only 3 possibilities: $\overline{X} > \mu_X$, $\overline{X} < \mu_X$ or $\overline{X} = \mu_X$. This is indeed the case; the two values will necessarily be in only one of these situations. However, the emphasis here is not on the two values themselves but on *our expectations about them*. If we think along these lines, there are clearly four possibilities. The first 3 correspond to the relationship between the two variables: we might expect $\overline{X}$ to be higher than, equal to or lower than μ_X. However, the 4th possibility has nothing to do with the variable but reflects our teleological position: we have no clue!! The mean measurement might be either higher or lower than the real mean: $\overline{X} \neq \mu$. This is the first important thing to remember: *hypothesis testing is not centered on the data; it is centered on our* a priori *beliefs about it*. After all, this is the meaning of the term hypothesis: an *a priori* belief, or assumption, that needs to be tested.

The second important fact in hypothesis testing has to do with the fundamental characteristic of the statistical thinking. Let us consider a normal distribution like the one in Fig. 5.1. Let us consider that this distribution represents the (normalized) distribution of the expression levels of the gene under study. This distribution has the mean $Z = 0$. This corresponds to $\overline{X} = \mu_X$. We know that if the gene is up-regulated in the condition under study, it will have an expression value higher than normal. We would like to set a threshold to the right, such that, if a gene is higher than the given threshold, we can call it up-regulated. This is our goal.

We have seen that 95% or so of the values of a normal distribution will be situated within $\pm 2\sigma$. The fundamental fact that needs to be understood here is that X can actually take *any value whatsoever* even if it follows strictly a normal distribution. Yes, it is unlikely that a variable coming from a normal distribution with zero mean and $\sigma = 1$ takes a value larger than 2. Yes, it is *very* unlikely that it takes a value larger than 3. And it is even more unlikely that it will take a value larger than 5. However, this is not impossible. In fact, the probability of this event can be easily calculated from the expression of the normal distribution. Once this fact is accepted, the conclusion follows: no matter where we set the threshold, a member of the original, normal distribution may end up there just by random chance. Therefore, *independently of the value of the threshold, we can never eliminate the possibility of making a mistake*.

Once we understand this fact, we have to adjust our goals: we should not aim at finding a threshold that allows us to identify correctly the up-regulated genes since this is futile; instead, we should aim at being able to calculate the exact probability of making a mistake for any given threshold. There is a big difference between a value situated 0.5 standard deviation larger than the mean and a value 3 standard deviations larger than the mean. The former will probably occur about 1 in 3 measurements while the latter will probably appear only once in 1,000 measurements. If we were to bet against this, we might feel uncomfortable to do so against the former while betting against the latter would be much safer. However, not all situations are suitable for taking the same amount of risk. For instance betting at roulette with a chance of losing of only 1 in 3 is not bad. This would mean winning 2 out of every 3 turns. Many people would do that[1]. However, jumping from a plane with a parachute that will open only 2 out of 3 jumps would probably not gather many volunteers. Sometimes, even very, very small probabilities of a negative outcome are unacceptable if the cost of such an occurrence is very high. As a real life example, Ford Motor Co. has recently recalled 13,000,000 Firestone tires because of 174 deaths and 700 injuries related to the tire tread separation. The empirical probability of being injured or dying because of this was only $874/13,000,000 = 0.00006$ or less than 1 in 10,000. However, Ford decided this was unacceptable and recalled the tires for a total cost of approximatively 3,000,000,000 dollars.

We saw earlier that it is not possible to choose a fixed threshold for the measurement values. The examples above illustrated the fact that choosing an a priori, fixed, threshold is not possible even for the probability of making a mistake. Therefore, we will focus on what is both possible and necessary, that is calculating the probability of making a mistake for any given situation. Let us consider the situation illustrated in Fig. 5.1. Let us consider that this is the distribution of the gene expression levels in the normal situation (on controls). Let us consider that we set the threshold for up-regulated genes at 2 standard deviations away from the mean. If a gene has a measurement of, for instance, $Z = 4$, we will call that gene up-regulated. However, a gene from the normal distribution of controls might have such a high expression value. This is unlikely, of course, but it may happen. If this happens and we call such a gene up-regulated because its value is larger than the threshold, we will be making a mistake: even if the gene has a high expression level, it comes from the normal distribution. The probability of this happening is called the **p-value**. The p-value provides information about the amount of trust we can place in a decision made using the given threshold. A more exact definition of the p-value will be given shortly.

Calculating the p-value for a given threshold is not difficult and can be accomplished as follows. The probability of a value being higher than a threshold x is 1 minus the probability of it being lower than or equal to the given threshold:

$$P(Z > x) = 1 - P(Z \leq x) \tag{5.1}$$

Furthermore, the probability of a value from a distribution being lower than a given threshold x is given directly by its cdf or by the area under its pdf. In Fig. 5.1, the shaded area gives the probability of a value from the given distribution being larger than 2. This is exactly the probability of making a mistake if we used 2 as the threshold for choosing up-regulated genes, or the p-value associated to this threshold.

---

[1]Heck, many people play roulette anyway, even if the chances of losing are much higher than 0.5

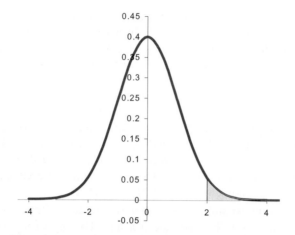

FIGURE 5.1:   A normal distribution with zero mean and standard deviation of one. The probability of a measurement being to the right of the marked value is given by the shaded area.

5.3   Hypothesis testing and significance

Hypothesis testing involve several important steps. The first step is to clearly define the problem. Such a problem might be stated as follows:

EXAMPLE 5.2

The expression level c of a gene is measured in a given condition. It is known from the literature that the mean expression level of the given gene in normal conditions is μ. We expect the gene to be up-regulated in the condition under study and we would like to test whether the data support this assumption.

The next step is to generate two hypotheses. These are statistical hypotheses and, unlike biological hypotheses, they have to take a certain, very rigid form. In particular, the two hypotheses must be **mutually exclusive** and **all inclusive**. Mutually exclusive means that the two hypotheses cannot be true both at the same time. All inclusive means that their union has to cover all possibilities. In other words, no matter what happens, the outcome has to be included in one or the other hypothesis. One hypothesis will be the null hypothesis. Traditionally, this is named H_0. The other hypothesis will be the alternate or research hypothesis, traditionally named H_a. The alternate hypothesis, or the research hypothesis, has to reflect our expectations. If we believe that the gene should be up-regulated, the research hypothesis will be $H_a : c > \mu$. The null hypothesis has to be mutually exclusive and also has to include all other possibilities. Therefore, the null hypothesis will be $H_0 : c \leq \mu$.

We have seen that the p-value is the probability of a measurement more extreme than a certain threshold occurring just by chance. If the measurement occurred by chance and we drew a conclusion based on this measurement, we would be making a mistake. We would erroneously conclude that the gene is up-regulated when in fact the measurement is only affected by chance. In the example considered previously (see Fig. 5.1), this corresponds to rejecting the null hypothesis $H_0 : c \leq \mu$ even if this null hypothesis is in fact true. In other words, the **p-value is the probability of drawing the wrong conclusion by rejecting a true null hypothesis**. Choosing a significance level means choosing a maximum acceptable level for this probability. The **significance level is the amount of uncertainty we are prepared to accept** in our studies.

For instance, when we choose to work at a significance level of 10% we accept that 1 in 10 cases our conclusion can be wrong. A significance level of 20% means that we can be wrong 1 out of 5 cases. Usual significance levels are 1%, 5%, 10% and 15% depending on the situation. At 15% significance level there might be a real phenomenon but it may also be just a random effect; a repetition of the experiment is necessary. At 10% not very many people will doubt our conclusions. At 5% significance level we can start betting money on it and at 1% other people will start betting money on our claims, as well. The 5% and 1% probability levels are thus standard critical levels.

Returning to the flow of hypothesis testing, once we have defined the problem and made explicit hypotheses, we have to choose a significance level. The next step is to calculate an appropriate statistic based on the data and calculate the p-value based on the chosen statistic. Finally, the last step is to either: i) reject the null hypothesis and accept the alternative hypothesis or ii) not reject the null hypothesis.

To summarize, the main steps of the hypothesis testing procedure are as follows :

1. Clearly define the problem.

2. Generate the null and research hypothesis. The two hypotheses have to be mutually exclusive and all inclusive.

3. Choose the significance level.

4. Calculate an appropriate statistic based on the data and calculate a p-value based on it.

5. Compare the calculated p-value with the significance level and either reject or not reject the null hypothesis.

This process will be now detailed and illustrated on some examples.

5.3.1 One-tail testing

In the simplest case, we expect the measurement to have a tendency. For instance, if we are working to improve an algorithm or a procedure, we might expect that the results of the improved procedure would be better than the results of the original. In gene expression experiments, we might understand or hypothesize the function of a gene and predict the direction of its changes.

EXAMPLE 5.3

The expression level of a gene is measured 4 times in a given condition. The 4 measurements are used to calculate a mean expression level of $\overline{X} = 90$. It is known from the literature that the mean expression level of the given gene, measured with the same technology in normal conditions, is $\mu = 100$ and the standard deviation is $\sigma = 10$. We expect the gene to be down-regulated in the condition under study and we would like to test whether the data support this assumption.

SOLUTION   We note that the sample mean is 90 which is lower than the normal 100. However, the question is whether this is meaningful since such a value may also be the result of some random factors. We need to choose a significance level at which we are comfortable for the given application. Let us say that we would like to publish these results in a peer-reviewed journal and, therefore, we would like to be fairly sure of our conclusions. We choose a significance level of 5%.

We have a clear problem and a given significance level. We need to define our hypotheses. The research hypothesis is "the gene is down-regulated" or:

$$H_a : \overline{X} < \mu \tag{5.2}$$

The null hypothesis has to be mutually exclusive and the two hypotheses together have to be mutually inclusive. Therefore:

$$H_0 : \overline{X} \geq \mu \tag{5.3}$$

Note that the two hypotheses do cover all possible situations. This is an example of a one-sided, or a one-tail, hypothesis in which we expect the values to be in one particular tail of the distribution.

From the sampling theorem, we know that the means of samples are distributed approximately as a normal distribution. Our sample has size $n = 4$ and mean $\overline{X} = 90$. We can calculate the value of Z as:

$$Z = \frac{\overline{X} - \mu}{\frac{\sigma}{\sqrt{n}}} = \frac{90 - 100}{\frac{10}{\sqrt{4}}} = \frac{-10}{10} \cdot 2 = -2 \tag{5.4}$$

The probability of having such a value just by chance, i.e. the p-value, is:

$$P(Z < -2) = F(-2) = 0.02275 \tag{5.5}$$

The computed p-value is lower than our significance threshold $0.022 < 0.05$. In the given circumstances, we can reject the null hypothesis. The situation is illustrated in Fig. 5.2. The shaded area represents the area corresponding to the critical value.

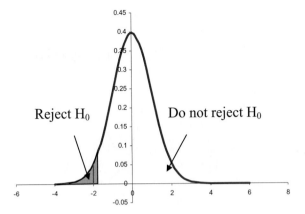

FIGURE 5.2: The computed statistic (-2) is more extreme than the critical value corresponding to the chosen level of significance. In this case, the null hypothesis can be rejected.

For any value of Z in this area, we will reject the null hypothesis. For values of Z anywhere in the remainder of the area under the graph of the pdf, we will not be able to reject the null hypothesis.

Since the hypotheses were designed to be mutually exclusive and all inclusive, we must now accept the research hypothesis. We state that "the gene is down-regulated at 5% significance level." This will be understood by the knowledgeable reader as a conclusion that is wrong in 5% of the cases or fewer. This is the classical approach to hypothesis testing. The p-value is simply compared to the critical value corresponding to the chosen significance level; the null hypothesis is rejected if the computed p-value is lower than the critical value. However, there is a difference between a p-value barely below the 0.05 threshold and a p-value several orders of magnitude lower. An alternative approach is to merely state the p-value and let the reader assess the extent to which the conclusion can be trusted. If we wanted to follow this approach, we would state that "the gene is down-regulated with a p-value of 0.0228."

⬚

EXAMPLE 5.4

The expression level of a gene is measured twice in a given condition. The 2 measurements are used to calculate a mean expression level of $\overline{X} = 90$. It is known from the literature that the mean expression level of the given gene, measured with the same technology in normal conditions, is $\mu = 100$ and the standard deviation is $\sigma = 10$. We expect the gene to be down-regulated in the condition under study and we would like to test whether the data support this assumption.

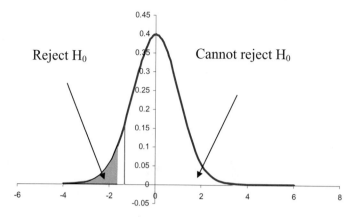

FIGURE 5.3:   The computed statistic (-1.414) is higher than the critical value corresponding to the chosen level of significance. In this case, the null hypothesis cannot be rejected.

SOLUTION   Note that this is exactly the same problem as before, with the only difference that now the gene was measured only twice. We will use the same null and research hypotheses:

$$H_a : \overline{X} < \mu \tag{5.6}$$

$$H_0 : \overline{X} \geq \mu \tag{5.7}$$

We are still dealing with the mean of a sample. Therefore, independently of the distribution of the gene expression level, we can use the normal distribution to test our hypotheses. We can calculate the value of Z as:

$$Z = \frac{\overline{X} - \mu}{\frac{\sigma}{\sqrt{n}}} = \frac{90 - 100}{\frac{10}{\sqrt{2}}} = \frac{-10}{10} \cdot 1.1414 = -1.414 \tag{5.8}$$

The probability of having such a value just by chance, i.e. the p-value, is:

$$P(Z < -1.41) = F(-1.41) = 0.0792 \tag{5.9}$$

This time, the computed value is higher than the 0.05 threshold and in this case, we will not be able to reject the null hypothesis. This is illustrated in Fig. 5.3. In this case, the measured difference may be the effect of a random event.   ⬜

The two examples above illustrated the essence of a general statistical phenomenon. If a variable seems different from its usual values, we can accept there is a real cause behind it if: i) either the difference is large enough (such as more than 3σ from the mean); or ii) the difference is small but this difference is obtained consistently, over sufficiently many measurements. In principle, the smaller the difference, the more measurements we will need in order to reject the null hypothesis for a given significance level.

The reasoning is the same even if the gene is expected to be up-regulated. This means that the measurement is in the other tail of the distribution. However, this is still a one-sided hypothesis. The following example illustrates this.

EXAMPLE 5.5

The BRCA1 mutation is associated with the over-expression of several genes [279]. HG2855A is one such gene. An experiment involving samples collected from a subject with the BRCA1 mutation measures the expression level of this gene 4 times. The mean of these measurements is 109. It is known from the literature that the standard deviation of the expression level of this gene is 10 and the mean expression level is 100. Does the data support the hypothesis that this gene is up-regulated?

SOLUTION We state the two hypotheses as:

$$H_a : \overline{X} > \mu \tag{5.10}$$

$$H_0 : \overline{X} \le \mu \tag{5.11}$$

Note that the alternative hypothesis reflects our expectations and that the null hypothesis includes the equality.

At this time, i.e. before doing any analysis, we also choose the significance level. We will work at a significance level of 5%.

The next step is to choose the appropriate statistical model. In this case, we are working with a mean of a sample and we know from the central limit theorem that sample means are normally distributed. Therefore, we can use the normal distribution. We normalize our variable by applying the Z transformation. We subtract the mean and divide by the standard deviation:

$$Z = \frac{\overline{X} - \mu}{\frac{\sigma}{\sqrt{n}}} = \frac{109 - 100}{\frac{10}{\sqrt{4}}} = \frac{9}{10} \cdot 2 = 1.8 \tag{5.12}$$

We calculate the probability of the Z variable taking a value greater than 1.8:

$$P(Z > 1.8) = 1 - P(Z \le 1.8) = 1 - F(1.8) = 1 - 0.96409 = 0.0359 \tag{5.13}$$

This value is lower than the chosen significance level $0.036 < 0.05$. Based on this, we can reject the null hypothesis and accept the research hypothesis: the gene is indeed up-regulated at the 5% significance level.

□

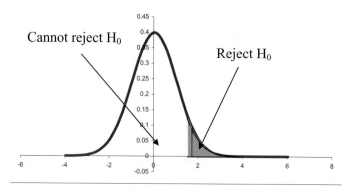

FIGURE 5.4: The computed statistic (1.8) is higher than the critical value corresponding to the chosen significance level. The probability of this happening by chance (0.036) is lower than the 0.05 significance chosen. In this case, the null hypothesis is rejected: the gene is up-regulated at the 5% significance level.

5.3.2 Two-tail testing

In some situations, however, we might not have any precise expectations about the outcome of the event. For instance, in a gene expression experiment we may not have any knowledge about the behavior of a given gene. Thus, the gene might be either up-regulated or down-regulated. In such cases, the reasoning has to reflect this. This is called a two-side or a two-tail test. The following example will illustrate such a situation.

EXAMPLE 5.6

A novel gene has just been discovered. A large number of expression experiments performed on controls revealed that the technology used together with consistent normalization techniques measured the mean expression level of this gene as 100 with a standard deviation of 10. Subsequently, the same gene is measured 4 times in 4 cancer patients. The mean of these 4 measurements is 109. Can we conclude that this gene is up-regulated in cancer?

SOLUTION   In this case, we do not know whether the gene will be up-regulated or down-regulated. The only research hypothesis we can formulate is that the measured mean in the condition under study is different from the mean in the control population. In this situation, the two hypotheses are:

$$H_a : \overline{X} \neq \mu \tag{5.14}$$

$$H_0 : \overline{X} = \mu \tag{5.15}$$

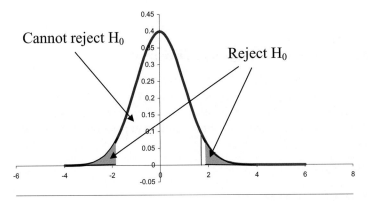

FIGURE 5.5:   In a two-tail test, the probability corresponding to the chosen significance level is divided equally between the two tails of the distribution.

We will work at the same significance level of 5%. The significance level corresponds to the probability of calling a gene differentially regulated when the gene has an unusual value in one of the tails of the histogram. In a one-sided test, we expected the gene to be in a specific tail and we set the threshold in such a way that the area in that tail beyond the threshold was equal to the chosen significance level. However, in this case, the gene can be in either one of the tails. In consequence, we have to divide the 5% probability into two equal halves: 2.5% for the left tail and 2.5% for the right tail. This situation is presented in Fig. 5.5.

We are dealing with a mean of a sample so we can use the normal distribution. We normalize our variable by applying the Z transformation. We subtract the mean and divide by the standard deviation:

$$Z = \frac{\overline{X} - \mu}{\frac{\sigma}{\sqrt{n}}} = \frac{109 - 100}{\frac{10}{\sqrt{4}}} = \frac{9}{10} \cdot 2 = 1.8 \qquad (5.16)$$

We calculate the probability of the Z variable taking a value greater than 1.8:

$$P(Z > 1.8) = 1 - P(Z \leq 1.8) = 1 - F(1.8) = 1 - 0.96409 = 0.0359 \quad (5.17)$$

This value is higher than the chosen significance level $0.036 > 0.025$. In this situation, we cannot reject the null hypothesis.   ☐

Note that all the numerical values in examples 5.5 and 5.6 were identical. However, in example 5.5, the null hypothesis was rejected and the gene was found to be differentially regulated at 5% significance while in example 5.6, the null hypothesis could not be rejected at the same significance level. This illustrates the importance of defining correctly the hypotheses to be tested *before the experiment is performed*.

5.4 "I do not believe God does not exist"

The Spanish philosopher Miguel de Unamuno has pointed out in one of his writings that one's position with respect to religion can take one of four alternatives. This is somewhat contrary to the common belief that one either believes in God or one does not. According to Miguel de Unamuno, the four different positions that one can adopt are as follows:

1. "I believe God exists"

2. "I do not believe God exists"

3. "I believe God does not exist"

4. "I do not believe God does not exist"

A careful consideration of the 4 statements above shows that all 4 are indeed distinct possibilities, very different from one another. Saying "I believe God exists" expresses a positive conviction about God's existence. This person knows something and that something happens to be about God's existence. "I do not believe God exists" expresses the lack of such a positive conviction. This person expressed their ignorance about something and that something happens to be again about God's existence. The third formulation, "I believe God does not exist" expresses another positive belief but this time it happens to be about the exact opposite: God's lack of existence. Finally, "I do not believe God does not exist" reflects the lack of a strong opinion regarding the same matter. Statements 1 and 3 reflect a positive attitude, the speaker feels certain about some fact. Statements 2 and 4 reflect a lack of knowledge about the facts of the matter. Let us substitute the "God exists" with x. Clearly, "I do not believe God does not exist" is not equivalent to saying "I believe God exist." The former merely states a lack of information about non-x whereas the latter makes a strong statement about x. These are not the same and cannot be substituted for each other. The lack of information, or proof, about something can in no circumstances be taken as proof about anything else. In particular, the lack of proof for a fact x cannot, under any circumstances, be taken as proof for the opposite of x.

This applies to the interpretation of the results of the hypothesis testing as described above. In particular, in the hypothesis testing there is a null hypothesis and a research hypothesis. As we have seen, sometimes the data do not allow us to reject the null hypothesis. This means that we do not have enough support to accept the research hypothesis. However, this does not necessarily mean that the research hypothesis is false. As an example, let us consider the null hypothesis: H_0 : "Genetically modified foods (GMOs) are safe" and the research hypothesis: H_a : "GMOs make the descendants of humans who consume them grow horns and tails 4 generations later." Since we have started to consume GMOs on a large scale only about 2 generations ago, no study will be able to reject the null hypothesis. The conclusion of such a study may be expressed as: "There is no data to support the idea that GMOs are harmful" or "There is no data to support the idea that GMOs are unsafe." However, and this is the big trap, *this does not mean that GMOs are safe.*

As a final, and hopefully convincing example, the tobacco industry claimed for many years that "there was no evidence showing that tobacco is harmful." Millions of people interpreted that as meaning "tobacco is safe" and jeopardized their health by smoking. When finally, enough data were gathered to prove that tobacco causes cancer and a plethora of other health problems, people got very upset. This shows that we should ask for positive evidence of the type "show us it is safe" instead of a lack of evidence of negative effects such as "there is no data showing it is harmful."

Going back to our gene expression experiments, let us assume that the research hypothesis is "the gene is up-regulated" and the null hypothesis is that "the gene is not up-regulated." In these circumstances, only two conclusions are possible:

1. The data are sufficient to reject the null hypothesis or

2. We cannot reject the null hypothesis.

If the null hypothesis is rejected, the research hypothesis has to be accepted because the two are mutually exclusive and all inclusive. Therefore, we can conclude the gene is up-regulated.

If the null hypothesis cannot be rejected, it means we cannot show it is false. However, the statistical argument can never conclude that the null hypothesis is true. In other words, we cannot say "the gene is NOT up-regulated." The gene may or may not be up-regulated. We simply do not have enough data to draw a conclusion at the chosen significance level. It may well happen that the very next experiment will produce more data (e.g. another replicate measurement with the same value) which, together with the already existing data, becomes conclusive proof that the gene is up-regulated (see examples 5.5 and 5.6 above).

5.5   An algorithm for hypothesis testing

We can now summarize an algorithm that can be used for hypothesis testing. Each step of the algorithm is illustrated with the corresponding step in Example 5.6.

1. State the problem: Is the gene up-regulated?

2. State the null and alternative hypothesis:

$$H_a : \overline{X} \neq \mu$$

$$H_0 : \overline{X} = \mu$$

Identify whether the situation corresponds to a one-tail or two-tail test; in this case, this is a two-tail test.

3. Choose the level of significance: 5%

4. Find the appropriate test statistic: we were using means of samples, therefore we used the Z transformation and normal distribution (from the central limit theorem).

5. Calculate the appropriate test statistic: $Z = 1.8$

6. Determine the p-value of the test statistic (the prob. of it occurring by chance):

$$P(Z > 1.8) = 1 - P(Z \le 1.8) = 1 - F(1.8) = 1 - 0.96409 = 0.0359$$

7. Compare the p-value with the chosen significance level: $0.0359 > 0.025$.

8. Reject or do not reject H_0 based on the test above: here we cannot reject H_0.

9. Answer the question in step 1: the data do not allow us to conclude the gene is up-regulated.

5.6   Errors in hypothesis testing

The hypothesis testing involved defining the H_0 (null) and H_a (research) hypotheses. So far, we have discussed the reasoning that allows us to pass judgement using such hypotheses. A very important issue is to assess the performance of the decisions taken with the algorithm presented in section 5.5. In order to do this, let us consider that the true situation is known. We will apply the algorithm as above then interpret the results in the context of the true situation.

Let us assume that H_0 is actually true and H_a is false. In this case:

- If we accept H_0, we have drawn the correct conclusion. We will call the instances in this category true negatives: they are negatives because they go against our research hypothesis and they are true because H_0 is indeed true and our H_a is false. If H_0 is "the gene is not regulated" and H_a is "the gene is either up or down-regulated," true negatives will be those genes which are not regulated and are reported as such by our algorithm.

- If we reject H_0, we have drawn an incorrect conclusion. We will call the instances in this category false positives. They are positives because they go with our research hypothesis and they are false because they are reported as such by our algorithm while, in fact, they are not. Non-regulated genes reported as regulated by our algorithm would be false positives.

Rejecting a null hypothesis when it is in fact true is called a **Type I error**. The probability of a Type I error is usually denoted by α. Let us consider a normal

Reported by the test	True (but unknown) situation	
	H_0 is true	H_0 is false
H_0 was not rejected	true negatives (correct decision) $1 - \alpha$	false negatives (Type II error) β
H_0 was rejected	false positives (Type I error) α	true positives (correct decision) $1 - \beta$

TABLE 5.1: The possible outcomes of hypothesis testing. The probability of making a Type I error corresponds directly to the significance level α. The probability of making a Type II error (β) corresponds to the number of false negatives. The term $1 - \beta$ is the power of the test and corresponds to the number of true positives that can be found.

distribution such as the one presented in Fig.5.4. Using a threshold such as the one shown in the figure will classify as up-regulated any gene expressed at a level higher than the threshold. However, this means all the genes in the shaded area of the graph will be false positives since H_0 will be rejected for them. Therefore, *the probability of a Type I error corresponds directly to the significance level chosen.*
Let us now assume H_0 is false and H_a is true. In this case:

- If we accept H_0, we have drawn an incorrect conclusion. The instances in this category will be false negatives (genes that are in fact regulated but are not reported as such by our algorithm).

- If we reject H_0, we have drawn the correct conclusion. The instances in this category are true positives.

The second type of mistake is called a **Type II error**. The probability of a Type II error is denoted by β. The probability of avoiding a Type II error corresponds to correctly picking the instances that do not belong to the distribution of reference. For our purposes, this corresponds to finding the differentially regulated genes (which do not belong to the distribution representing the normal variation of expression levels). This is exactly the purpose of the hypothesis testing algorithm. The higher this probability, the better will the algorithm be at finding such genes. This probability is called the **power of the test** and can be calculated as $1 - \beta$.
The possible outcomes of hypothesis testing are summarized in Table 5.1. The columns correspond to the true but unknown situation whereas the rows correspond to the two possible outcomes of the hypothesis testing. Row 1 corresponds to a situation in which H_0 cannot be rejected; row 2 corresponds to a situation in which H_0 can be rejected and, therefore, H_a can be accepted.

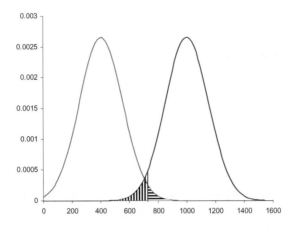

FIGURE 5.6: A problem with measurements from two populations. The distribution to the left represents genes that are not expressed, the distribution to the right corresponds to expressed genes. Assume we are using the null hypothesis "the gene is not expressed." We choose the threshold corresponding to a significance level of 5%. For any gene with a measured value to the right of this threshold we will reject the null hypothesis and conclude the gene is expressed. The area to the right of the threshold and belonging to the distribution on the left (horizontal dashing) is equal to the significance level and the probability of a Type I error. The area to the left of the threshold and belonging to the distribution on the right (vertical dashing) is equal to the probability of a Type II error.

A natural tendency is to try to minimize the probability of making an error. The probability of making a Type II error is not directly controllable by the user.[2] However, the probability of making a Type I error, α, is exactly the significance level. Therefore, one might be tempted to use very high standards and choose very low values for this probability. Unfortunately, this is not a good idea since α and β are closely related and using a lower value for α has the immediate consequence of reducing the power of the test. An example will illustrate this phenomenon.

Let us assume that our measurement actually come from two distributions. Fig. 5.6 illustrates this situation. The genes that are not expressed have the distribution to the left (towards lower intensities) while the gene that are expressed will come from the distribution to the right (towards higher intensities). Assume we are using the null hypothesis "the gene is not expressed." We choose the threshold corresponding to a significance level of 5%. For any gene with a measured value to the right of this threshold we will reject the null hypothesis and conclude the gene is expressed. If a

---

[2]This probability can be controlled only indirectly, through a careful choice of the number of measurements (sample size).

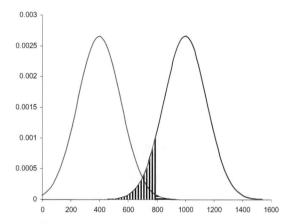

FIGURE 5.7: The same problem as in Fig. 5.6 but with a threshold corresponding to a lower α. The area corresponding to a Type I error is decreased but the area corresponding to a Type II error (β) is increased. In consequence, the power of the test, $1 - \beta$, will be diminished.

gene from the distribution to the left (not expressed) happens to be in this area, we will be making a Type I error (false positive). If a gene from the distribution to the right is in this area, we will correctly categorize it as expressed. The area to the right of the threshold and belonging to the distribution on the left (horizontal dashing) is equal to the significance level and the probability of a Type I error. The area to the left of the threshold and belonging to the distribution on the right (vertical dashing) is equal to the probability of a Type II error. In this area, we will not be able to reject the null hypothesis. For a gene from the left distribution, this is the correct decision. For a gene from the right distribution, we will be making a Type II error (false negative).

Fig. 5.7 shows what happens if a lower α is used and the threshold is moved to the right. The area corresponding to the Type I error will be reduced but, at the same time, the area corresponding to a Type II error will be increased. This reduces the power of the test. The genes from the right distribution that fall between the old and the new position of the threshold will now be false negatives while before they were correctly identified as true positives. The optimal threshold corresponds to the value for which the two distributions yield the same probability (same height on this graph).

5.7   Summary

This chapter presented the framework of the classical statistical hypothesis testing. The key elements introduced here include how to formulate the research and null hypothesis, p-value and its link to the significance level, types of errors and the link between significance and power.
Important aspects discussed in this chapter include:

- The framework of the statistical testing (hypotheses, significance levels, etc.) has to be defined before the experiment is done.

- When using two-tail testing, the probability corresponding to the significance level is equally distributed between the tails of the histogram. Thus, the computed p-value is compared with a probability of $\alpha/2$ instead of α.

- Not being able to reject the null hypothesis does not mean that the null hypothesis is false. For instance, the fact that there is no evidence that x is harmful does not mean that x is safe. Similarly, not being able to conclude that a gene is differentially regulated does not mean that the gene is not so.

- There are two types of errors in classical hypothesis testing. The Type I error corresponds to a situation in which a true null hypothesis is rejected. The Type II error corresponds to not rejecting a false null hypothesis. The probability of a Type I error is the significance level chosen. The probability of a Type II error is denoted by β. The quantity $1 - \beta$ is known as the power of the test and is directly proportional to the ability of the test to identify correctly those instances for which the research hypothesis is true (e.g. differentially regulated genes).

- The requirements of decreasing the probability of Type I (α) and Type II (β) errors are contradictory. Lowering α increases β and reduces the power of the test. This is to say that lowering the probability of false positives also reduces the ability of detecting true positives. Conversely, lowering β increases α. This is to say that increasing the ability of the test to detect true positives also increases the probability of introducing false positives.

5.8   Solved problems

1. The expression level of a gene is measured in a particular experiment and found to be 90. A large number of control experiments have been done with the same technology and it is known that:

(a) The expression level of this gene follows a normal distribution in the control population.

(b) The standard deviation of the expression level of this gene in controls is 10.

(c) The mean expression level in controls is 100.

The experiment is expected to suppress the activity of this gene. Can one conclude that the gene is indeed expressed at a lower level in the experiment performed?

SOLUTION   We choose a significance level of 5%. We expect the gene to be lower so this is a one-tail test. We formulate the null and research hypotheses.

$$H_a : \overline{X} < \mu \tag{5.18}$$

$$H_0 : \overline{X} \geq \mu \tag{5.19}$$

This time we have a single measurement as opposed to a mean of a sample. Therefore, we cannot use the central limit theorem to assume normality. Fortunately, in this case, it is known that the expression level of the given gene follows a normal distribution. If this had not been a known fact, one would have had to study first how the gene is distributed in the control population. However, in the given circumstances we can calculate Z as follows:

$$Z = \frac{X - \mu}{\sigma} = \frac{90 - 100}{10} = -1 \tag{5.20}$$

The probability of obtaining a value this small or smaller just by chance is:

$$P(Z < -1) = 0.158 \tag{5.21}$$

This value is higher than 0.05 (5%) and, therefore, we cannot reject the null hypothesis. Note that the null hypothesis cannot be rejected even if we lower our standards to 10% or even 15%. Also note that the values involved are exactly the same as in example 5.3: the measured value was 90 and the problem was modelled by a normal distribution with $\mu = 100$ and $\sigma = 10$. However, in example 5.3 there was enough evidence to reject the null hypothesis due to the fact that the measured value was obtained from 4 replicates.   ▯

2. A particular gene is expected to be up-regulated in a given experiment. The gene is measured 4 times and a mean expression level of 109 is obtained. It is known from the literature that the standard deviation of the expression level of this gene is 10 and the mean expression level is 100. Can one conclude that the gene is indeed up-regulated?

SOLUTION

We choose a significance level of 5%. We expect the gene to be expressed higher so this is a one-tail test.

We formulate the null and research hypotheses as follows:

$$H_a : \overline{X} > \mu \qquad (5.22)$$

$$H_0 : \overline{X} \leq \mu \qquad (5.23)$$

In this problem, we do not know whether the gene is distributed normally. However, the gene has been measured several times and we know the mean of those measurements. According to the central limit theorem, the distribution of such sample mean values approximates well a normal distribution. We can calculate the value of Z as:

$$Z = \frac{\overline{X} - \mu}{\frac{\sigma}{\sqrt{n}}} = \frac{109 - 100}{\frac{10}{\sqrt{4}}} = 1.8 \qquad (5.24)$$

The probability of obtaining such a value just by chance is (p-value) :

$$P(Z \geq 1.8) = 1 - P(Z < 1.8) = 1 - 0.9647 = 0.0359 \qquad (5.25)$$

We compare the p-value with the significance level $0.0359 < 0.05$ and conclude that we can reject the null hypothesis at the chosen 5% level of significance. We can indeed conclude that, at this significance level, the gene is up-regulated.

□

Chapter 6

Classical approaches to data analysis

> *A knowledge of statistics is like a knowledge of foreign languages or of algebra;*
> *it may prove of use at any time under any circumstances.*

> — Arthur L. Bowley, Elements of Statistics, Part I, Chap. 1, p. 4

6.1   Introduction

This chapter will illustrate the use of classical methods in analyzing microarray data. The reader should be familiar with the basic statistical terms defined in Chapter 4 as well as with the hypothesis testing issues discussed in Chapter 5.

There are two different criteria that are used to distinguish between different types of problems and therefore hypothesis testing situations:

1. The number of samples involved

2. Whether we assume that the data comes from a known distribution or not.

According to the first criterion, there are three different types of problem that can arise. There are:

1. Problems involving one sample

2. Problems involving two samples

3. Problems involving more than two samples.

The second criterion divides hypothesis testing into:

1. Parametric testing – where the data are known or assumed to follow a certain distribution (e.g. normal distribution)

2. Non-parametric testing – where no a priori knowledge is available and no such assumptions are made.

The two criteria are orthogonal in the sense that each distinct class according to the first criterion can fall into any of the classes according to the second criterion. Thus, there are 6 possible combinations and therefore 6 different approaches to hypothesis testing:

1. Parametric testing

   (a) Involving one sample

   (b) Involving two samples

   (c) Involving more than two samples

2. Non-parametric testing

   (a) Involving one sample

   (b) Involving two samples

   (c) Involving more than two samples

In this text, we will focus on parametric testing. Non-parametric testing does not rely on any particular assumptions about an underlying distribution and often work with the order of the measurements instead of their values. This may be problematic for a small number of measurements as it is often the case in microarray experiments. However, some non-parametric tests (e.g. Wilcoxon's test) are used in the analysis of Affymetrix data.

6.2   Tests involving a single sample

6.2.1   Tests involving the mean. The t distribution.

Tests involving a single sample may focus on the *mean* or *variance* of the sample. The following hypotheses may be formulated if the testing regards the mean of the sample:

$$H_0 : \mu = c, \ H_a : \mu \neq c \tag{6.1}$$

$$H_0 : \mu \geq c, \ H_a : \mu < c \tag{6.2}$$

$$H_0 : \mu \leq c, \ H_a : \mu > c \tag{6.3}$$

The hypotheses in 6.1 corresponds to a two-tail testing in which no a priori knowledge is available while 6.2 and 6.3 correspond to a one-tail testing in which the measured value c is expected to be higher and lower than the population mean, respectively. A typical situation of a test involving a single sample is illustrated by the following example.

Example 6.1

The expression level of a gene, measured with a given technology and normalized consistently, is known to have a mean of 1.5 in the normal human population. A researcher measures the expression level of this gene on several arrays using the same technology and normalization procedure. The following values have been obtained: 1.9, 2.5, 1.3, 2.1, 1.5, 2.7, 1.7, 1.2 and 2.0. Is this data consistent with the published mean of 1.5?

This problem involves a single sample including 8 measurements. It is easy to establish that a two-tail test is needed since there is no particular expectation for the measured mean: $H_0 : \mu = c$, $H_a : \mu \neq c$. However, we do face an obstacle: the population variance is not known. One way of dealing with the lack of knowledge about the population variances is to estimate it from the sample itself. We have seen that estimating population parameters from a sample, especially a rather small sample like this one, is imprecise since the estimates may be well off the real values. It should be expected that at some point we will probably need to give up something but this is the best we can do. Thus, instead of:

$$Z = \frac{\overline{X} - \mu}{\frac{\sigma}{\sqrt{n}}} \tag{6.4}$$

we will use a similar expression in which the population standard deviation σ is substituted by the sample standard deviation s:

$$t = \frac{\overline{X} - \mu}{\frac{s}{\sqrt{n}}} \tag{6.5}$$

However, the Z variable had a special property. If we took many samples from a given population and we calculated the Z value for each sample, the distribution of those Z values would follow a normal distribution. It was because of this reason that we could use the tables or calculated values of the normal distribution in order to calculate p values. If we took again many samples from a given population and calculated the values of the t variable above for all such samples, we would notice that the distribution thus obtained is not a normal distribution. Therefore, we will not be able to use the tables of the normal distribution to solve the given problem. However, the t variable in 6.5 does follow another classical and well known distribution called *t* **distribution** or **Student's** *t* **distribution**.

The name of this distribution has an interesting story. The distribution was discovered by William S. Gossett, a 32-year old research chemist employed by the famous Irish brewery Guinness. When Gossett arrived in Dublin he found that there was a mass of data concerning brewing which called for some statistical analysis. Gossett's main problem was to estimate the mean value of a characteristic of various brews on the basis of very small samples. Confronted with this problem, Gossett found the t distribution and developed a method allowing the computation of confidence limits for such small samples. Gossett's findings were of exceptional importance for the scientific community, but Guinness would not allow him to publish the results of his work under his name. Thus, he chose the name "Student" and published his report under this signature in the journal Biometrika in 1908 [111, 128]. Since then, this test has been known as Student's t-test. Gossett's name was released much after the publication of his work. He died in 1937 leaving us one of the most useful, simple

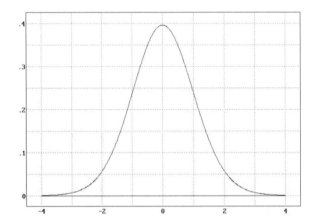

FIGURE 6.1: An example of a t distribution. The shape is very similar to that of a normal distribution. For small samples, the t distribution has more values in its tails compared to the normal distribution. For larger sample sizes (degrees of freedom), the t distribution tends to the normal distribution.

and elegant statistical tools as a proof that beer research can lead to true advances in scientific knowledge.

The t distribution has zero mean and a standard deviation that depends on the **degrees of freedom**, or **d.f.**, of the data. The notion of degrees of freedom is a perennial cause of confusion and misunderstandings as well as a source of inspiration for many jokes.[1] The number of d.f. can be defined as the number of independent pieces of information that go into the estimation of a statistical parameter. In general, the number of degrees of freedom of an estimate is equal to the number of independent measurements that go into the estimate minus the number of parameters estimated as intermediate steps in the estimation of the parameter itself.

For instance, if the variance, s^2, is to be estimated from a random sample of n independent scores, then the number of degrees of freedom is equal to the number of independent measurements n minus the number of parameters estimated as intermediate steps. In this case, there is one parameter, μ, estimated by $\overline{X}$ and therefore the d.f. is $n-1$. In a similar way, we can conclude that the number of degrees of freedom in our problem is $n - 1$ where n is the sample size.

The shape of the t distribution is similar to the shape of the normal distribution, especially as n grows larger (see Fig. 6.1). For lower values of n, the t distribution has more values in its tails than the normal distribution. In practical terms, this means that for a given percentage of the distribution, we will need to go further into the tail. For example, given a t distribution with 4 degrees of freedom, if we wanted to set thresholds on the tails such as the central portion includes 95% of the measure-

---

[1] For instance, it has been said that "the number of d.f. is usually considered self-evident – except for the analysis of data that have not appeared in a textbook" [100].

ments, we need to go approximately 2.78 standard deviations away from the mean. In comparison, for a normal distribution, the interval extends only ± 1.96 standard deviations from the mean. Intuitively, this is consistent to the fact that we are estimating a parameter (population standard deviation) and, therefore, the data have to show more extreme values, further from the mean, for us to be able to reject the null hypothesis at any given significance level (5% in the example above). For large values of n, the t distribution tends to the normal distribution. This reflects the fact that as the sample size grows larger, the sample standard deviation will become a more and more accurate estimate of the population standard deviation. For values of n sufficiently large, the t distribution will be indistinguishable from the normal distribution. The t distribution is widely available like any other classical distribution. Its values can be obtained from tables or computer software.

Armed with this new distribution, we can now return to solving our problem.

Example 6.1

The expression level of a gene measured with a given technology and normalized consistently, is known to have a mean expression level of 1.5 in the normal human population. A researcher measures the expression level of this gene on several arrays using the same technology and normalization procedure. The following values have been obtained: 1.9, 2.5, 1.3, 2.1, 1.5, 2.7, 1.7, 1.2 and 2.0. Is this data consistent with the published mean of 1.5? □

SOLUTION

1. State the null and alternative hypothesis:

$$H_0 : \mu = 1.5$$

$$H_a : \mu \neq 1.5$$

Identify whether the situation corresponds to a one-tail or two-tail test; in this case, this is a two-tail test.

2. Choose the level of significance: 5%.

3. Find the appropriate statistical model and test statistic: as we discussed, the appropriate model is the t distribution.

4. Calculate the appropriate test statistic:

$$t = \frac{\overline{X} - \mu}{\frac{s}{\sqrt{n}}} = \frac{1.87 - 1.5}{\frac{0.51}{\sqrt{8}}} = 2.21$$

5. Calculate the p-value corresponding to the calculated statistic:[2] p=0.057

---

[2] We used the spreadsheet function TDIST.

6. Compare the p-value with the chosen significance level: $0.057 > 0.05$. Alternatively, we could have compared the calculated value of the t statistic to the critical value of the same corresponding to the 5% significance level: $2.21 < 2.306$.

7. Reject or do not reject H_0 based on the test above: here we cannot reject H_0.

8. Draw the conclusion: the data seem consistent with the published mean.

$$\square$$

6.2.2   Choosing the number of replicates

The size of the sample has a clear and direct influence on the results. Once a sample is collected, the statistic calculated on it together with the sample size will decide whether the measured change is significant or not. Many times, genes may not vary that much in absolute value while the variation can still be very meaningful from a biological point of view. Taking this into consideration, one could ask the converse question. Given a certain size of the expected variation, what would be the number of measurements that we need to collect in order to be able to draw a statistically meaningful conclusion from the data? Let us consider the following example.

Example 6.2

The expression level of a gene measured with a specific technology and normalized in a standard way has a mean of 100 and a standard deviation of 10, in the normal population. A researcher is conducting an experiment in which the expression level of this gene is expected to increase. However, the change is expected to be relatively small, of only about 20%. Assuming that each array provides a single measurement of the given gene, how many arrays does the researcher need to run in order to be able to detect such a change 95% of the time? $\square$

The null hypothesis is $H_0 : \mu \leq 100$. We will work at the 1% significance level. Once again, we can use the Z statistic:

$$Z = \frac{\overline{X} - \mu}{\frac{\sigma}{\sqrt{n}}} \tag{6.6}$$

Usually, we know the values of all variables on the right hand side and use them to calculate the value of Z. In this case, we do not know the value of n, the number of replicate measurements but we can calculate the critical value of Z that would correspond to the threshold beyond which we can reject the null hypothesis. From a table of the normal distribution, the value of z that corresponds to $\alpha = 0.01$ is $z_\alpha = 2.33$. With this, we can calculate the minimal value of the sample mean that would be significant at this significance level:

$$z_\alpha = \frac{\overline{X}_t - \mu}{\frac{\sigma}{\sqrt{n}}} \tag{6.7}$$

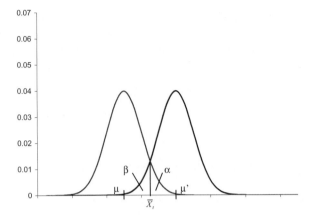

FIGURE 6.2: The distribution to the left represents the expression of the given gene in normal individuals ($\mu = 100$, $\sigma = 10$). The distribution to the right represents the expression of the given gene in the condition under study. We assume this distribution has a mean of $\mu = 120$ and the same standard deviation $\sigma = 10$. We would like to detect when a gene comes from the right distribution 95% of the time. The $\overline{X}_t$ has to be chosen such that it corresponds at the same time to: i) $\alpha = 0.01$ on the distribution to the left and ii) $1 - \beta = 0.95$ on the distribution to the right.

The threshold value of $\overline{X}$ can be extracted as:

$$\overline{X}_t = \mu + z_\alpha \frac{\sigma}{\sqrt{n}} = 100 + 2.33 \frac{10}{\sqrt{n}} \tag{6.8}$$

The problem is that we cannot actually calculate $\overline{X}_t$ because we do not know n yet. However, we have a supplementary requirement, that of being able to detect a change of a given magnitude a certain percentage of times. This refers to the power of the test. If the true mean is $\mu + 20\% = 120$, then the Z value for the distribution centered at 120 could be written as:

$$z_\beta = \frac{\mu' - \overline{X}_t}{\frac{\sigma}{\sqrt{n}}} = \frac{120 - \overline{X}_t}{\frac{10}{\sqrt{n}}} \tag{6.9}$$

We want to reject the null hypothesis with a probability of $1 - \beta = 0.95$. The Z value that corresponds to $\beta = 0.05$ is -1.645. From this second distribution:

$$\overline{X}_t = \mu' - z_\beta \frac{\sigma}{\sqrt{n}} = 120 - 1.645 \frac{10}{\sqrt{n}} \tag{6.10}$$

Once again, we cannot calculate the value of $\overline{X}_t$ since we do not know the number of replicated measurements n. However, the two conditions ($\alpha = 0.05$ and $\beta = 0.95$) have to be satisfied simultaneously by the same $\overline{X}_t$. This is illustrated in Fig. 6.2.

The distribution to the left represents the expression of the given gene in normal individuals ($\mu = 100$, $\sigma = 10$). The distribution to the right represents the expression of the given gene in the condition under study. We assume this distribution has a mean of $\mu = 120$ and the same standard deviation $\sigma = 10$. We would like to detect when a gene comes from the right distribution 95% of the time. The $\overline{X}_t$ has to be chosen such that both conditions on α and β are satisfied at the same time. This condition can be written as:

$$\mu + z_\alpha \frac{\sigma}{\sqrt{n}} = \mu' - z_\beta \frac{\sigma}{\sqrt{n}} \tag{6.11}$$

From this, we can extract the number of replicates as follows:

$$\mu - \mu' = (z_\alpha + z_\beta) \frac{\sigma}{\sqrt{n}} \tag{6.12}$$

and finally:

$$n = \frac{(z_\alpha + z_\beta)^2 \sigma^2}{(\mu - \mu')^2} \tag{6.13}$$

This formula allows us to calculate the number of replicates needed in order to detect a specific change in the gene expression assessed as a mean of several replicate measurements, when the standard deviation is known and specific thresholds have been chosen for the probability α of making a type I error (false positives) and the power of the test $1 - \beta$ (the ability to detect true positives).

Equation 6.13 is very meaningful and illustrates several important phenomena. Firstly, the number of replicate measurements, or the sample size, depends on the chosen significance level, α, the chosen power of the test, β, the variability of the measurements as reflected by their standard deviation σ and the difference between means that we want to be able to detect. For fixed α and β, the number of replicates is directly proportional to the standard deviation and inversely proportional to the difference between means that we want to detect. A large difference in means can be detected with a smaller number of replicate measurements and a larger variability will require more replicates. Furthermore, the dependence is **quadratic**. In other words, if the standard deviation increases from $\sigma = 1$ to $\sigma = 2$, we will need 4 times more replicates in order to be able to detect the same change $\mu - \mu'$.

6.2.3 Tests involving the variance (σ^2). The chi-square distribution

Example 6.2 illustrated the need for information regarding variance. However, so far, we do not have any tools able to test hypotheses regarding the variance. Let us consider the following example.

Example 6.3

The expression level of a gene is measured many times with a certain array technology and found to have a mean expression level of 40 with a standard deviation of 3. A colleague scientist proposes a novel array technology. The same gene is measured 30 times and found to have a standard deviation of 2. Is there evidence that the new technology offers significantly more uniform measurements of the gene expression levels? ▯

This is a typical example of a *one sample test regarding variance*. The null and research hypotheses are as follows:

$$H_0 : \sigma^2 \geq c$$

$$H_a : \sigma^2 < c \qquad (6.14)$$

Unfortunately, none of the distributions discussed so far allows us to test these hypotheses because none of them is concerned with the standard deviation of the sample. We would need a variable similar to the t variable that behaves in a certain standard way. Fortunately, the quantity:

$$\frac{(n-1)s^2}{\sigma^2} \qquad (6.15)$$

behaves like a random variable with the interesting and useful property that if all possible samples of size n are drawn from a normal population with a variance σ^2 and for each such sample the quantity $\frac{(n-1)s^2}{\sigma^2}$ is computed, these values will always form the same distribution. This distribution will be a sampling distribution called a χ^2 (chi-square) distribution. The parametric aspect of the testing using a chi-square distribution comes from the fact that the population is assumed to be normal. The χ^2 distribution is similar to the t distribution inasmuch its particular shape depends on the number of degrees of freedom. If samples of size n are considered, the number of degrees of freedom of the chi-square distribution will be $n - 1$. The χ^2 distribution is shown in Fig. 6.3.

The expected value of the chi-square distribution with ν degrees of freedom is:

$$\mu = E\left(\chi^2\right) = \nu \qquad (6.16)$$

and its variance is:

$$Var\left(\chi^2\right) = 2\nu \qquad (6.17)$$

Given that we now know that the chi-square variable behaves in a certain way, we can use this to test our hypotheses. Given a certain sample, we can calculate the value of the chi-square statistic on this sample :

$$\chi^2 = \frac{(n-1)s^2}{\sigma^2} \qquad (6.18)$$

If the value of the sample standard deviation s is close to the value of the population standard deviation σ, the value of χ^2 will be close to $n - 1$. If the value of the sample standard deviation s is very different from the value of the population standard deviation σ, the value of χ^2 will be very different from $n - 1$. Given that the

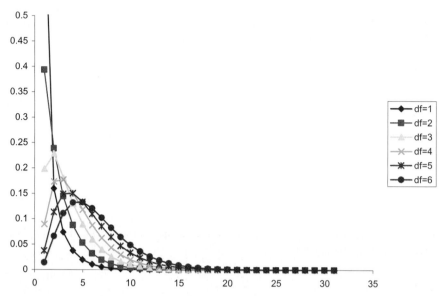

FIGURE 6.3:   The shape of the χ^2 (chi-square) distribution depends on the number of degrees of freedom (df).

shape of the χ^2 distribution is known, we can use it to calculate a precise value for the probability of the measured sample having a variance different from the known population variance.

Let us use the χ^2 distribution in order to solve the problem in Example 6.3. In order to do this, we will calculate the value of the χ^2 statistic for our sample and see how unlikely it is for such value to occur if the null hypothesis is true:

$$\chi^2 = \frac{(n-1)\,s^2}{\sigma^2} = \frac{(30-1)2^2}{3^2} = \frac{29 \cdot 4}{9} = 12.88 \qquad (6.19)$$

The value of the χ^2 statistic is expected to be near its mean of $\nu = n - 1 = 29$ if H_0 is true, and different from 29 if H_a is true. The one-tail probability that corresponds to a χ^2 value of 12.88 for a distribution with 29 degrees of freedom is 0.995. In other words, if H_0 is true, the measured value of the χ^2 statistic will be less or equal to 12.88 in 99.5% of the cases. In these conditions, we can reject the null hypothesis at the 5% significance level.

As we discussed in Chapter 5, an alternative approach is to compare the calculated value of the χ^2 statistic with the critical values of the χ^2 from a distribution with 29 degrees of freedom. For instance the critical values for probabilities of 0.95 and 0.05 are 17.70 and 42.55, respectively. Note that this corresponds to a two-tail testing situation and the sum of the probabilities in both tails is 10% so this corresponds to a 10% significance level. In other words, we will not be able to reject the null hypothesis at 10% significance as long as the calculated χ^2 statistic is in the interval $(17.70, 42.55)$ for a two-tail test. If the calculated value of the χ^2 statistic is more extreme than these values, the H_0 hypothesis will be rejected.

However, our research hypothesis stated that we expect the new technology to be better and therefore we are in a one-tail testing situation. In this case, the only interesting critical value is 17.70 and this corresponds to a 5% significance. We note that $12.88 < 17.70$ and we reject the null hypothesis: there is enough data to support the hypothesis that the new microarray technology is significantly better.

We can summarize the testing involving the variance of one sample as follows. The hypothesis can take one of three forms:

$$H_0 : \sigma^2 \geq c \quad H_a : \sigma^2 < c$$

$$H_0 : \sigma^2 \leq c \quad H_a : \sigma^2 > c$$

$$H_0 : \sigma^2 = c \quad H_a : \sigma^2 \neq c \tag{6.20}$$

The first two correspond to a one-tail test in which the data are expected to be on a specific side of the value c; the last corresponds to a two-tail test. The value c is chosen or known before the data are gathered. The value c is never calculated from the data itself. In any given experiment, only one set of hypotheses is appropriate.

The testing for variance is done using a χ^2 distribution. The χ^2 distribution is a family of distributions with a shape determined by the number of degrees of freedom. There are two basic alternatives for hypothesis testing. The first alternative is to calculate the p-value corresponding to the calculated value of the χ^2 statistic and compare it with the significance level α. If the p value is smaller than the predetermined alpha level, H_0 is rejected. The second alternative is to compare the calculated value of the χ^2 statistic with the critical value corresponding to the chosen significance level. If the calculated χ^2 value is more extreme than the appropriate critical χ^2 value, the H_0 will be rejected.

6.2.4 Confidence intervals for standard deviation

The χ^2 distribution can be used to calculate confidence intervals for the standard deviation. Let us consider the following example.

Example 6.4
The expression level of a specific gene is measured 16 times. The sample variance is found to be 4. What are the 95% confidence limits for the population variance? □

We can use the definition of the χ^2 variable:

$$\chi^2 = \frac{(n-1)s^2}{\sigma^2} \tag{6.21}$$

to extract the σ^2:

$$\sigma^2 = \frac{(n-1)s^2}{\chi^2} \tag{6.22}$$

Gene ID	Cancer patients			Control subjects		
	P 1	P 2	$\cdots$ P n	C 1	C 2	$\cdots$ C n
AC002115	5.2	6.1	$\cdots$ 5.9	0.9	0.7	$\cdots$ 0.8
AB006782	2.1	1.6	$\cdots$ 1.9	-1.3	0.6	$\cdots$ 0.2
AB001325	0.5	-2.8	$\cdots$ 1.5	1.7	3.8	$\cdots$ 0.2
AB001527	0.5	-2.8	$\cdots$ 1.5	1.7	3.8	$\cdots$ 0.2
AB006190	-2.1	-3.9	$\cdots$ -2.1	-1.9	-1.3	$\cdots$ 0.2
AB002086	4.1	3.6	$\cdots$ 1.5	-1.3	0.6	$\cdots$ 0.2
$\cdots$						
AB000450	-1.4	1.3	$\cdots$ -2.5	2.1	-3.2	$\cdots$ 0.8

TABLE 6.1: An experiment comparing the expression levels of some genes in two groups: cancer patients (P) and control subjects (C).

and then use this to set the conditions regarding the probability of making a Type I error, α:

$$C\left[\frac{(n-1)s^2}{\chi^2_{1-\frac{\alpha}{2}}} \leq \sigma^2 \leq \frac{(n-1)s^2}{\chi^2_{\frac{\alpha}{2}}}\right] = 1 - \alpha \tag{6.23}$$

We can now extract the square root in order to obtain σ:

$$C\left[\sqrt{\frac{(n-1)s^2}{\chi^2_{1-\frac{\alpha}{2}}}} \leq \sigma \leq \sqrt{\frac{(n-1)s^2}{\chi^2_{\frac{\alpha}{2}}}}\right] = 1 - \alpha \tag{6.24}$$

These are the confidence intervals for the standard deviation.

6.3   Tests involving two samples

6.3.1   Comparing variances. The F distribution.

Let us consider a typical experiment whose goal is to compare the gene expression levels of some cancer patients to the expression levels of the same genes in a group of healthy people. An example of such data is shown in Table 6.1.

Probably, the most interesting question that one could ask refers to whether a given gene is expressed differently between cancer patients and healthy subjects. This is a question that involves the mean of the two samples. However, as we shall see, in order to answer this question we must first know whether the two samples have the same variance. Let us consider the following example which will help us describe the method used to compare variances of two samples.

Example 6.5
  The expression level of a gene is measured in a number of control subjects and

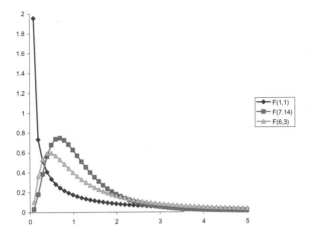

FIGURE 6.4: The F distribution models the behavior of the random variable $\frac{s_1^2}{s_2^2}$ where s_1^2 and s_2^2 are the sample variances of two samples drawn from two normal populations with the same variance σ^2. The shape of the F distribution depends on the degrees of freedom ν_1 and ν_2 of the two samples considered. The figure shows the shapes of the F distributions for $F_{\nu_1=\nu_2=1}$, $F_{\nu_1=7,\nu_2=14}$ and $F_{\nu_1=6,\nu_2=3}$.

patients. The values measured in controls are: 10, 12, 11, 15, 13, 11, 12 and the values measured in patients are: 12, 13, 13, 15, 12, 18, 17, 16, 16, 12, 15, 10, 12. Is the variance different between controls and patients?

The problem can be used to state the following pair of hypotheses:

$$H_0 : \sigma_1 = \sigma_2, \quad H_a : \sigma_1 \neq \sigma_2 \tag{6.25}$$

At this point, none of the test statistics used so far is appropriate to study this problem. In order to develop one, let us consider the following situation. Let us consider a normal population with variance σ^2. Let us draw a random sample of size n_1 of measurements from this population and calculate the variance of this sample s_1^2. Let us then draw another sample of size n_2 from the same population and calculate its sample variance s_2^2. Most likely, these two variances will not be the same. We would like to study their relationship but following two variables at the same time is inconvenient. A good way to combine the two numbers into a single one is to take their ratio $\frac{s_1^2}{s_2^2}$. Since the two sample standard deviations will be, in most cases, slightly different from each other, the ratio $\frac{s_1^2}{s_2^2}$ will be different from one. However, it turns out that this ratio is a random variable that follows a sampling distribution known as the F distribution. Once again, the actual shape of this distribution depends on the degrees of freedom ν_1 and ν_2 of the two samples (see Fig. 6.4).

Since the two samples are coming from populations with the same variance, it is intuitive that the expected value of the ratio $\frac{s_1^2}{s_2^2}$ is 1. Values of this statistic that deviate considerably from 1 will indicate that H_0 is unlikely. This distribution also has the interesting property that the left tail for an F with ν_1 and ν_2 degrees of freedom is the reciprocal of the right tail for an F with the degrees of freedom reversed:

$$F_{\alpha(\nu_1,\nu_2)} = \frac{1}{F_{1-\alpha(\nu_2,\nu_1)}} \tag{6.26}$$

We can now address the problem in Example 6.5. In this case, $\nu_1 = n_1 - 1 = 7 - 1 = 6$ and $\nu_2 = n_2 - 1 = 13 - 1 = 12$. We can use a spreadsheet to calculate the critical values of F. We choose $\alpha = 0.05$ but we work in a two-tail framework so we calculate two values, one for the right tail:

$$F_{0.975(6,12)} = 0.186 \tag{6.27}$$

and one for the left tail:

$$F_{0.025(6,12)} = 3.72 \tag{6.28}$$

Thus, if the ratio of the variances (s_1^2/s_2^2) is less than 0.19 or more than 3.72, H_0 can be rejected. In these circumstances, a type I error of rejecting a true null hypothesis H_0 will happen at most 5% of the time.

The results of a hypothesis testing using an F statistic should (obviously) be independent on the choice of the order of the samples (which one is the numerator and which one is the denominator). The following example will illustrate this.

Example 6.6

The expression level of a gene is measured in a number of control subjects and patients. The values measured in controls are: 10, 12, 11, 15, 13, 11, 12 and the values measured in patients are: 12, 13, 13, 15, 12, 18, 17, 16, 16, 12, 15, 10, 12. Is the variance different between controls and patients? ⬜

SOLUTION The null hypothesis is: $H_0 : \sigma_A^2 = \sigma_B^2$ and the research hypothesis is: $H_a : \sigma_A^2 \neq \sigma_B^2$. We know that the ratio of the sample variances of two samples drawn from two normal populations with the same population variance will follow an F distribution.

We consider sample 1 coming from population A and sample 2 coming from population B: $\nu_A = 6$, $s_A^2 = 2.66$, $\nu_B = 12$, $s_B^2 = 5.74$.

We choose to consider the ratio s_A^2/s_B^2 first. Therefore, we use:

$$F_{0.975(6,12)} = 3.73 \tag{6.29}$$

$$F_{0.025(6,12)} = \frac{1}{F_{0.975(12,6)}} = \frac{1}{5.37} = 0.1862 \tag{6.30}$$

If the statistic s_A^2/s_B^2 is lower than 0.186 or higher than 3.73 we will reject the null hypothesis. If the statistic s_A^2/s_B^2 is in between the two values, we will not be able to reject the null hypothesis:

$$\frac{1}{5.37} \leq \frac{s_A^2}{s_B^2} \leq 3.73 \tag{6.31}$$

Now, let us consider the ratio s_B^2/s_A^2.

$$F_{0.975(12,6)} = 5.37 \tag{6.32}$$

$$F_{0.025(12,6)} = \frac{1}{F_{0.975(6,12)}} = \frac{1}{3.73} = 0.268 \tag{6.33}$$

If the statistic s_B^2/s_A^2 is lower than 0.268 or higher than 5.37 we will reject the null hypothesis. If the statistic s_B^2/s_A^2 is in between the two values, we will not be able to reject the null hypothesis:

$$\frac{1}{3.73} \leq \frac{s_B^2}{s_A^2} \leq 5.37 \tag{6.34}$$

An examination of equations 6.31 and 6.34 reveals that the same conditions are used in both cases, independently of the choice of the populations. For instance, if s_A^2 is to be significantly higher than s_B^2, the statistic s_A^2/s_B^2 has to satisfy:

$$\frac{s_A^2}{s_B^2} > F_{0.975(6,12)} = 3.73$$

or, alternatively, the statistic s_B^2/s_A^2 has to satisfy:

$$\frac{s_B^2}{s_A^2} < F_{0.025(12,6)} = \frac{1}{F_{0.975(6,12)}} = \frac{1}{3.73}$$

$$\square$$

Thus, it is clear that the two different choices lead to exactly the same numerical comparison which is the basis for rejecting or not rejecting the null hypothesis.

In summary, there are three possible sets of hypotheses regarding variances of two samples:

$$H_0 : \sigma_1^2 \geq \sigma_2^2 \quad H_a : \sigma^2 < \sigma_2^2$$
$$H_0 : \sigma_1^2 \leq \sigma_2^2 \quad H_a : \sigma^2 > \sigma_2^2$$
$$H_0 : \sigma_1^2 = \sigma_2^2 \quad H_a : \sigma^2 \neq \sigma_2^2$$

The first two correspond to a one-tail test in which one of the samples is expected to have a variance lower than the other. As always, in any experiment only one set of hypotheses is appropriate. The testing for variance is done using an F statistic. The value of the F statistic calculated from the two samples is compared with the critical values corresponding to the chosen α calculated from the F distribution. The F distribution is a family of distributions with a shape determined by the degrees of freedom of the two samples. Alternatively, if the p-value calculated for the F statistic is lower than the chosen significance level, the null hypothesis H_0 can be rejected.

6.3.2   Comparing means

As mentioned in the previous section, given two groups of gene expression measurements, e.g. measurements in cancer patients vs. measurements in a sample of healthy individuals, probably the most interesting question that one could ask is whether a given gene is expressed differently between cancer patients and healthy subjects. This question refers to the mean of the measurements: we would like to know whether the mean expression level of a gene is different between the two samples. Note that the question refers to *a given gene*. This apparent trivial detail is absolutely crucial for the choice of the approach to follow and the correctness of the final result.

Let us consider for instance the following gene, measured in 6 cancer patients and 6 controls:

Gene ID	P1	P2	P3	P4	P5	P6	C1	C2	C3	C4	C5	C6
AC04219	-5.09	-6.46	-7.46	-4.57	-5.67	-6.84	5.06	7.25	4.17	6.10	5.06	5.66

In this case, the answer seems clear: all cancer patients have negative values while all control subjects have positive values. We can be pretty confident that there is a difference between the expression level of this gene in cancer vs. healthy subjects. However, in most cases the numbers are not so very different. For instance, a conclusion is really difficult to draw from the data collected about the following gene:

Gene ID	P1	P2	P3	P4	P5	P6	C1	C2	C3	C4	C5	C6
AC002378	0.66	0.51	1.12	0.83	0.91	0.50	0.41	0.57	-0.17	0.50	0.22	0.71

In this case, the range of values for both cancer and controls is much narrower. All but one values are positive and although the values for cancer patients seem a bit larger, there are control cases (e.g. C6) that are larger than some cancer cases (e.g. P2). Furthermore, "pretty confident" may not be sufficiently accurate. We would like to have a better way to characterize the degree of confidence in our conclusion. As we have seen in Chapter 5, the probability of drawing the wrong conclusion, or the p-value, might be a more useful and more accurate way to characterize the amount of trust that we are willing to put into our conclusion.

Let us approach this problem using the framework defined in Chapter 5:

1. State the problem.

   Given a single gene (e.g. AC002378), is this gene expressed differently between cancer patients and healthy subjects?

2. State the null and alternative hypothesis.

   The null hypothesis, H_0, is that all measurements originate from a single distribution (see Fig. 6.5). The research hypothesis, H_a, is that there are two distributions: the measurements of the cancer patients come from one distribution while the measurements of the healthy patients come from a different

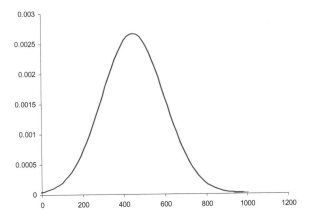

FIGURE 6.5: The null hypothesis H_0: there is no difference in the expression of this gene in cancer patients vs. control subjects and all measurements come from a single distribution.

distribution (see Fig. 6.6). We summarize the hypotheses as:

$$H_0 : \mu_1 = \mu_2, \ H_a : \mu_1 \neq \mu_2$$

In this case, we do not have any expectation regarding in which of the two situations the gene is expressed higher or lower so this is a two-tail test situation.

Note that a cursory examination of the data reveals that the measurements corresponding to the cancer patients are in general higher than the measurements corresponding to the controls. However, it would be incorrect to use this in order to justify the use of a one-tail test by saying that we expect the cancer measurements to be higher. The hypotheses must be stated before examining the data and the choice between a one-tail and a two-tail test must be made based on our a priori knowledge and expectations and not based on the collected data itself.

3. Choose the level of significance.

   We choose to work at 5% significance level.

4. Find the appropriate statistical model and test statistic.

Returning to the problem at hand, the t distribution is the appropriate statistical model. However, there are several versions of this test as follows.

1. **One-tail vs. two-tails.** The one-tail test is to be used when there is an *a priori* knowledge or expectation that one set of measurements will be either higher or lower than the other. If there is no such knowledge, one should use a two-tail test.

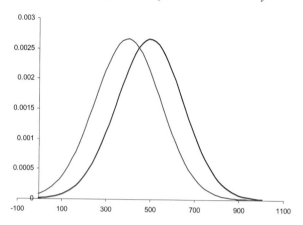

FIGURE 6.6: The research hypothesis H_a: there are two distributions, one that describes the expression of the given gene in cancer patients and one that describes the expression of the same gene in control subjects.

2. **Paired vs. unpaired.** A paired test is used when there is a natural pairing between experiments. A typical example is a situation in which a set of patients are treated with a given drug. A certain parameter is measured several times before and after the drug is administered and the task is to decide whether the drug has a statistical significant effect. In this case, the data collected from one patient before and after the administration of the drug can be tested with a paired t-test. If there are no known reasons to associate measurements in the two samples, the unpaired test is to be used.

3. **Equal variance vs. unequal variance.** As suggested by the names, the equal-variance t-test is appropriate when the two samples are known, or presumed, to come from distributions with equal variances while the unequal variance test is appropriate when the two distributions have different variances.

In our problem, there is no *a priori* expectation that the cancer measurements be either higher or lower than the control measurements. Furthermore, there is no special connection between cancer patients and control subjects. The choice between the equal variance and the unequal variance can be decided by testing the variances as discussed in Section 6.3.1.

6.3.2.1 Equal variances

If the two samples are drawn from normal populations,[3] $(\overline{X}_1 - \overline{X}_2)$ will be distributed normally with an expected value equal to $\mu_1 - \mu_2$. It can be shown that the

---

[3]This assumption makes this test parametric.

variance of a difference of two variables is equal to the sum of variances of the two variables. We can write the variance of $\overline{X}_1 - \overline{X}_2$ as:

$$\frac{\sigma_1^2}{n_1} + \frac{\sigma_2^2}{n_2} \tag{6.35}$$

The t statistic used in the tests for mean involving one sample was (see Eq. 6.5):

$$t = \frac{\overline{X} - \mu}{\sqrt{\frac{s^2}{n}}} \tag{6.36}$$

We can re-write this t-statistic using the new variable $X_1 - X_2$ with the new mean $\mu_1 - \mu_2$ and the new sample variance $\frac{s_1^2}{n_1} + \frac{s_2^2}{n_2}$:

$$t = \frac{(\overline{X}_1 - \overline{X}_2) - (\mu_1 - \mu_2)}{\sqrt{\frac{s_1^2}{n_1} + \frac{s_2^2}{n_2}}} \tag{6.37}$$

This is the case in which the two variances are assumed to be equal ($\sigma_1^2 = \sigma_2^2$), so the best estimate of the overall variability, and hence the new sample variance, will be given by a *pooled sample variance*:

$$s_1 = s_2 = s_p^2 = \frac{(n_1 - 1) \cdot s_1^2 + (n_2 - 1) \cdot s_2^2}{n_1 + n_2 - 2} \tag{6.38}$$

Substituting this in Equation. 6.37, we obtain the formula for the t-statistic of two independent samples with equal variances:

$$t = \frac{(\overline{X}_1 - \overline{X}_2) - (\mu_1 - \mu_2)}{\sqrt{s_p^2 \left(\frac{1}{n_1} + \frac{1}{n_2}\right)}} \tag{6.39}$$

Formulae 6.38 and 6.39 can be used to test hypotheses regarding the means of two samples of equal variance. The degrees of freedom are given by the number of measurements minus the number of intermediate values we need to calculate. In this case, there are $n_1 + n_2$ measurements and two intermediate values s_1^2 and s_2^2, so $\nu = n_1 + n_2 - 2$.

We can now return to the example given at the beginning of this section and answer the question regarding gene AC002378:

Gene ID	P1	P2	P3	P4	P5	P6	C1	C2	C3	C4	C5	C6
AC002378	0.66	0.51	1.12	0.83	0.91	0.50	0.41	0.57	-0.17	0.50	0.22	0.71

For this gene, we calculate the following:

$$\overline{X}_1 = 0.755$$

$$\overline{X}_2 = 0.373$$

$$s_1^2 = 0.059$$

$$s_2^2 = 0.097$$

At this point, we have to decide whether we can assume equal variances or not. For this purpose, we take $H_0 : \sigma_1^2 = \sigma_2^2$ and $H_a : \sigma_1^2 \neq \sigma_2^2$. We choose to work at the 5% significance level and we calculate the F statistic:

$$F = \frac{s_1^2}{s_2^2} = \frac{0.059}{0.097} = 0.60$$

We calculate the probability of obtaining such a value for the F statistic just by chance[4] as 0.70 and we decide that the data do not allow us to reject H_0 at the chosen significance level. We then assume that the two samples have equal variance and calculate the pooled variance:

$$s_p^2 = \frac{(n_1 - 1) \cdot s_1^2 + (n_2 - 1) \cdot s_2^2}{n_1 + n_2 - 2} = \frac{(6-1) \cdot 0.059 + (6-1) \cdot 0.097}{6 + 6 - 2} = 0.078$$
(6.40)

Finally, we use formula 6.39 to calculate the value of the t-statistic:

$$t = \frac{(\overline{X}_1 - \overline{X}_2) - (\mu_1 - \mu_2)}{\sqrt{s_p^2 \left(\frac{1}{n_1} + \frac{1}{n_2}\right)}} = \frac{(0.755 - 0.373) - 0}{\sqrt{0.078 \left(\frac{1}{6} + \frac{1}{6}\right)}} = 2.359 \qquad (6.41)$$

and calculate the p-value, or the probability of having such a value by chance,[5] as 0.04. This value is lower than the chosen significance level of 0.05 and therefore we reject the null hypothesis: the gene AC002378 is expressed differently between cancer patients and healthy subjects.

6.3.2.2   Unequal variances

If we can assume that the two sets of measurements come from independent normal populations but we cannot assume that those populations have the same variance, a modified t-test is necessary. This situation arises if an F test undertaken on the ratio of the sample variances $\frac{s_1^2}{s_2^2}$ provides sufficient evidence to reject the hypothesis that the two variances are equal $H_0 : \sigma_1^2 = \sigma_2^2$. In this case, the t-statistic can still be calculated as:

$$t = \frac{(\overline{X}_1 - \overline{X}_2) - (\mu_1 - \mu_2)}{\sqrt{\frac{s_1^2}{n_1} + \frac{s_2^2}{n_2}}} \qquad (6.42)$$

---

[4]This can be calculated using the function FDIST in Excel.
[5]Using for instance the function TDIST in Excel.

where s_1 and s_2 are the respective sample variances. However, the degrees of freedom need to be adjusted as:

$$\nu = \frac{\left(\frac{s_1^2}{n_1} + \frac{s_2^2}{n_2}\right)^2}{\frac{\left(\frac{s_1^2}{n_1}\right)^2}{n_1-1} + \frac{\left(\frac{s_2^2}{n_2}\right)^2}{n_2-1}} \tag{6.43}$$

Usually this value is not an integer and needs to be *rounded down*.

A close inspection of formulae 6.39 and 6.42 shows that if $n_1 = n_2$ or $s_1^2 = s_2^2$, the two formulae will yield exactly the same value for the t-statistic.

6.3.2.3   Paired testing

In some situations, the measurements are naturally paired. Examples include:

1. Simultaneous tests – e.g. cells from the same culture are split into two groups and each group is given a different treatment.

2. Before and after tests – e.g. a certain parameter is measured in a given animal before and after the treatment with a given substance.

3. Matched tests – e.g. the subjects are matched in pairs that have similar characteristics (e.g. same age, height, weight, diet, etc.).

In all these cases, the variables measured are dependent. Let us assume that there are n measurements performed before the treatment and n measurements after the treatment. Since these measurements are paired, we can think that what we want to actually follow is not the values themselves but the differences "after treatment – before treatment." These differences are independent so the t-test developed earlier will be appropriate:

$$t = \frac{\overline{X} - \mu_d}{\frac{s_d}{\sqrt{n}}} \tag{6.44}$$

where μ_d is the mean difference, s_d is the sample standard deviation of the differences, etc. Clearly, the degrees of freedom are $n-1$ where n is *the number of differences*, or paired measurements. Note that in an unpaired t-test there would be $2n-1$ degrees of freedom since there would be $2n$ *independent* measurements.

In conclusion, the only difference between paired and unpaired testing is the computation of the degrees of freedom. The rest of the statistical argument follows the same lines used in all other hypothesis testing situations.

6.3.3   Confidence intervals for the difference of means $\mu_1 - \mu_2$

If the null hypothesis $H_0 : \mu_1 - \mu_2$ can be rejected, a natural question is to ask how different the two means are? This can be translated directly into the goal of establishing confidence intervals for the difference $\mu_1 - \mu_2$. Such intervals can be calculated using the same method used in Section 6.2.4. The intervals are:

$$C\left[\left(\overline{X}_1 - \overline{X}_2\right) - t_0 \cdot s_{\overline{X}_1 - \overline{X}_2} \le \mu_1 - \mu_2 \le \left(\overline{X}_1 - \overline{X}_2\right) - t_0 \cdot s_{\overline{X}_1 - \overline{X}_2}\right] = 1 - \alpha \tag{6.45}$$

where the number of degrees of freedom is $\nu = n_1 + n_2 - 2$ and the standard error $s_{\overline{X}_1 - \overline{X}_2}$ is:

$$s_{\overline{X}_1 - \overline{X}_2} = \sqrt{s_p^2 \left(\frac{1}{n_1} + \frac{1}{n_2}\right)}$$

6.4 Summary

This chapter discussed a number of statistical techniques that can be used directly to answer questions that arise in the analysis of microarray data. The chapter addressed only parametric testing involving one or two samples. A sample in this context is a set of measurements. Such measurements can be:

- Repeated measurements of the expression level of a gene in various individuals of a biological population (biological replicates)

- Repeated measurements of the expression level of a gene using several spots on a given microarray (spot replicates)

- Repeated measurements of the expression level of a gene using several arrays (array replicates).

Each of these levels allows answering specific questions at that particular level. For instance, spot replicates can provide quality control information for a given array, array replicates can provide quality control information for a given technology, etc. However, the results can be combined. For instance, an analysis at the spot replicate level can be used in order to decide how many replicates should be printed on a custom array and what the confidence limits for the mean of such measurements are. Once this has been decided, the mean of such measurements can be chosen to represent the expression level of a gene on a given array.

The caveats include the fact that the techniques discussed in this chapter do not allow us to test hypotheses about interactions of different levels. ANalysis Of VAriance (ANOVA) methods must be used for such purposes. Also, the techniques discussed here assume that the data was normalized in a consistent way. Applying these techniques on raw data or trying to compare data that was normalized using different techniques will, most probably, produce erroneous results.

The following list includes questions that arise in the analysis of microarray data that can be addressed with the techniques discussed in this chapter.

1. A gene is known to be expressed at level c when measured with a given technology in a given population under given circumstances. A new set of measurements produce the values: $m_1, m_2, \ldots m_k$. Are these data consistent with the accepted value c?

   - Hypothesis: $H_0 : \mu = c$
   - Suitable distribution: t distribution.
   - Assumptions: The t distribution assumes normality. However, if the variable is a mean of a sample $(\overline{X})$, the central limit theorem allows us to infer valid conclusions even if the original population of X is not normal.
   - Reasoning: classical hypothesis testing.

2. Given that the measurements of a gene in normal conditions has a mean μ and a standard deviation σ, how many replicate measurements are needed in order to detect a change of δ % at least 95% of the time?

   - Hypothesis: there are two null hypotheses involved, one for the initial population and one for the population with an assumed shift in mean that we want to detect.
   - Suitable distribution: Z distribution (normal)
   - Assumptions: The Z distribution assumes normality but if the variable is a mean of a sample $(\overline{X})$, the central limit theorem allows us to infer valid conclusions even if the original population of X is not normal.
   - Reasoning: calculate a Z_α value by using the initial distribution and setting conditions for Type I error (α); then use a distribution with a shift in mean equal to the minimum change to be detected, set conditions for the Type II error and calculate a Z_β value. The number of replicates can be calculated by setting the condition: $Z_\alpha = Z_\beta$.

3. The expression level of a gene measured with technology A in a given set of conditions has a standard deviation of σ. The expression level of the same gene in the same conditions is measured n times with technology B and has a sample variance s^2. Does technology B offer a more consistent way of measuring gene expression?

   - Hypothesis: $H_0 : \sigma^2 = c$
   - Suitable distribution: χ^2 distribution
   - Assumptions: normal distribution
   - Reasoning: classical hypothesis testing.

4. The expression level of a specific gene is measured n times. The sample variance is found to be s. What are the 95% confidence limits for the population variance?

   - Hypothesis: not applicable (not a hypothesis testing situation)

- Suitable distribution: χ^2 distribution
- Assumptions: normal distribution
- Reasoning: Extract the variance from the expression of the χ^2 distribution and set conditions for Type I error (α).

5. Given two sets of measurements of the expression level of a gene (perhaps in two different conditions), are the variances of the two sets equal?

   - Hypothesis: $\sigma_1 = \sigma_2$
   - Suitable distribution: F distribution
   - Assumptions: the two populations are normal
   - Reasoning: classical hypothesis testing.

6. Given two sets of measurements of the expression level of a gene (perhaps in two different conditions), what is the pooled variance of the combined set of measurements?

   - Hypothesis: not applicable (not a hypothesis testing situation)
   - Suitable distribution: not applicable (not a hypothesis testing situation)
   - Assumptions: the two populations are normal
   - Reasoning: see Eq. 6.38.

7. Given two sets of measurements of the expression level of a gene in two different conditions, is this gene expressed differently between the two conditions?

   - Hypothesis: $\mu_1 = \mu_2$
   - Suitable distribution: t distribution
   - Assumptions: the two populations are normal or the random variable is a mean of a sample (central limit theorem)
   - Reasoning: classical hypothesis testing but care must be taken in using the appropriate formulae for sample variance and degrees of freedom (equal variance vs. unequal variance, paired vs. unpaired, etc.).

8. If a gene has been found to have different mean measurements between two different conditions, how different are these means? What are the 95% confidence intervals for the difference of the means?

   - Hypothesis: not applicable (not a hypothesis testing situation)
   - Suitable distribution: t distribution
   - Assumptions: normality
   - Reasoning: see Eq. 6.45

Note that very often such questions need to be combined. For instance, in order to choose the appropriate test to see whether a gene is expressed differently between two conditions, one usually needs to test whether the two sets of measurements have the same variance. Another example is measuring a gene several times and calculating the sample variance and mean as well as the confidence intervals for population variance and mean. When sufficient data are collected, these can be used as good estimates of the population variance and mean, which in turn can be used to answer questions about the reliability of a new technology or about the expression value in a new condition.

6.5 Exercises

1. Given two independent populations with variances σ_1 and σ_2. Consider the variables X_1 and X_2 drawn from the two populations, respectively. Calculate the variance of the variables:

   (a) $X_1 + X_2$

   (b) $X_1 - X_2$

   (c) $X_1 \cdot X_2$

2. Consider two independent samples for which the null hypothesis of having the same means can be rejected. Calculate the confidence intervals for $\mu_1 - \mu_2$.

Chapter 7

Analysis of Variance – ANOVA

In a nutshell, understanding any system means engineering it.

—Lynn Conway

7.1 Introduction

7.1.1 Problem definition and model assumptions

Let us consider an experiment measuring the expression level of a given gene in a number of k conditions. Each gene i is measured n_i times for a total of $\sum_{i=1}^{k} n_i$ measurements as follows:

		Condition			
1	2	$\cdots$	i	$\cdots$	k
X_{11}	X_{21}	$\cdots$	X_{i1}	$\cdots$	X_{k1}
X_{12}	X_{22}	$\cdots$	X_{i2}	$\cdots$	X_{k2}
$\vdots$	$\vdots$	$\vdots$	$\vdots$	$\vdots$	$\vdots$
X_{1n_1}	X_{2n_2}	$\cdots$	X_{in_i}	$\cdots$	X_{kn_k}

Note that this may look like a matrix with the symbolic notation used but, in reality, is more general because each column can have a different number of measurements n_i.

The typical question asked here is to decide whether there are any differences between the expression level of the given gene between the k conditions. However, this is a rather imprecise formulation. We need to formulate clearly the null and research hypotheses. The null hypothesis seems to be pretty easy to formulate. Under the null hypothesis, the different conditions are not really different and, therefore, all measurements actually come from a single distribution. In these conditions, all means would be the same:

$$H_0 : \mu_1 = \mu_2 = \cdots = \mu_k \tag{7.1}$$

However, from the initial formulation, it is not clear whether the alternate or research hypothesis would require for *all conditions* or only *a subset of conditions* to be different from each other. In other words, there seem to be three possibilities for H_a:

1. H_a: All means are different from each other.

2. H_a: Several but not all means are different from each other.

3. H_a: There is at least one pair of means that are different from each other.

Let us consider the first choice above as our research hypothesis. The problem would then be formulated as the following pair of hypotheses:

H_0: $\mu_1 = \mu_2 = \cdots = \mu_k$
H_a: All means are different from each other.

It is clear that this pair of hypotheses does not include all possible situations. In other words, it is not all inclusive. For instance, if $\mu_1 \neq \mu_2$ and $\mu_2 \neq \mu_3$ but $\mu_4 = \mu_5 = \cdots = \mu_k$, both hypotheses above would be false which shows they are not defined correctly. A similar situation happens if the second alternative above is chosen as the research hypothesis since it is always possible to find a situation in which neither hypothesis is true. The only correct combination is to use H_0 as defined above together with its logical negation as the research alternative. From elementary logics, if a statement is of the form "for all x it is true that $P(x)$," its negation will be of the form "there exists at least one x for which $P(x)$ is false." Therefore, the *only* correct problem formulation is:

H_0: $\mu_1 = \mu_2 = \cdots = \mu_k$
H_a: There is at least one pair of means that are different from each other.

This particular data layout and set of hypotheses is characteristic to a **Model I**, or **fixed effects**, **ANOVA**. In Model I ANOVA, the researcher has a specific interest in the conditions or treatments under study. In particular, the question regards whether there are differences between any pair of the specific conditions considered. If such differences exist (at least one), it is of interest to identify which ones are different. Furthermore, in this data layout, each measurement belongs to a single group. In other words, the data was factored one-way. This analysis is called a **one-way ANOVA**. The other assumptions made here are as follows:

1. The k samples are independent random samples drawn from k specific populations with means $\mu_1, \mu_2, \ldots, \mu_k$ (constant but unknown).

2. All k populations have the same variance σ^2.

3. All k populations are normal.

Since the framework used to develop the ANOVA methodology assumes the populations are normal, ANOVA is a parametric test. ANOVA stands for ANalysis Of

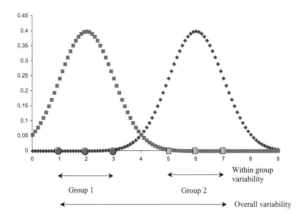

FIGURE 7.1: Two samples with means $\mu_1 = 2$ and $\mu_2 = 6$. The variability within groups is much smaller than the overall variability; this may allow us to reject the hypothesis that the two samples were drawn from the same distribution.

VAriance which may seem a bit counterintuitive since the problem was formulated as a test for means. However, the reasoning behind this is still intuitive. It is very likely that two different random samples will have two different means whether they come from the same population or not. Given a certain difference in means, one can reject a null hypothesis regarding the means if the variance of each sample is sufficiently small in comparison to the overall variance. This is illustrated in Fig. 7.1 and Fig. 7.2.

Fig. 7.1 shows two samples with means $\mu_1 = 2$ and $\mu_2 = 6$. In this case, the variability within groups is much smaller than the overall variability. This may allow us to reject the hypothesis that the two samples were drawn from the same distribution and accept that there is a significant difference between the two samples. Fig. 7.2 illustrates an example with two samples in which the variability within groups is comparable to the overall variability. Note that the two sample means are still $\mu_1 = 2$ and $\mu_2 = 6$, as in Fig. 7.1. However, in this situation we are unlikely to reject the null hypothesis that the two samples were drawn from the same distribution. These two examples show how an analysis of variance can provide information that allow us to draw conclusions about hypotheses involving means.

Another interesting observation is that both the set of hypotheses used here and the examples in Fig. 7.1 and Fig. 7.2 look very similar to hypotheses and examples given in Chapter 6 when the t-test was discussed. This is no coincidence. If there are only two samples, the ANOVA will provide the same results as a t-test. However, the t-test can be applied to comparisons involving only two samples while ANOVA can be applied to any number of samples. An astute reader could argue that this is not a problem since one could use a divide-and-conquer approach and split the null hypothesis $H_0 : \mu_1 = \mu_2 = \cdots = \mu_k$ into a number of hypotheses involving only two samples $H_0^i : \mu_i = \mu_j$. Subsequently, each of these hypotheses could be tested with

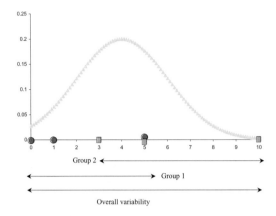

FIGURE 7.2:   Two samples with means $\mu_1 = 2$ and $\mu_2 = 6$. The variability within groups is comparable to the overall variability. In this situation, we are unlikely to reject the null hypothesis that the two samples were drawn from the same distribution.

an individual t-test. Thus the null given hypothesis $H_0 : \mu_1 = \mu_2 = \cdots = \mu_k$ will be rejected if at least one of the $H_0^i : \mu_i = \mu_j$ is rejected. Although this may seem as a valid reasoning from a logical point of view, it is invalid from a statistical point of view. The main problem is that the statistical hypothesis testing does not provide absolute conclusions. Instead, each time a null hypothesis is rejected, there is a non-zero probability of the null hypothesis being actually true. This is the probability of a Type I error or the significance level. When many such tests are carried out for the purpose of drawing a single conclusion, as above, a single mistake in each one of the individual tests is sufficient to invalidate the conclusion. Thus, the probability of a Type I error increases with the number of tests even if the probability of a Type I error in each test is bounded by the chosen level of significance α. This will be explained fully in the section dealing with the correction for multiple experiments.

7.1.2   The "dot" notation

A notation very useful when many sums are involved is the "dot" notation. This notation uses a dot instead of the index for which the sum is calculated. Using this notation, the sum of all measurements of a gene in a given condition i can be written as:

$$\sum_{j=1}^{n_i} X_{ij} = C_{i.} \qquad (7.2)$$

	Condition 1	$\cdots$	Condition k	
measurement 1	X_{11}	$\cdots$	X_{k1}	$X_{.1} = \sum_{i=1}^{k} X_{i1}$
measurement 2	X_{12}	$\cdots$	X_{k2}	$X_{.2} = \sum_{i=1}^{k} X_{i2}$
$\vdots$	$\vdots$	$\cdots$	$\vdots$	$\vdots$
measurement n	X_{1n}	$\cdots$	X_{kn}	$X_{.n} = \sum_{i=1}^{k} X_{in}$
	$X_{1.} = \sum_{j=1}^{n} X_{1j}$	$\cdots$	$X_{k.} = \sum_{j=1}^{n} X_{kj}$	$X_{..} = \sum_{i,j} X_{ij}$

FIGURE 7.3: The "dot" notation. Summation on an index is represented by a dot substituting the given index. In this example, there are k conditions and each condition is measured n times for a total number of $n \times k$ measurements.

The same notation can be used for means and other quantities. For instance, the mean of the i-th sample can be written as:

$$\frac{\sum_{j=1}^{n_i} X_{ij}}{n_i} = \frac{C_{i.}}{n_i} = \overline{X}_{i.} \tag{7.3}$$

Furthermore, a summation over two indices can be represented by using two dots as follows:

$$\sum_{i=1}^{k} \sum_{j=1}^{n_i} X_{ij} = C_{..} \tag{7.4}$$

More examples using the dot notation are given in Fig. 7.3.

7.2 One-way ANOVA

7.2.1 One-way Model I ANOVA

The general idea behind one-way Model I ANOVA is very simple. The measurements of each condition vary around their mean. This is an *within group* variability and will be characterized by a corresponding *within group variance*. At the same time, the means of each treatment will vary around an overall mean. This is due to a *inter-group* variability. Finally, as a result of the two above, each individual measurement varies around the overall mean. The idea behind ANOVA is to study the relationship between the inter-group and the within-group variabilities (or variances). We can start by considering the difference between an individual measurement X_{ij} and the global mean:

$$X_{ij} - \overline{X}_{..} \tag{7.5}$$

Let us add and subtract the mean of the condition i and regroup the terms in a convenient way:

$$X_{ij} - \overline{X}_{..} = X_{ij} - \overline{X}_{..} + \overline{X}_{i.} - \overline{X}_{i.} = \left(\overline{X}_{i.} - \overline{X}_{..}\right) + \left(X_{ij} - \overline{X}_{i.}\right) \tag{7.6}$$

This is a *very* interesting observation. Equation 7.6 tells us that the deviation of an individual measurement from the overall mean can be seen as the sum of the deviations of that individual measurement from the mean of its condition and the deviation of that mean from the overall mean. This is most interesting because the first term $\left(\overline{X}_{i.} - \overline{X}_{..}\right)$ characterizes the deviation of an individual condition mean from the overall mean or the deviations *between groups* while the term $\left(X_{ij} - \overline{X}_{i.}\right)$ characterizes the deviation of an individual measurement from the mean of its condition or the deviation *within group*. Unfortunately, we have seen in Chapter 4 that the deviations are not good estimators of the sample variance because their sum cancels out. The sample variability is better characterized by the variance of the sample which was defined as the sum of squared deviations of individual measurements from the mean $\sum(X_i - \overline{X})^2$ divided by the sample size minus 1. For samples of equal size, the denominator is constant and the variance will be determined by the sum of (deviation) squares. Such sums of squares (SS) turn out to be most important quantities in the ANOVA approach and we will focus on them.

We can characterize the total variability (using the dot notation defined above) as:

$$\sum_{i=1}^{k}\sum_{j=1}^{n_i}(X_{ij} - X_{..})^2 \tag{7.7}$$

This expression is the sum of the squares of all differences from individual observation to the grand mean of the data set and is called the **Total Sum of Squares** of SS_{Total}. This is a measure of the scatter of all data considered, as a single group, around its mean. Note that if we divide this expression by $N - 1$ where N is the total number of measurements, we obtain exactly the global sample variance.

We need to work out a formula relating the overall variance with the variance between conditions and the variance within conditions. This translates to working out a formula between their respective sum of squares: SS_{Total}, $SS_{between\ groups}$ and $SS_{within\ group}$.

All measurements within a group are measurements of the same gene in the same condition. Any differences between such measurements are due to the experimental error. In consequence, the sum of squares within a group is also called the **Error Sum of Squares** $SS_{within\ group} = SS_{Error}$ or **Residual Sum of Squares**. The sum of squares between groups characterizes the variance of the groups around the overall mean. Therefore, one individual such sum of square will characterize the variance of its condition (or treatment) $SS_{between\ groups} = SS_{Cond}$. This is also called the **Among Treatment Sum of Squares**.

7.2.1.1 Partitioning the Sum of Squares

The **Total Sum of Squares**:

$$SS_{Total} = \sum_{i=1}^{k}\sum_{j=1}^{n_i}(X_{ij} - X_{..})^2 \tag{7.8}$$

can be re-written using equation 7.6 as:

$$SS_{Total} = \sum_{i=1}^{k} \sum_{j=1}^{n_i} \left[\left(\overline{X}_{i.} - \overline{X}_{..} \right) + \left(X_{ij} - \overline{X}_{i.} \right) \right]^2$$

$$= \sum_{i=1}^{k} \sum_{j=1}^{n_i} \left[\left(\overline{X}_{i.} - \overline{X}_{..} \right)^2 + 2 \left(\overline{X}_{i.} - \overline{X}_{..} \right) \left(X_{ij} - \overline{X}_{i.} \right) + \left(X_{ij} - \overline{X}_{i.} \right)^2 \right]$$

$$= \sum_{i=1}^{k} \sum_{j=1}^{n_i} \left(\overline{X}_{i.} - \overline{X}_{..} \right)^2 + \sum_{i=1}^{k} \sum_{j=1}^{n_i} 2 \left(\overline{X}_{i.} - \overline{X}_{..} \right) \left(X_{ij} - \overline{X}_{i.} \right) +$$

$$+ \sum_{i=1}^{k} \sum_{j=1}^{n_i} \left(X_{ij} - \overline{X}_{i.} \right)^2$$

$$= \sum_{i=1}^{k} \sum_{j=1}^{n_i} \left(\overline{X}_{i.} - \overline{X}_{..} \right)^2 + \sum_{i=1}^{k} \sum_{j=1}^{n_i} \left(X_{ij} - \overline{X}_{i.} \right)^2 +$$

$$+ 2 \sum_{i=1}^{k} \left(\overline{X}_{i.} - \overline{X}_{..} \right) \sum_{j=1}^{n_i} \left(X_{ij} - \overline{X}_{i.} \right) \tag{7.9}$$

The very last term contains the sum:

$$\sum_{j=1}^{n_i} \left(X_{ij} - \overline{X}_{i.} \right)$$

which is the sum of all deviations of measurements in sample i with respect to its mean. We have seen in Chapter 4, equation 4.8 that this sum is zero. In fact, this was an obstacle against using the sum of deviates as a measure of variance. Using this, the total sum of squares becomes:

$$SS_{Total} = \sum_{i=1}^{k} \sum_{j=1}^{n_i} \left(\overline{X}_{i.} - \overline{X}_{..} \right)^2 + \sum_{i=1}^{k} \sum_{j=1}^{n_i} \left(X_{ij} - \overline{X}_{i.} \right)^2 \tag{7.10}$$

or

$$SS_{Total} = SS_{Cond} + SS_{Error} \tag{7.11}$$

This is the fundamental result that constitutes the basis of the ANOVA approach. The result states that the total sum of squares SS_{Total} or the overall variability can be partitioned into the variability SS_{Cond} due to the difference between conditions (treatments) and the variability SS_{Error} within treatments. This is essentially the same partitioning that can be done on the deviations (Eq. 7.6) only that the SS partitioning is more useful since the SS is directly related to the variance.

7.2.1.2   Degrees of freedom

It is interesting to calculate the number of degrees of freedom for each of the sum of squares. The overall variance is reflected by SS_{Total}. For this variance, there are $N - 1$ degrees of freedom where $N = \sum_{i=1}^{k} n_i$ is the total number of measurements in the data set. The condition variance is reflected by SS_{Cond} which has $k - 1$ degrees of freedom if there are k different conditions. Finally, there are k error variances, each of them having $n_i - 1$ degrees of freedom. We can calculate the number of degrees of freedom for the errors as:

$$\sum_{i=1}^{k} (n_i - 1) = \sum_{i=1}^{k} n_i - \sum_{i=1}^{k} 1 = N - k \tag{7.12}$$

Now, one can appreciate that the partitioning that is true for the sum of squares:

$$SS_{Total} = SS_{Cond} + SS_{Error} \tag{7.13}$$

also carries over to the degrees of freedom:

$$N - 1 = (k - 1) + (N - k) \tag{7.14}$$

7.2.1.3   Testing the hypotheses

We have seen that the sums of squares are closely related to the variances. The following quantities can be defined:

1. **Error Mean Squares** (MS_E)

$$\frac{SS_{Error}}{\sum_{i=1}^{k}(n_i - 1)} = \frac{SS_{Error}}{N - k} \tag{7.15}$$

   It can be shown that the expected value of this quantity is σ^2.

2. **Condition Mean Squares** (or Treatment Mean Squares):

$$MS_{Cond} = \frac{SS_{Cond}}{k - 1} \tag{7.16}$$

   It can be shown that the expected value of this quantity is $\sigma^2 + \sum_i \frac{n_i(\mu_i - \mu)^2}{k-1}$. The proof of this statement is beyond the scope of this book. However, it can be seen that if the null hypothesis is true and all means are equal, the second term of this expression becomes zero and the expected value of the Condition Mean Squares MS_{Cond} becomes equal to σ^2, as well.

The two statistics above allow us to test the null hypothesis using the techniques developed to test for hypotheses on variance. In particular, the statistic:

$$\frac{MS_{Cond}}{MS_E} \tag{7.17}$$

is distributed as an F distribution with $\nu_1 = k - 1$ and $\nu_2 = N - k$.

We have seen that the expected values for both the denominator and numerator are σ^2 when the null hypothesis is true. Therefore, the F statistic above will have a value of 1 if the null hypothesis is true and a value larger than 1 when some of the means are different. Since the value of this statistic can only be larger or equal to one, the F test used to possibly reject the null hypothesis will always be one-tailed.

Most statistical software packages allow the user to perform an ANOVA analysis very easily. Furthermore, ANOVA is also available in the Data Analysis Pack that is an option to Microsoft Excel. Let us consider the following example.

Example 7.1

The expression level of a gene is measured in two different conditions: control subjects and cancer patients. The measurements in control subjects produced the following values: 2, 3 and 1. The measurements in cancer patients produced the values 6, 7 and 5. Is there a significant difference between the expression level of this gene in the two conditions? □

SOLUTION   We can organize the data as follows:

	Control	Cancer
Measurement 1	2	6
Measurement 2	3	7
Measurement 3	1	5

The results of the ANOVA analysis can be summarized as shown in the following table. This is the summary as it is presented by Excel.

Source of variation	SS	df	MS	F	p-value	F critical
Between groups	24	1	24	24	0.00805	7.708
Within groups	4	4	1			
Total	28	5				

Note that the partition of the sum of squares is noticeable in the first column ($24 + 4 = 28$) and the partitioning of the degrees of freedoms can be observed in the second column ($1 + 4 = 5$). The third column shows the condition and error mean squares and the fourth column shows the value of the F statistic. The p-value of this observed statistic is 0.008 (column 4) which is smaller than the critical value for the 5% significance level (column 5). In conclusion, the data allow us to reject the null hypothesis and conclude that the given gene is expressed differently in the two conditions. □

As an observation, a two-tail t-test assuming equal variances.[1] will produce exactly the same p-value of 0.008 in this example. As noted above, a comparison that involves *only* two conditions can be analyzed with either ANOVA or a t-test. The following example is a situation in which more than two samples are involved and the t-test cannot be applied directly anymore.

Example 7.2

The expression level of a gene is measured in three different conditions: mesothelioma patients who survived more than 1 year after they were operated on (long term survivors), mesothelioma patients who survived less than 1 year (short term survivors) and control subjects. Is there a significant difference between the expression level of this gene in the three conditions?

	Control	Short-term	Long-term
Patient 1	2	6	2
Patient 2	3	7	2
Patient 3	1	5	1

▯

SOLUTION

The results of the ANOVA analysis can be summarized as shown in the following table. The critical value for the F statistic corresponds to a significance level of 5% for $F_{2,6}$.

Source of variation	SS	df	MS	F	p-value	F critical
Between groups	34.88	2	17.44	22.42	0.001	5.14
Within groups	4.66	6	0.77			
Total	39.55	8				

Once again, the data allow us to reject the null hypothesis. However, this time there are 3 different conditions/treatments and the only information provided by the model I ANOVA that we have performed is that at least one pair of means is different. Unfortunately, ANOVA does not provide information as to *which* of the three possible pairs of means contradicts the null hypothesis. The three possibilities are: $\mu_1 \neq \mu_2$, $\mu_1 \neq \mu_3$ and $\mu_2 \neq \mu_3$. This can be done with a series of pair-wise t-tests with corrections for multiple experiments as discussed in Chapter 9. ▯

Example 7.3

There is a conjecture that a certain gene might be linked to ovarian cancer. The ovarian cancer is sub-classified into 3 categories: stage I, stage II and stage III-IV. There are 3 samples available, one from each stage. These samples are labelled with

---

[1] This is also known as a homoscedastic test.

3 colors and hybridized on a 4 channel cDNA array (one channel remains unused). The experiment is repeated 5 times and the following measurements are collected:

Array	mRNA 1	mRNA 2	mRNA 3
1	100	95	70
2	90	93	72
3	105	79	81
4	83	85	74
5	78	90	75

Is there any difference between the 3 mRNA samples?

SOLUTION   We will use a one-way Model I ANOVA analysis at the 5% significance level. We start by calculating the sums and averages for the 3 samples.

Array	mRNA 1	mRNA 2	mRNA 3
1	100	95	70
2	90	93	72
3	105	79	81
4	83	85	74
5	78	90	75
$T_{i.}$	456	442	372
$\overline{X}_i$	91.2	88.4	74.4
$T_{i.}^2$	207936	195364	138384

We calculate the sums of squares:

$$SS_{Treat} = \sum_i \frac{T_{i.}^2}{n_i} - \frac{T_{..}^2}{N} = 810.13 \tag{7.18}$$

$$SS_{Total} = \sum_i \sum_j X_{ij}^2 - \frac{T_{..}^2}{N} = 1557.33 \tag{7.19}$$

$$SS_{Error} = SS_{Total} - SS_{Treat} = 1557.33 - 810.13 = 747.2 \tag{7.20}$$

We calculate the mean squares by dividing the sums of squares by their respective degrees of freedom. There are $k - 1 = 3 - 1 = 2$ degrees of freedom for the treatments and $N - k = 15 - 3 = 12$ degrees of freedom for the error.

$$MS_{Treat} = \frac{SS_{Treat}}{k-1} = \frac{810.13}{2} = 405.06 \tag{7.21}$$

$$MS_{Error} = \frac{SS_{Error}}{N-k} = \frac{747.2}{12} = 62.26666667 \tag{7.22}$$

And finally, we calculate the value of the F statistic:

$$F = \frac{MS_{Treat}}{MS_{Error}} = \frac{405.06}{62.26} = 6.50 \tag{7.23}$$

These results can be summarized in the usual ANOVA table:

	SS	df	MS	F	p-value	F critical
Treatments	810.13	2	405.06	6.50	0.012	3.885
Error	747.2	12	62.26			
Total	1557.33					

We see that the p-value is lower than the chosen significance level and we reject the null hypothesis. Implicitly, we accept the research hypothesis that at least one pair of means are different. We now have to test which means are different. We will use a t-test with Bonferroni correction for multiple experiments (see Chap. 9). Our chosen significance level is 0.05 and there are 3 parallel experiments. The value of α adjusted according to Bonferroni is:

$$\alpha_{Bonferroni} = \frac{0.05}{3} = 0.017 \tag{7.24}$$

We will use the t-test with the assumption of equal variances (the ANOVA itself assumes that all distributions involved are normally distributed with equal variance). There are $N - k = 15 - 3 = 12$ degrees of freedom. The t statistic is calculated as:

$$t = \frac{\overline{X}_{i.} - \overline{X}_{j.}}{\sqrt{MS_{Error}\left(\frac{1}{n_i} + \frac{1}{n_j}\right)}} \tag{7.25}$$

The t-values and the corresponding p-values are as follows:

	1 vs. 2	1 vs. 3	2 vs. 3
t	0.561048269	3.687585516	3.07298793
p-value	0.585089205	0.00310545	0.009662999

The null hypothesis can be rejected for the comparisons between 1 vs. 3 and 2 vs. 3. The null hypothesis cannot be rejected for the comparison 1 vs. 2. The conclusion is that 3 is different from both 1 and 2. In other words, this gene seems to be expressed differently in the later stages (III and IV) of ovarian cancer with respect to the earlier stages (stages I and II).

7.2.2   One-way Model II ANOVA

We have seen that in Model I ANOVA, the researcher has a specific interest in the conditions or treatment under study. In particular, the question regards whether there are differences between any pair of the specific conditions considered. If such differences exist (at least one), it is of interest to identify which ones are different. In Model II, or random effects, ANOVA, the particular conditions studied are not of interest and the question focuses on whether there is a significant variability among the conditions. The data layout looks similar to the layout for Model I ANOVA:

Condition

1	2	$\cdots$	i	$\cdots$	k
X_{11}	X_{21}	$\cdots$	X_{i1}	$\cdots$	X_{k1}
X_{12}	X_{22}	$\cdots$	X_{i2}	$\cdots$	X_{k2}
$\vdots$	$\vdots$	$\vdots$	$\vdots$	$\vdots$	$\vdots$
X_{1n_1}	X_{2n_2}	$\cdots$	X_{in_i}	$\cdots$	X_{kn_k}

but often the conditions or treatments involved in Model II ANOVA are chosen at random from a larger population. The hypotheses for a Model II can be formulated as follows:

$$H_0: \quad \sigma_{Cond}^2 = 0$$
$$H_a: \quad \sigma_{Cond}^2 > 0$$

The null hypothesis states that there is no significant difference between conditions (columns) and the research hypothesis states that there is a significant difference between conditions. Once again, in the Model II analysis there is no interest towards the individual conditions which may have been drawn at random from a larger population.

The partitioning of the sums of squares still holds:

$$SS_{Total} = SS_{Cond} + SS_{Error} \tag{7.26}$$

but the expected value for the condition mean square MS_{Cond} is:

$$E\left(MS_{Cond}\right) = \sigma^2 + n_0\sigma_{Cond}^2 \tag{7.27}$$

where n_0 is:

$$n_0 = \frac{N - \sum_{i=1}^{k}\frac{n_i^2}{N}}{k-1} \tag{7.28}$$

and N is the total number of measurements $N = \sum_{i=1}^{k} n_i$.
The error mean square MS_{Error} for this test is:

$$E\left(MS_E\right) = \sigma^2 \tag{7.29}$$

Once again, if H_0 is true, the factor n_0 above is 0, both numerator and denominator are σ^2 and the expected value for the F statistic:

$$F = \frac{MS_{Cond}}{MS_{Error}} \tag{7.30}$$

will be 1. If H_0 is not true, the F statistic will be greater than 1. Again a one-tail testing can be performed using the F distribution.

Example 7.4

A set of 5 genes is proposed as a set of house-keeping genes to be used in the normalization. These genes are assumed to be expressed at approximately constant levels throughout the body. A number of 5 tissues have been randomly chosen to investigate the expression levels of this set of genes. The measurements of the gene expression levels provided the following data:

	Liver	Brain	Lung	Muscle	Pancreas
Gene 1	12	14	17	10	32
Gene 2	15	20	31	15	10
Gene 3	17	23	19	20	12
Gene 4	20	21	14	25	13
Gene 5	12	19	26	30	41

Can this set of genes be used as housekeeping genes for normalization purposes?

SOLUTION

In this example, the particular tissues studied are not of interest. These tissues have been picked at random from a larger population. The question of interest is whether there is a significant variability among the tissues for this particular set of genes. This is a Model II ANOVA situation. The null and research hypotheses are:

$$H_0: \quad \sigma^2_{Tissue} = 0$$
$$H_a: \quad \sigma^2_{Tissue} > 0$$

We can calculate the $T_{i.}$ and $\overline{X}_i$ values as follows:

	12	14	17	10	32
	15	20	31	15	10
	17	23	19	20	12
	20	21	14	25	13
	12	19	26	30	41
T_i	76	97	107	100	108
$\overline{X}_i$	15.2	19.4	21.4	20	21.6

We then calculate the basic summary statistics:

$$\sum_i \sum_j X_{ij}^2 = 10980$$

$$T_{..} = \sum_i \sum_j X_{ij} = 488$$

$$N = 25$$

The sum of squares can be calculated using the more convenient formulae:

$$SS_{Total} = \sum_i \sum_j X_{ij}^2 - \frac{T_{..}^2}{N} = 10980 - \frac{488^2}{25} = 1454.24$$

and

$$SS_{Tissue} = \sum_i \frac{T_{i.}^2}{n_i} - \frac{T_{..}^2}{N} = 133.84$$

We can now extract SS_{Error} from the partitioning formula 7.11:

$$SS_{Error} = SS_{Total} - SS_{Tissue} = 1454.24 - 133.84 = 1320.4$$

There are 4 degrees of freedom for the treatments ($k - 1$ where $k = 5$) and 20 degrees of freedom for the error ($N - k = 25 - 5$). In consequence the mean squares will be:

$$MS_{Tissue} = \frac{SS_{Tissue}}{k - 1} = \frac{133.84}{4} = 33.46$$

$$MS_{Error} = \frac{SS_{Error}}{N - k} = \frac{1320.4}{20} = 66.02$$

yielding an F statistic of:

$$F = \frac{33.46}{66.02} = 0.507$$

The p-value corresponding to this F value can be obtained from an F distribution with 4 and 20 degrees of freedom and is $p = 0.73$. Under these circumstances, we cannot reject the null hypothesis. In conclusion, there is no evidence that the chosen genes are expressed differently between tissues and it is probably[2] safe to use them for normalization purposes.

7.3 Two-way ANOVA

The one-way ANOVA methods allow us to investigate data in which only one factor is considered. However, in many problems the data might be influenced by more than one factor. In fact, even in the data used in the one-way ANOVA there is a second dimension, that of the individual measurements within the same group. These appeared as rows in the data matrix. This section will discuss the methods used to analyze data when several factors might contribute to the variability of the data.

---

[2]Note that strictly speaking, our hypothesis testing did not prove that the genes are expressed uniformly throughout the given tissue but merely showed that there are no obvious inter-tissue differences so far. See Section 5.4 for a more detailed discussion of how to interpret the conclusions of hypothesis testing.

7.3.1   Randomized complete block design ANOVA

Let us consider a microarray experiment in which three mRNA samples are hybridized on a number of arrays. For the sake of the presentation, we will assume this was done with 3-channel arrays. On such arrays, the mRNA can be labelled with 3 colors and 3 samples can be tested simultaneously on the same array. Each gene will be measured 6×3 times (6 arrays $\times$ 3 mRNA samples). Let us consider the data gathered for one gene:

	Treatments		
	mRNA 1	mRNA 2	mRNA 3
Array 1	100	95	88
Array 2	90	93	85
Array 3	105	79	87
Array 4	83	85	83
Array 5	78	90	89
Array 6	93	75	75

Each of the expression values in this array can be looked at in two ways. Firstly, each value comes from a given mRNA sample. The mRNA sample is the first factor that affects the measurements. Secondly, each value was obtained using a given microarray. Therefore, the microarray on which the measurement was performed is the second factor that can influence the measurements. All values in any given row share the fact that they have been obtained on the same microarray. Such values represent **a block** of data. The goal of this experiment is to compare the given mRNA samples. However, each microarray will be slightly different from every other microarray and we would like to remove the variability introduced by the arrays such that we can make a better decision about the mRNA samples. In this situation, we are not interested in the differences between microarrays. We had to measure the 3 mRNAs at the same time because this is how the microarrays work and we would like to distinguish between the variability introduced by them and the variability due to the different mRNA samples. This experiment design is called **randomized complete block design**. The design is called **randomized** because each mRNA sample was assigned randomly within each block. The term block denotes the fact that the data can be partitioned into chunks that are expected to share certain features (e.g. all values obtained from a given microarray will share all characteristics determined by washing or drying). Finally, the term **complete** denotes the fact that each mRNA sample is measured exactly once in each block.

The hypotheses will be:

1. H_0: $\mu_{mRNA_1} = \mu_{mRNA_2} = \mu_{mRNA_3}$

2. H_a: There is at least one pair of means that are different from each other.

The data can be organized as follows:

Block		mRNA sample (Treatment)				Totals	Mean
		1	2	$\cdots$	k		
	1	X_{11}	X_{21}	$\cdots$	X_{k1}	$T_{.1.}$	$\overline{X}_{.1}$
	2	X_{12}	X_{22}	$\cdots$	X_{k2}	$T_{.2.}$	$\overline{X}_{.2}$
	$\vdots$	$\vdots$	$\vdots$	$\vdots$	$\vdots$	$\vdots$	$\vdots$
	b	X_{1b}	X_{2b}	$\cdots$	X_{kb}	$T_{.b.}$	$\overline{X}_{.b}$
	Totals	$T_{1.}$	$T_{2.}$	$\cdots$	$T_{k.}$	$T_{..}$	
	Means	$\overline{X}_{1.}$	$\overline{X}_{2.}$	$\cdots$	$\overline{X}_{k.}$	$\overline{X}_{..}$	

The notations remain the same as in Section 7.1.2: a dot denotes a summation on the corresponding index, T is the simple sum of the terms and $\overline{X}$ stands for a mean. We will make the following assumptions:

1. Each measurement is a random, independent sample from a population. The measurement located in the cell (i, j) is assumed to have mean μ_{ij}. The input data represents samples from $b \times k$ populations.

2. All these populations are normally distributed with a variance σ^2.

3. There is no interaction between the effects of the arrays and those of the samples.

The randomized complete block model assumes that each individual datum is a result of a superposition of various effects:

$$X_{ij} = \mu + B_j + V_i + \epsilon_{ij} \qquad (7.31)$$

where X_{ij} is the value measured for sample i and array j, μ is an overall mean effect, B_j is the effect of the array j, V_i is the effect of the sample i and ϵ_{ij} is the random error. In our example, an array is a block and a sample is a variety of mRNA. The effects above can be quantified as:

1. B_j is the effect of the j-th block. This can be assessed by:

$$B_j = \mu_{.j} - \mu \qquad (7.32)$$

which is the difference between the array mean and the overall mean. This is very intuitive since if we were to describe how different array j is from the overall mean, this is exactly the quantity that we would want to use.

2. V_i is the effect of the i-th variety (or mRNA sample):

$$V_i = \mu_{i.} - \mu \qquad (7.33)$$

3. ϵ_{ij} is the effect of the random error that affects measurement X_{ij} in particular. Therefore, this can be quantified as:

$$\epsilon_{ij} = X_{ij} - \mu_{ij} \qquad (7.34)$$

The rest of the process is exactly as in the one-way ANOVA. The magic happens again with the sum of squares (see Eqs. 7.9 through 7.11) and all the inconvenient terms disappear to leave a very simple and intuitive relation:

$$SS_{Total} = SS_{Treat} + SS_{Blocks} + SS_{Error} \tag{7.35}$$

This partitioning also happens at the level of degrees of freedom. There are k samples so SS_{Treat} will have $k - 1$ degrees of freedom, there are b blocks so SS_{Blocks} will have $b - 1$ degrees of freedom and there are $b \times k$ individual measurements so SS_{Error} will have $(b - 1) \cdot (k - 1)$ degrees of freedom. The partition of the degrees of freedom will be:

$$\begin{aligned} SS_{Total} &= SS_{Treat} + SS_{Blocks} + & SS_{Error} \\ bk - 1 &= (k - 1) + (b - 1) + (b - 1) \cdot (k - 1) \end{aligned} \tag{7.36}$$

Following the same steps used in the one-way ANOVA, we then calculate the mean squares MS_{Treat}, MS_{Blocks} and MS_{Error}:

$$MS_{Treat} = \frac{SS_{Treat}}{k - 1} \tag{7.37}$$

$$MS_{Block} = \frac{SS_{Block}}{b - 1} \tag{7.38}$$

$$MS_{Error} = \frac{SS_E}{(b - 1) \cdot (k - 1)} \tag{7.39}$$

and the F statistics for both samples and arrays:

$$F = \frac{MS_{Treat}}{MS_{Error}} \tag{7.40}$$

$$F = \frac{MS_{Block}}{MS_{Error}} \tag{7.41}$$

It can be shown that the F statistics above are 1 when the null hypothesis is true and larger than 1 when the distributions are different. This means that we can test our hypotheses using one-tail of the F distributions with the correspondent number of degrees of freedom.

7.3.2 Comparison between one-way ANOVA and randomized block design ANOVA

Before we solve some examples, let us try to identify the differences between the one-way ANOVA and the randomized block design ANOVA. In both cases the data seem to be the same rectangular matrix so it is important to clarify when to use which method. Let us reconsider the partitioning of the degrees of freedom for the randomized block design:

$$\begin{aligned} SS_{Total} &= SS_{Treat} + SS_{Blocks} + & SS_{Error} \\ bk - 1 &= (k - 1) + (b - 1) + (b - 1) \cdot (k - 1) \end{aligned} \tag{7.42}$$

This expression is an invitation to some reflections. Let us extract the degrees of freedom corresponding to the error:

$$(b-1) \cdot (k-1) = bk - 1 - (k-1) - (b-1) \tag{7.43}$$

We can now perform the computations and also consider that $bk = N$ which is the total number of measurements. We obtain:

$$(b-1) \cdot (k-1) = bk - k - b + 1 = N - k - (b-1) \tag{7.44}$$

This is a very interesting and meaningful result if we compare it with the number of degrees of freedom of the error in one-way ANOVA (see Eq. 7.14). The degrees of freedom of the error have been reduced by $b-1$ with respect to the one-way ANOVA. It turns out that if everything else is kept constant, reducing the degrees of freedom of the denominator increases the critical values of the F distribution. For any given set of data, it will be harder to produce an F statistic more extreme than the critical value. In other words, we will be able to reject the null hypothesis less often which means we have decreased the power of the test (see Section 5.6 for the definition of the power of a statistical test). If this were the only difference between the two approaches, the randomized block design would never be used. Fortunately, there is one other important difference. Let us go back to the partitioning of the sum of squares for the randomized complete block design:

$$SS_{Total} = SS_{Treat} + SS_{Blocks} + SS_{Error} \tag{7.45}$$

and let us extract from here the term corresponding to the error:

$$SS_{Error} = SS_{Total} - (SS_{Treat} + SS_{Blocks}) \tag{7.46}$$

This equation can be directly compared to the one for the one-way analysis:

$$SS_{Error} = SS_{Total} - SS_{Treat} \tag{7.47}$$

For any given data set, the sum of squares for the total and the error are exactly the same. We can now see what the randomized blocking does: **it extracts the variability due to the blocks from the variability due to the treatments**. This allows us to test for differences between treatments using only the "interesting" variability given by the different samples and ignore, if we choose to do so, the variability due to the arrays. The SS_{Error} goes to the denominator of the F statistic ($F = \frac{MS_{Treat}}{MS_{Error}}$) and therefore decreasing it will increase the value of F. If the variability due to the array is large, being able to remove it from the error variability may allow us to reject the null hypothesis in situations in which a one-way ANOVA might not.

Furthermore, doing an analysis using blocking can reassure us that the effects of the arrays are not falsifying our conclusions. This should appear as a low value of SS_{Blocks} and a high p-value for the corresponding F statistic. In this case, one can safely conclude that blocking is not necessary and a completely randomized design together with a one-way ANOVA is sufficient. Alternatively, a significant p-value for the blocks in a design as above might signal that the laboratory process that produced the microarrays is inconsistent and the arrays are significantly different from each other.

7.3.3 Some examples

Armed with this new methodology, let us consider the following examples.

Example 7.5

A microarray experiment measured the expression level of the genes in the liver of mice exposed to ricin vs. control mice. Two mRNA samples are collected from mice exposed to ricin and one is collected from a control mouse. The 3 samples are labelled and hybridized using a 4-channel technology on 6 arrays. One of the channels is used for normalization purposes so the experiment provides 3 measurements for each gene. The data gathered for one gene of particular interest are given below. Are there any significant differences between the mRNA samples coming from the 3 animals?

	Treatments		
	ricin 1	ricin 2	control
Array 1	100	95	88
Array 2	90	93	85
Array 3	105	79	87
Array 4	83	85	83
Array 5	78	90	89
Array 6	93	102	75

☐

SOLUTION

A quick inspection of the data reveals that the last column has consistently lower values than the rest. This column corresponds to the control mouse and it is very likely from a biological point of view to expect the control mouse to be different than the experiment. At this point, it is very tempting to conclude without further ado that this gene does play a role in the defense mechanism triggered by ricin. However, we know better by now so we will analyze the data using ANOVA.

We start by calculating the sums and means for the columns and rows:

		mRNA sample				
		1	2	3	Totals	Means
	Array 1	100	95	88	283	94.33
	Array 2	90	93	85	268	89.33
	Array 3	105	79	87	271	90.33
Block	Array 4	83	85	83	251	83.66
	Array 5	78	90	89	257	85.66
	Array 6	93	102	75	270	90
	Totals	549	544	507	1600	
	Means	91.5	90.66	84.5		88.88

We then calculate the sums of squares:

$$SS_{Total} = \sum_i \sum_j X_{ij}^2 - \frac{T_{..}^2}{N} = 1141.77$$

$$SS_{Treat} = \sum_i \frac{T_{i.}^2}{b} - \frac{T_{..}^2}{N} = 175.44$$

$$SS_{Blocks} = \sum_k \frac{T_{.j}^2}{} - \frac{T_{..}^2}{N} = 212.44$$

$$SS_{Error} = SS_{Total} - (SS_{Treat} + SS_{Blocks}) = 1141.77 - (175.44 + 212.44) = 753.88$$

$$MS_{Treat} = \frac{SS_{Treat}}{k-1} = \frac{175.44}{2} = 87.72$$

$$MS_{Block} = \frac{SS_{Block}}{b-1} = \frac{212.44}{5} = 42.48$$

$$MS_{Error} = \frac{SS_E}{(b-1) \cdot (k-1)} = \frac{753.88}{10} = 75.38$$

and the F statistics for both samples and arrays:

$$F = \frac{MS_{Treat}}{MS_{Error}} = \frac{87.72}{75.38} = 1.16 \tag{7.48}$$

$$F = \frac{MS_{Block}}{MS_{Error}} = \frac{42.48}{75.38} = 0.56 \tag{7.49}$$

The p-value for the treatments is obtained from the F distribution with 2 and 10 degrees of freedom, respectively, and the p-value for the blocks is obtained from the F distribution with 5 and 10 degrees of freedom, respectively. The two values are:

$$F_{2,10}(1.16) = 0.35$$

$$F_{5,10}(0.56) = 0.72$$

The conclusion is that the null hypothesis stating that the samples are drawn from the same distribution cannot be rejected. In other words, there are no statistically significant differences between the mice. Furthermore, the null hypothesis that the blocks are the same cannot be rejected either. Therefore, we can conclude that the data offer no indication that there are significant differences between the arrays.

This is usually summarized in an ANOVA results table. The following table has been constructed in Excel and includes both the p-values and the critical values of the F distribution:

	SS	df	MS	F	p-value	F critical
Rows	212.444	5	42.488	0.563	0.726	3.325
Columns	175.444	2	87.722	1.163	0.351	4.102
Error	753.888	10	75.388			
Total	1141.777	17				

Example 7.6

A microarray experiment measured the expression level of the genes in the liver of mice exposed to ricin vs. control mice. Two mRNA samples are collected from mice exposed to ricin and one is collected from a control mouse. The 3 samples are labelled and hybridized using a 4-channel technology on 6 arrays. One of the channels is used for normalization purposes so the experiment provides 3 measurements for each gene. The data gathered for one gene of particular interest are given below. Are there any significant differences between the mRNA samples coming from the 3 animals?

		Treatments	
Array	ricin 1	ricin 2	control
1	110	105	95
2	90	95	83
3	105	79	82
4	85	85	73
5	82	80	74
6	93	102	82

SOLUTION

We will analyze this data as a complete random block design. We start by calculating the sums and means for the columns and rows:

		mRNA sample			Totals	Means
		1	2	3		
	Array 1	110	105	95	310	103.33
	Array 2	90	95	83	268	89.33
	Array 3	105	79	82	266	88.66
Block	Array 4	85	85	73	243	81
	Array 5	82	80	74	236	78.66
	Array 6	93	102	82	277	92.33
	Totals	565	546	489	1600	
	Means	94.16	91	81.5	88.88	

We then calculate the sums of squares:

$$SS_{Total} = \sum_i \sum_j X_{ij}^2 - \frac{T_{..}^2}{N} = 2087.77$$

$$SS_{Treat} = \sum_i \frac{T_{i.}^2}{b} - \frac{T_{..}^2}{N} = 521.44$$

$$SS_{Blocks} = \sum_k \frac{T_{\cdot j}^2}{k} - \frac{T_{\cdot\cdot}^2}{N} = 1162.44$$

$$SS_{Error} = SS_{Total} - (SS_{Treat} + SS_{Blocks}) = 2087.77 - (521.44 + 1162.44) = 403.88$$

$$MS_{Treat} = \frac{SS_{Treat}}{k-1} = \frac{521.44}{2} = 260.72$$

$$MS_{Block} = \frac{SS_{Block}}{b-1} = \frac{1162.44}{5} = 232.48$$

$$MS_{Error} = \frac{SS_E}{(b-1) \cdot (k-1)} = \frac{403.88}{10} = 40.38$$

and the F statistics for both samples and arrays:

$$F = \frac{MS_{Treat}}{MS_{Error}} = \frac{260.72}{40.38} = 6.45 \qquad (7.50)$$

$$F = \frac{MS_{Block}}{MS_{Error}} = \frac{232.48}{40.38} = 5.75 \qquad (7.51)$$

The p-value for the treatments is obtained from the F distribution with 2 and 10 degrees of freedom, respectively and the p-value for the blocks is obtained from the F distribution with 5 and 10 degrees of freedom, respectively. The two values are:

$$F_{2,10}(1.16) = 0.009$$

$$F_{5,10}(0.56) = 0.015$$

The ANOVA summary will look like this:

	SS	df	MS	F	p-value	F critical
Rows	1162.44	5	232.48	5.75	0.009	3.325837383
Columns	521.44	2	260.72	6.45	0.015	4.10
Error	403.88	10	40.38			
Total	2087.77	17				

The conclusion here is that null hypothesis can be rejected for the treatments. There is sufficient evidence that the treatments are different at the 5% significance level.
Let us now analyze the same data using a one-way Model I ANOVA approach. Here we do not consider the rows as blocks and we will disregard their variance. We focus on the inter-group variance as it compares with the within group variance.
The summary for the one-way Model I is:

	SS	df	MS	F	p-value	F critical
Treatments	521.44	2	260.72	2.49	0.11	3.68
Error	1566.33	15	104.42			
Total	2087.77					

The p-value resulted from the one-way Model I analysis does not allow us to reject the null hypothesis. In this situation, the variability due to the blocks masked the variability due to the treatments. This shows that the two way analysis in which the blocks were considered as a source of variance was more powerful in this case than the Model I analysis which lumps together the variance due to the blocks and the variance due to the treatments. The two-way analysis showed not only that the treatments are different but also that there is a considerable variation between arrays.
□

7.3.4 Factorial design two-way ANOVA

In the randomized block design, it was assumed that the array effect and the sample effect are independent and there are no interactions between the two besides the simple addition of their effects. In principle, two factors can have a **synergetic interaction** in which the result of both factors applied at the same time will be larger than the sum of the results of the two factors applied independently or an **antagonistic interaction**, or interference, in which the two factors applied at the same time have an effect less than the sum of the two factors applied independently. These types of interactions are taken into consideration by the two-way factorial design. A general equation for a factorial design involving two factors would be:

$$X_{ijk} = \mu + \alpha_i + \beta_j + (\alpha\beta)_{ij} + \epsilon_{ijk} \tag{7.52}$$

In this equation, α_i is the effect of the i-th unit of factor α, β_j is the effect of the j-th unit of factor β, $(\alpha\beta)_{ij}$ is the effect of the interaction between the i-th factor α and the j-th β factor and ϵ_{ijk} is the random noise.

The data for this design will be laid out using 3 indexes and can be imagined as a 3-dimensional matrix. The first index will represent the factor A or the first dimension, the second index the factor B or the second dimension and the third index will represent the position of the measurement within the ij cell or the third dimension. This layout is shown in Fig. 7.4. The data can also be laid out as shown in Fig. 7.5. This layout emphasizes a cell at the intersection of every row and column. This cell represents the set of measurements influenced by factor A_i and B_j. We can calculate totals and means for each such cell as well as some totals and means for each row and column of cells. This is shown in Fig. 7.6.

Each measurement observed can be thought of as the sum of the effects of factors A, B as well as their interactions:

$$X_{ijk} = \mu + \alpha_i + \beta_j + (\alpha\beta)_{ij} + \epsilon_{ijk} \tag{7.53}$$

These effects can be quantified as :

1. μ is the overall mean effect,

2. $\alpha_i = \mu_{i..} - \mu$ is the effect of the i-th factor A,

3. $\beta_j = \mu_{.j.} - \mu$ is the effect of the j-th factor B,

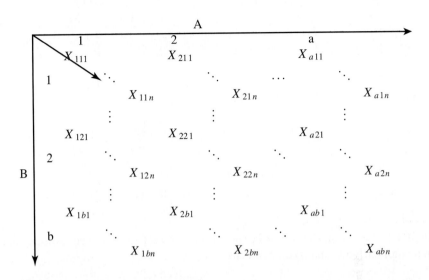

FIGURE 7.4: A factorial design with two factors. The first index represents factor A or the first dimension, the second index represents factor B or the second dimension and the third index will represent the position of the measurement within the ij cell or the third dimension.

B	A 1		2		a	
1	X_{111}		X_{211}		X_{a11}	
		X_{11n}		X_{21n}		X_{a1n}
		$\vdots$		$\vdots$		$\vdots$
2	X_{121}		X_{221}		X_{a21}	
		X_{12n}		X_{22n}		X_{a2n}
		$\vdots$		$\vdots$		$\vdots$
b	X_{1b1}		X_{2b1}		X_{ab1}	
		X_{1bn}		X_{2bn}		X_{abn}

FIGURE 7.5: The layout of the factorial design two-way ANOVA.

Factor B	1	2	3	$\cdots$	a	Level B Totals and Means
1	$T_{11.}$ $\overline{X}_{11.}$	$T_{21.}$ $\overline{X}_{21.}$	$T_{31.}$ $\overline{X}_{31.}$	$\cdots$	$T_{a1.}$ $\overline{X}_{a1.}$	$T_{.1.}$ $\overline{X}_{.1.}$
2	$T_{12.}$ $\overline{X}_{12.}$	$T_{22.}$ $\overline{X}_{22.}$	$T_{32.}$ $\overline{X}_{32.}$	$\cdots$	$T_{a2.}$ $\overline{X}_{a2.}$	$T_{.2.}$ $\overline{X}_{.2.}$
$\vdots$	$\vdots$	$\vdots$	$\vdots$	$\vdots$	$\vdots$	$\vdots$
b	$T_{1b.}$ $\overline{X}_{1b.}$	$T_{2b.}$ $\overline{X}_{2b.}$	$T_{3b.}$ $\overline{X}_{3b.}$	$\cdots$	$T_{ab.}$ $\overline{X}_{ab.}$	$T_{.b.}$ $\overline{X}_{.b.}$
Level A Totals and Means	$T_{1..}$ $\overline{X}_{1..}$	$T_{2..}$ $\overline{X}_{2..}$	$T_{3..}$ $\overline{X}_{3..}$	$\cdots$	$T_{a..}$ $\overline{X}_{a..}$	$T_{...}$ $\overline{X}_{...}$

(Factor A spans the columns 1, 2, 3, ..., a)

FIGURE 7.6: Calculating totals and means for each cell as well as totals and means for each row and column of cells

4. $(\alpha\beta)_{ij} = \mu_{ij.} - \mu_{i..} - \mu_{.j.} + \mu$ is the effect of the interaction between the i-th factor A and the j-th factor B,

5. $\epsilon_{ijk} = X_{ijk} - \mu_{ij.}$ is the effect of the random noise.

Using these expressions, Equation 7.53 can be re-written as:

$$(X_{ijk}-\mu) = (\mu_{i..}-\mu) + (\mu_{.j.}-\mu) + (\mu_{ij.}-\mu_{i..}-\mu_{.j.}+\mu) + (X_{ijk}-\mu_{ij.})$$

Deviation of individual measurement from the grand mean	=	deviation of the i-th A factor from the grand mean	+	deviation of the j-th B factor from the grand mean	+	interaction of the i-th factor A and j-th factor B	+	unexplained variability within a cell

deviation of cell mean from the grand mean

The assumptions of this model are as follows:

1. Each cell constitutes an independent random sample of size n from a population with the mean μ_{ij}.

2. Each such population is normally distributed with the same variance σ^2.

The partitioning performed on the model can be also performed on the estimates of the corresponding terms. Each population mean μ can be estimated by the mean of the sample drawn from the corresponding population $\overline{X}$:

Population mean	Sample mean
μ	$\overline{X}_{...}$
$\mu_{i..}$	$\overline{X}_{i..}$
$\mu_{.j.}$	$\overline{X}_{.j.}$
$\mu_{ij.}$	$\overline{X}_{ij.}$

This yields:

$$(X_{ijk}-\overline{X}_{...}) = (\overline{X}_{i..}-\overline{X}_{...})+(\overline{X}_{.j.}-\overline{X}_{...})+(\overline{X}_{ij.}-\overline{X}_{i..}-\overline{X}_{.j.}+\overline{X}_{...})+(X_{ijk}-\overline{X}_{ij.}) \quad (7.54)$$

The sums of squares can be partitioned in:

$$SS_{Total} = \underbrace{SS_A + SS_B + SS_{A\times B}}_{SS_{Cells}} + SS_{Error} \quad (7.55)$$

where the first three terms correspond to the cell sums of squares:

$$SS_{Cells} = SS_A + SS_B + SS_{A\times B} \quad (7.56)$$

from which we can extract $SS_{A\times B}$. The same partitioning will hold true for the degrees of freedom, as well:

$$
\begin{array}{ccccccccc}
SS_{Total} & = & SS_A & + & SS_B & + & SS_{A\times B} & + & SS_{Error} \\
abn - 1 & = & (a-1) & + & (b-1) & + & (a-1)(b-1) & + & ab(n-1)
\end{array}
$$

$$\underbrace{}_{N-1} \qquad \underbrace{}_{ab-1}$$

Although the details of the formulae are different, the ideas behind the factorial design ANOVA are the same as for every other ANOVA. Sums of squares will be calculated for each of the factors above. Then mean squares will be obtained by dividing each sum of squares by its degrees of freedom. This will produce a variance-like quantity for which an expected value can be calculated if the null hypothesis is true. These mean squares can be compared the same way variances can: using their ratio and an F distribution. Each ratio will have the MS_E as the denominator. Essentially, each individual test asks the question whether a certain component MS_x has the variance significantly different from the variance of the noise MS_E. For instance, in order to test whether there is interaction between the two factors, we can use the ratio:

$$F_{A\times B} = \frac{MS_{A\times B}}{MS_E} \quad (7.57)$$

with the degrees of freedom

$$\nu_1 = (a-1)(b-1) \quad (7.58)$$

and

$$\nu_2 = ab(n-1) \quad (7.59)$$

7.3.5   Data analysis plan for factorial design ANOVA

The data analysis for factorial design ANOVA can be summarized as follows:

1. **Test for interaction between factors.**

   The hypotheses here are:

   (a) H_0: $(\alpha\beta)_{ij} = 0$ for all i, j

   (b) H_a: There is at least one pair i, j for which $(\alpha\beta)_{ij} \neq 0$

   This is tested with by calculating the statistic:

   $$F_{A \times B} = \frac{MS_{A \times B}}{MS_E} \qquad (7.60)$$

   with the $\nu_1 = (a-1)(b-1)$ and $\nu_2 = ab(n-1)$ degrees of freedom, respectively. The null hypothesis is rejected if the numerator is significantly larger than the denominator.

2. If there are significant interactions, the next step is to **separate the means** by using pair-wise t-tests with correction for multiple experiments (e.g. Bonferroni, see Chapter 9) or Duncan's multiple range test [125].

3. If there are no significant interactions, we need to **test whether there are significant differences in means for each of the factors considered**:

   (a) Test for difference between the A factors. Here, the hypotheses are:

       i. H_0: $\alpha_i = 0$ for all i (or all $\mu_{i..}$ are equal)
       ii. H_a: There exists at least a pair of factors A for which the two corresponding means are not equal.

   This is tested with by calculating the statistic:

   $$F_A = \frac{MS_A}{MS_E} \qquad (7.61)$$

   with $a - 1$ and $ab(n-1)$ degrees of freedom, respectively.

   (b) Test for difference between the B factors. Here, the hypotheses are:

       i. H_0: $\beta_i = 0$ for all i (or all $\mu_{i..}$ are equal)
       ii. H_a: There exists at least a pair of factors A for which the two corresponding means are not equal.

   This is tested with by calculating the statistic:

   $$F_B = \frac{MS_B}{MS_E} \qquad (7.62)$$

   with $b - 1$ and $ab(n-1)$ degrees of freedom, respectively.

Source of Variation	SS	df	MS	E(MS)
Cells	SS_{Cells}	$ab - 1$	MS_{Cells}	$\sigma^2 + n \sum_i \sum_j \frac{(\mu_{ij.} - \mu_{...})^2}{ab-1}$
A factors	SS_A	$a - 1$	MS_A	$\sigma^2 + nb \sum_i \frac{(\mu_{i..} - \mu_{...})^2}{a-1}$
B factors	SS_B	$b - 1$	MS_B	$\sigma^2 + na \sum_j \frac{(\mu_{.j.} - \mu_{...})^2}{b-1}$
A × B	$SS_{A \times B}$	$(a-1)(b-1)$	$MS_{A \times B}$	$\sigma^2 + n \sum_i \sum_j \frac{(\alpha\beta)^2_{ij}}{(a-1)(b-1)}$
Error	SS_{Error}	$ab(n-1)$	$MS_{A \times B}$	σ^2
Total	SS_{Total}	$abn - 1$		

TABLE 7.1: The main quantities involved in a factorial design ANOVA: sums of squares, degrees of freedom and mean squares. The table also shows the expected values for each mean square MS.

7.3.6 Reference formulae for factorial design ANOVA

The various sums of squares involved can be calculated as follows:

$$SS_{Total} = \sum_i \sum_j \sum_k \left(X_{ijk} - \overline{X}_{...} \right)^2 = \sum_i \sum_j \sum_k X_{ijk}^2 - \frac{T_{...}^2}{abn} \qquad (7.63)$$

$$SS_A = \sum_i \sum_j \sum_k \left(\overline{X}_{i..} - \overline{X}_{...} \right)^2 = \sum_i \left(\frac{T_{i..}^2}{bn} \right) - \frac{T_{...}^2}{abn} \qquad (7.64)$$

$$SS_B = \sum_i \sum_j \sum_k \left(\overline{X}_{.j.} - \overline{X}_{...} \right)^2 = \sum_i \left(\frac{T_{.j.}^2}{an} \right) - \frac{T_{...}^2}{abn} \qquad (7.65)$$

$$SS_{Cells} = \sum_i \sum_j \sum_k \left(\overline{X}_{ij.} - \overline{X}_{...} \right)^2 = \sum_i \sum_j \left(\frac{T_{ij.}^2}{n} \right) - \frac{T_{...}^2}{abn} \qquad (7.66)$$

The expected values of the various mean squares involved are shown in Table 7.1. These formulae are meaningful and instructive from a theoretical point of view but they will probably never be used in practice in the analysis of microarray data since most of the software available make these computations transparent to the user.

7.4 Quality control

Note that the same data can be analyzed using different models. An example of such a situation was discussed in Example 7.6 in which the same data was analyzed using a complete random block approach and an Model I approach. If this is the case, a natural question is how do we know which model to use? Furthermore, assuming that we have performed the analysis using some model, how can we assess the quality of our work? Recall that each specific ANOVA model makes some specific assumptions (e.g. that the effects can be added, that there are no other effects, etc.) Furthermore, ANOVA itself is a parametric approach since it makes the assumption that the data are distributed normally, the genes are independent, etc. Many of these assumptions are clearly not satisfied in microarray data analysis (e.g. genes do interact with each other) so the results of the ANOVA analysis may be inaccurate. However, ANOVA is so elegant that it provides us with the means to estimate the quality of the results. The key observation is that all ANOVA models make one other common assumption: **all models assume that the noise is random** (and normally distributed like everything else). Furthermore, ANOVA estimates all factors involved in the model use. Thus, at the end of the analysis one can go back and use these estimates to calculate an estimate of the random error. For instance, if the model:

$$\log\left(y_{ijkg}\right) = \mu + A_i + D_j + G_g + (AD)_{ij} + (AG)_{ig} + (VG)_{kg} + (DG)_{jg} + \epsilon_{ijkg}$$
$$(7.67)$$

was used, then the random error can be extracted as:

$$\epsilon_{ijkg} = \mu + A_i + D_j + G_g + (AD)_{ij} + (AG)_{ig} + (VG)_{kg} + (DG)_{jg} - \log\left(y_{ijkg}\right)$$
$$(7.68)$$

In this equation, all terms on the right hand side can be estimated by using partial sums (see for instance Eq. 7.54). We can then look at the distribution of the residuals and assess the quality of our analysis. If the residuals do not show any particular trends, the conclusion is that the analysis has accounted for all systematic effects and the results are credible. If, on the other hand, the residuals show any trends or substantial deviation from a random normal distribution, the conclusion is that the model did not capture the effects of all factors and interactions. In this case, an alternative model might be sought. If several models show no tendencies in their residual distributions, the model with the lowest MS_E is to be preferred since for that model, the factors considered managed to explain more of the variability exhibited by the data.

The following example from [179] will illustrate the usage of two ANOVA models to analyze the same data.

Example 7.7

An experiment is studying the effect of 2,3,7,8-tetrachlordibenzo-p-dioxin (TCDD) on cells from the human hepatoma cell line HepG2. The experiment used a two channel cDNA technology. Instead of using replicated spots (the same gene spotted several times on every array), the experiment involved performing 6 different labelling and hybridization experiments. The experimental design is shown in the following table:

Array	Cy3	Cy5
1	variety 2	variety 1
2	variety 2	variety 1
3	variety 1	variety 2
4	variety 2	variety 1
5	variety 1	variety 2
6	variety 1	variety 2

There were 1920 genes spotted on the array but 13 of them were consistently below the detectable level and are discarded from consideration. Calculate the degrees of freedom available to estimate the error for an analysis using the ANOVA models:

$$\log\left(y_{ijkg}\right) = \mu + A_i + D_j + V_k + G_g + (AG)_{ig} + (VG)_{kg} + (DG)_{jg} + \epsilon_{ijkg} \quad (7.69)$$

and

$$\log\left(y_{ijkg}\right) = \mu + A_i + D_j + G_g + (AD)_{ij} + (AG)_{ig} + (VG)_{kg} + (DG)_{jg} + \epsilon_{ijkg} \quad (7.70)$$

$\square$

SOLUTION We will consider there are a arrays, d dyes, v varieties and n genes. Using this notation the degrees of freedom for each term in the first model are:

Term	Formula	df
Arrays	$a-1$	5
Dyes	$d-1$	1
Varieties	$v-1$	1
Genes	$n-1$	$(1920-13)-1 = 1906$
Array $\times$ Gene	$(a-1)(n-1)$	$5 \times 1906 = 9530$
Variety $\times$ Gene	$(v-1)(n-1)$	$(2-1)(1907-1) = 1906$
Dye $\times$ Gene	$(d-1)(n-1)$	$(2-1)(1907-1) = 1906$
Total	$N-1 = a \times d \times n - 1$	$6 \times 2 \times 1907 - 1 = 22883$

The degrees of freedom for the error can be calculated as the difference between the total degrees of freedom and the sum of degrees of freedom of all known terms:

$$\nu_{Error} = \nu_{Total} - \left[\nu_{arrays} + \nu_{dyes} + \cdots + \nu_{Dye \times Gene}\right] \quad (7.71)$$

In this case:

$$\nu_{Error} = 22883 - (5+1+1+1906+9530+1906+1906) = 22883 - 15225 = 7628$$

For the second model, the degrees of freedom are as follows:

Term	Formula	df
Arrays	$a - 1$	5
Dyes	$d - 1$	1
Genes	$n - 1$	$(1920 - 13) - 1 = 1906$
Array $\times$ Dye	$(a - 1)(d - 1)$	$(6 - 1) \cdot 1 = 5$
Array $\times$ Gene	$(a - 1)(n - 1)$	$5 \times 1906 = 9530$
Variety $\times$ Gene	$(v - 1)(n - 1)$	$(2 - 1)(1907 - 1) = 1906$
Dye $\times$ Gene	$(d - 1)(n - 1)$	$(2 - 1)(1907 - 1) = 1906$
Total	$N - 1 = a \times d \times n - 1$	$6 \times 2 \times 1907 - 1 = 22883$

The degrees of freedom of the error for the second model are:

$$\nu_{Error} = 22883 - (5+1+1906+5+9530+1906+1906) = 22883 - 15259 = 7624$$

The conclusion is that both models allow a good estimate of the quality of the analysis because there are sufficient degrees of freedom for the error. These models are fully discussed in [179].

7.5 Summary

This chapter discussed a set of techniques known as ANalysis Of Variance (ANOVA). In spite of the name, in most cases, ANOVA is used to test hypotheses about the means of several groups of measurements. One-way ANOVA analyzes the data considering only one way of partitioning the data into groups, i.e. taking into consideration only one factor. Model I ANOVA focuses on the means of the specific groups of data collected. Model I asks the question whether the means of these groups in particular are different from each other. The null hypothesis is that all groups have the same mean; the research hypothesis is that there is at least a pair of groups whose means are different. If the answer is affirmative, the Model I approach proceeds to separate the means and decide which groups in particular are different from each other. Model II ANOVA is not concerned with the specific groups. The null hypothesis is that there is no significant variability between the groups considered and the research hypothesis is that such variability is significant. The specific groups are not of interest in Model II ANOVA and the analysis stops by rejecting or not the null hypothesis that the overall variance is zero. Two-way ANOVA considers data that can be grouped according to at least two factors. This chapter discussed the randomized block design and the factorial design ANOVA. The general idea behind ANOVA is to identify individual sources of variance in such a way that all un-interesting variability can be removed from the test that addresses the research problem. ANOVA is a parametric approach that assumes normality. Moreover, each specific ANOVA model assumes the data are influenced by a certain number of factors interacting in a given way. These assumptions are important and they may not be true in all situations. Once an ANOVA analysis was performed, an inspection of the model's residuals can provide quality control information. If the residuals are normally distributed, it is likely that the model used was able to capture well the phenomenon under study. If the residuals show any noticeable trends, this is an indication that the model was not able to explain properly the data. A crucial issue for ANOVA is the experimental design. If the experimental design does not provide enough degrees of freedom for a specific model, ANOVA will not be able to provide an accurate analysis. More details about the design of experiments will be provided in Chapter 8.

7.6 Exercises

1. Construct an Excel spreadsheet performing an ANOVA analysis for a randomized complete block design involving 4 samples hybridized simultaneously (using a 4 channel technology) on 6 arrays. Use random numbers as experimental values. Then change the numbers corresponding to one mRNA sample such that the mean of that sample is lower. See how much the mean can go down without the F test showing a significant difference between the different samples. Settle on a data set such that the p-value is barely below 0.05.

2. Repeat the experiment above by changing the values corresponding to one array. See how much the array mean can go down without the F test showing a significant difference.

3. Perform the Model I test on the two data sets obtained above. How do the conclusions of the Model I and Model II (randomized block design) compare? Compare the values of the F statistic, MS_E, the degrees of freedom and the power of the two tests.

Chapter 8

Experiment design

One day when I was a junior medical student, a very important Boston surgeon visited the school and delivered a great treatise on a large number of patients who had undergone successful operations for vascular reconstructions. At the end of the lecture, a young student at the back of the room timidly asked, "Do you have any controls?" Well, the great surgeon drew himself up to his full height, hit the desk, and said, "Do you mean did I not operate on half the patients?" The hall grew very quiet then. The voice at the back of the room very hesitantly replied, "Yes, that's what I had in mind." Then the visitor's fist really came down as he thundered, "Of course, not. That would have doomed half of them to their death." It was absolutely silent then, and one could scarcely hear the small voice ask, "Which half?"

—E.E. Peacock, Medical World News, Sept. 1, 1972

It is often said that experiments must be made without preconceived ideas. That is impossible. Not only would it make all experiments barren, but that would be attempted which could not be done.

— Henri Poincare: The Foundations of Science, Science and Hypothesis

To consult a statistician after an experiment is finished is often merely to ask him to conduct a post-mortem examination. He can perhaps say what the experiment died of.

—Ronald A. Fisher: Indian Statistical Congress, 1938, vol. 4, p. 17

8.1 The concept of experiment design

Chapter 7 discussed in detail some of the classical statistical analysis methods able to process data from experiments influenced by several factors. Very conveniently, in Chapter 7 the data happened to be such that these methods could be applied. Measurements were available for all needed factor interactions. However, this does not happen automatically in every experiment involving data collection. In order for the experiment to provide the data necessary for the analysis, the experiment needs to be *designed*. The design of the experiment is a crucial but often neglected phase in microarray experiments. A designed experiment is a test or a series of tests in which a researcher makes purposeful changes to the input variables of a process or a system such that one may observe and identify the reasons for changes in the output response [214]. If the experiments are not designed properly, no analysis method will be able to obtain valid conclusions. It is very important to provide data for a proper comparison for every major source of variation.

Unfortunately, many a data analyst has been called to analyze the data only after the experiment was carried out and the data collected. It is only by sheer luck that an experiment will provide all required data if it was not designed to do so. Therefore, in many such instances, the statistician or data analyst will have no other choice but to require that a new set of data be collected.

In a designed experiment, the factors that are thought to contribute to the noise of the system are identified, some of them controlled, and the statistical methods for analysis are chosen from the very beginning of the experiment. A laboratory experiment in which these elements have not been identified from the very beginning is a data collection study rather than an experiment.

8.2 Comparing varieties

Ronald Fisher was a statistician who pioneered the field of experiment design. He is currently considered the "grandfather of statistics." His focus at that time was on agricultural experiments. The development of new varieties of crops is a long term and tedious work. The evaluation of several varieties in a study may need a vast area of land. Fisher's classical experiment involves comparing the yield of two strains of corn x and y. In order to compare the two strains, the researcher plans to seed one acre of land with each strain and compare the amount of corn produced from each strain. Unfortunately, the land available for this is divided into two lots A and B situated a certain distance away from each other. The researcher decides to seed lot A with strain x and lot B with strain y. One crop season later, the researcher will be able to compare the two yields only to realize there is no way to assign the difference to a specific cause. Assuming that the crop on lot A, planted with seed x yielded more corn, the researcher will never know whether this was because x is indeed more productive or perhaps because lot A had a more fertile soil. This is a typical example of a confounding experiment design. In most cases, the data provided by a confounded experiment design simply do not allow the researcher to answer the question posed and no data analysis method or approach can change this.[1] This situation is illustrated in Fig. 8.1.

Let us recall how the number of degrees of freedom is partitioned in ANOVA:

$$\nu_{Total} = \sum \nu_{factors} + \nu_{error}$$

---

[1]However, in some situations, an experiment can be designed in such a way that certain uninteresting variables are confounded.

| | Corn variety | | |
Field		x	y
	A	M_{Ax}	
	B		M_{By}

FIGURE 8.1: A confounding experiment design. The data are influenced by two factors. As performed, this experiment does not allow us to decide whether the difference between the two measurements is due to the different strains of corn or to the different soil in the two lots.

Here the total number of degrees of freedom is $N - 1 = 2 - 1 = 1$. Facing this hardship, we can reduce our demands and be willing to perhaps ignore the influence of the different lots and consider the two measurements as two measurements of the two strains in the same conditions. This would collapse the rows of the matrix in Fig. 8.1 producing a matrix with a single row and two columns. Even in this case, the degrees of freedom for the variety would then be $2 - 1 = 1$ and there are no degrees of freedom available in order to estimate the error.

$$\nu_{Total} = \nu_{Corn} + \nu_{error}$$
$$1 = 1 + 0$$

Alternatively, one can ignore the difference between the varieties and calculate the difference between the lots. This would collapse the columns in Fig. 8.1 producing a matrix with a single column and two rows. Even in this case, the number of measurements is insufficient:

$$\nu_{Total} = \nu_{Lots} + \nu_{error}$$
$$1 = 1 + 0$$

A better experiment design would seed each strain on both lots available. This time the matrix of the experiment design would look like the one in Fig. 8.2. There are $4 - 1 = 3$ total degrees of freedom, $2 - 1 = 1$ block (lot) degrees of freedom and $2 - 1 = 1$ variety (corn) degrees of freedom. The error can now be estimated using $3 - (1+1) = 1$ degree of freedom. The ANOVA approach described in Chapter 7 can now be used to test the hypothesis that the two strains are different while adjusting for the variability introduced by the different soil in the two lots. Furthermore, the same approach can even test whether the two lots are significantly different. Fisher's landmark contribution was the development of the ANOVA methods discussed in Chapter 7 that simultaneously estimate the relative yield of the crop varieties and the relative effects of the blocks of land.

Corn variety

Field		x	y
	A	M_{Ax}	M_{Ay}
	B	M_{Bx}	M_{By}

FIGURE 8.2: A better experiment design. There are $4 - 1 = 3$ total degrees of freedom, $2 - 1 = 1$ block (lot) degrees of freedom and $2 - 1 = 1$ variety (corn) degrees of freedom. The error can be estimated using $3 - (1 + 1) = 1$ degree of freedom.

8.3 Improving the production process

Experiment design methods have been used extensively to improve performance in production plants. In this context, the objective of the experiment design is to troubleshoot a process and transform it into a "robust" process with minimal influence of external sources of variability [214].

Let us consider the example of a microarray production process in a microarray core facility. A major problem in a microarray facility is establishing the correct protocols for printing the arrays, hybridization and scanning. The arrays produced in such a facility need to be reproducible and meet high quality standards. The factorial design described in Chapter 7 can be used to ensure the robustness of the process and assess the effects of different experimental factors [285]. In general, the target preparation step is a major contributor to the overall experimental variability. The target preparation involves two factors: i) reverse transcription of mRNA (which is dependent on the reverse transcriptase enzyme), and ii) incorporation of fluorescently labelled nucleotides during this reverse transcription. Other factors that add to the variability of the targets include the RNA type used in the experiment, age and type of the fluorescent label, dNTP age and the enzyme type.

Printing the arrays is a process that involves many sources of variance: different protocols, different people performing various steps, different consumable materials such as glass slides, buffers, enzymes, etc., instrumentation, etc. This combination of factors transforms the input into an output that has one or more observable response. This is illustrated in Fig. 8.3. There are many variables that have an influence on the output. Some of these are controllable $(x_1, x_2, \ldots, x_n)$ and some uncontrollable $(z_1, z_2, z_3, \ldots, z_n)$. The questions posed by the experimenter may include [214]:

1. Which variables are most influential on the response, y?

2. Where can we set the influential x-s so that y is almost always near the desired nominal value? Some of the sources of noise are understandingly unavoidable. What can we do to minimize them?

3. Where to set the influential x-s so that variability in y is as small as possible?

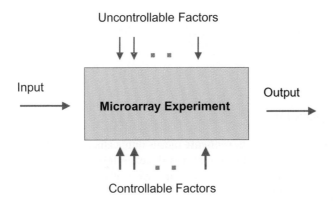

FIGURE 8.3: The output of a process is affected by the input of that process as well as by a number of controllable and uncontrollable factors.

4. Where to set the influential x-s so that the effects of the uncontrollable variables $z_1, z_2, \ldots, z_3$ are minimized?

For example, let us consider an experiment studying the effect of a drug on the expression levels of certain genes. In this case, the input is the amount of drug administered and the output is the expression level of the genes, or more precisely, the difference between the expression levels of the genes without the drug and the expression levels of the genes when the drug is administered. There are situations in which the variance introduced by the various sources is so great that the interesting effect, the gene regulation due to the drug, is completely covered by the variation in the output determined by the other sources of variance. In such circumstances, a preliminary study can be undertaken in order to assess the relative importance of the various sources of variance as well as find ways of minimizing the undesired variability [285]. Once the levels of these factors have been optimized, the experiment can then proceed to vary the concentrations of the drug and measure the differences of the output signals.

8.4 Principles of experimental design

There are three basic principles of experimental design:

1. replication

2. randomization

3. blocking

8.4.1   Replication

In a strict linguistic sense, to **replicate** means duplicate, repeat, or perform the same task more than once. Replication allows the experimenter to obtain an estimate of the experimental error. This estimate of error can become the basis for drawing conclusions whether the observed differences in the data are significant.

Replication is a widely misunderstood term in the microarray field. Often, the misunderstanding is related to the definition of the task to be performed. Thus, if the purpose is to understand and control the noise introduced by the location of the spot on the slide, one can replicate spots by printing exactly the same DNA at different locations on the same slide. If the purpose is to understand and control the noise introduced by the hybridization stage for instance, one can print several exact copies of a given slide (with all other parameters and DNA sources exactly the same) and hybridize several times with exactly the same mRNA in exactly the same conditions. Finally, if the purpose is to control the biological variability, different mRNA samples can be collected from similar specimens and the microarray should be used in exactly the same conditions from all other points of view. The common misunderstanding is related to the fact that often researchers refer to replicates without specifying which one factor was varied while keeping everything else constant. Even more misleading, sometimes the term "replicates" is used to describe results obtained by varying several factors at the same time. For instance, combining two different expression values of the same gene obtained on different arrays, in different hybridization conditions and perhaps with different mRNA, will certainly contain more information than a single expression value, but many people would not think of these two measurements as being replicates. Strictly speaking, these measurements *can* be considered replicates only that the unexplained variability (or noise) will be so large that, in most cases, these data will not provide sufficient support to reject a null hypothesis. An alternative point of view could give up the strict hypothesis testing approach used so far in favor of a Bayesian approach that make inferences based on prior knowledge and accumulated experience.

Spot replication of DNA sequences by printing them adjacent to one another has been used in the field, but is not always the best choice. The very purpose of spotting the same sequence more than once is to have several independent pieces of information. If the spots are printed next to each other, any defect that affects the slide locally, any time during the whole process, is likely to affect all such spots and thus defeat the purpose of having independent measurements. For filter microarrays using radioactive labelling, if a gene is expressed at a high level, its spot can be extremely large[2] and can cover part or all of its neighbors. Since overlapping spots can not be distinguished from each other, this is a source of experimental error. As another example, printing all replicates of a gene in a limited area on the array means that if that particular area is affected by a local washing or drying problem, all spots corresponding to the given gene will be affected and no reliable information about this gene will be available. A better choice is to distribute the spots randomly on the entire surface of the array. Clearly, spots that are replicated on the same array share all sources of variability related to the array and the process that the array was subject to. Thus, the only remaining source of variability between such measurements is the location of the spots on the slide.

For spotted cDNA arrays, there is a non-negligible probability that the hybridization of any single spot containing complementary DNA will not reflect the presence of the mRNA. This probability is about 5% [193]. This means that if a cDNA array contains 5000 spots corresponding to expressed genes, approximately 250 such spots will not appear to have a high signal. The converse situation is also true. It turns out that a spot can provide a substantial signal even if the corresponding mRNA is not present. This probability is even higher, of about 10% [193]. These probabilities are rather large and they should make us think twice before deciding to print single spots.

If we accept that printing each spot just once is not a very good idea, the next question is how many replicates do we need? Would 2 spot replicates be sufficient? How about 3? And what if we wanted to have really good data and are prepared to pay more for it? Shall we go for 5? Several approaches are possible here in order to answer this question. An analytical approach would try to estimate the population mean and variance, choose a minimum detectable change and calculate the number of replicates as in Example 6.2. However, this approach is not always feasible for a large variety of reasons. The fundamental phenomenon in action here is that the variance of the mean of a set of measurements is reduced by the number of replicates as follows (see Theorem 4.1 in Chapter 4):

$$\sigma_{\overline{X}}^2 = \frac{\sigma^2}{n} \tag{8.1}$$

where $\sigma_{\overline{X}}^2$ is the variance of the sample mean, σ^2 is the variance of the measurements and n is the number of replicates.

An empirical approach would just choose a "reasonable" number knowing that any number of replicates is better than not having replicates at all. If this approach is chosen, a good minimum for the number of spot replicates is 3. This is because if two replicate spots yield two very different values such as 10 and 100, one would not know which value is more likely to be the outlier. However, if there are three spot replicates and the values read are 10, 90 and 100, it is more likely that the value 10 is the outlier. In such cases, the software used for the analysis can be instructed to disregard the outliers. Of course, it is still possible to have ambiguous situations such as 3 measurements of 10, 50 and 100 but such a situation is far less probable than

---

[2]This phenomenon is known as bleeding.

having an ambiguous situation when using only two measurements. Indeed, Lee and colleagues [193] used 3 replicates and showed that both false positives (spots showing signal in the absence of the correspondent mRNA) and false negatives (spots not showing signal in the presence of the correcpondent mRNA) can be reduced considerably by combining the data provided by only 3 replicate spots.

Another perennial question related to replication regards the practice of pooling biological samples in microarray experiments. This method has been proposed as a way to reduce costs and time in such studies. Let us consider, for example, a situation in which a certain treatment is applied to a number of 5 animal subjects. There are two approaches. A first approach would combine, or pool, the mRNA coming from the 5 animals and use the pooled sample to hybridize a number, let us say, 5 arrays. The second approach would use each individual sample to hybridize a different array. One of the advantages of pooling is that it can reduce the data collection effort. In this case, the pooled sample could have been hybridized only on 3 arrays instead of 5. However, the classical question is: assuming that the number of arrays is the same, which way is better: hybridize a pooled sample 5 times or hybridize each of the 5 individual samples separately?

Arguments can be brought for both choices but, overall, pooling has the drawback of averaging without control. As Claude Bernard put it: *"If we collect a man's urine during twenty-four hours and mix all this urine to analyze the average, we get an analysis of a urine which simply does not exist; for urine, when fasting, is different from urine during digestion. A startling instance of this kind was invented by a physiologist who took urine from a railroad station urinal where people of all nations passed, and who believed he could thus present an analysis of* average *European urine!"* [34]. Conversely, the advantage of not pooling is that an average can always be calculated from the individual values whereas the individual values cannot be extracted from the average or pooled sample.

As another example, consider the case in which 4 of the animals above have a certain gene expressed consistently at a low level. However, the 5th animal, due to some individual characteristics (e.g. some illness), has a very high level of expression of this gene. When pooling the samples, the large amount of mRNA coming from this last animal can increase the overall level such that the gene appears expressed higher in any hybridization with the pooled sample. In this situation, individual hybridizations would have made us aware of the fact that 4 out of 5 measurements were consistently low while still giving us the opportunity to calculate an average should we wish to do so.

There are several arguments in favor of pooling, as well. The most compelling of them all is that sometimes it is not possible to extract enough mRNA from a single individual. In such cases, of course, having some measurement is better than having no measurement at all and pooling is the only way to go.

8.4.2   Randomization

Randomization has been proclaimed to be the cornerstone underlying statistical methods [214]. **Randomization** requires the experimenter to use a random choice for every factor that is not of interest but might influence the outcome of the experiment. Such factors are called **nuisance factors**. The simplest example is the printing of replicate spots on the array. If such replicates are printed next to each other, a localized defect of the array will affect all of them making it impossible to distinguish the interesting gene effect from the un-interesting effect of the defect. Randomization requires that the replicate spots be printed at random locations throughout the array. Another example is the use of microarray slides from different batches in an experiment comparing a treatment group vs. a control group. If all control animals are tested using slides from one batch and all treated animals are tested using slides from a different batch, it will be impossible to distinguish between the un-interesting variability introduced by the slides and the interesting variability introduced by the treatment. These two factors would be confounded in such an experiment design. However, if the slides are assigned randomly between the controls and the treated animals, the bias is eliminated and the influence of this nuisance factor reduced. Randomization may not always be possible but should always be attempted.

8.4.3 Blocking

Blocking is a design technique used to increase the accuracy with which the influence of the various factors is assessed in a given experiment. A **block** is a subset of experimental conditions which are expected to be more homogeneous than the rest. **Blocking** refers to the method of creating homogeneous blocks of data in which the nuisance factor is kept constant and the factor of interest is allowed to vary. Blocking is used to eliminate the variability due to the difference between blocks (see section 7.3.1 in Chapter 7 for more details). A typical example of a block in microarrays is the microarray slide itself. Since all spots on a given slide are subject to the same factors during the slide processing (hybridization, washing, drying, etc.) it is expected that the measurements of the spots coming from a single slide will be more homogeneous (have a lower variance) than the measurements across the whole experiment. This can be observed particularly well on the control spots if such spots are being used. The two channel cDNA process deals with this very elegantly by hybridizing both control and treatment samples on the same array. Unfortunately, this is not currently possible with the Affymetrix technology which requires each sample to be hybridized on its own array. However, the cDNA process introduces the supplementary nuisance factor of the dyes, which, in turn, would require blocking.

Both blocking and randomization deal with nuisance factors. The difference is that blocking can only be used when the nuisance factor is under our control. Examples include any choice of materials or substances. If the nuisance factor considered is not under our control (e.g. drying marks on the surface of the microarray), randomization remains the only tool available. This is summarized by the general rule: "**block what you can, randomize what you cannot.**"

8.5 Guidelines for experimental design

The following guidelines for designing experiments can be taken into account when planning an experiment:

1. **Describe exactly what your research problem is.** Write the questions you want answered in a laboratory book or your LIMS system. Spell out every detail. The fact is that you may ask many questions from a single experiment. It is necessary to make an effort and foresee what the problems in the experiment might be, what can affect it in a negative or positive way, what the sources of variation might be, what the goals are and what the final results could be. There is an abundance of literature on every subject and sometimes a researcher can gather a lot of information a long time before the actual experiment starts. Stating the objectives is extremely important for going through the whole process of designing the experiment, planning it, implementing it and analyzing the data.

2. **Choose the technology to be used.** Make the choices between cDNA/oligo arrays, commercial/custom arrays, specific brand and type of microarray if commercial arrays are to be used. A very important issue is to choose an array that is appropriate for the biological question being asked. In an exploratory research, in which no hypothesis has been yet formulated, large, comprehensive arrays may be best. In a hypothesis driven research, in which phenomena involving certain specific gene regulatory pathways are hypothesized, more focused arrays may be more useful and convenient. An important issue is to choose a commercial array that has a good representation of the genes conjectured to be relevant. The same issue applies to custom arrays printed in house. These issues are discussed in more detail in Chap. 15.

3. **Involve a collaborator who has experience in experiment design and data analysis**. Try to obtain a firm commitment from them for the data analysis part since experiment design and data analysis are parts of the same thought process and it is not a good idea to change collaborators in-between. Be aware of the fact that communication across field boundaries can be very challenging. In spite of a growing number of people with interdisciplinary interests, it was not too long ago that the word "mitochondria" seemed offensive to some statisticians, and even today, "heteroscedasticity" will sound a bit scary to most biologists! The goal of this chapter is not to allow the life scientist to design their own experiments but help them understand the issues involved, such that they can communicate effectively with a statistician or computer scientist.

4. **Choose the factors that can influence your output in a significant way and their corresponding levels of interest.** Identify which are the major factors you want to follow. In most cases, the major inputs will be the ones directly

related to your scientific question: the effect of the drug, differences between illness and healthy, etc. In most case, the interesting outputs of a microarray experiment will include the expression of various genes or, in ANOVA terms, the variety-gene interaction (see Chapter 7).

5. **Identify the nuisance factors that you would like to consider.** The nuisance factors could be the types of enzyme, types of dyes, nucleotides, who prepared the mRNA sample, etc. Each nuisance factor you choose to follow will require extra work; so the rule of thumb is to keep the number of factors as low as possible by choosing only the ones that are expected to influence considerably the outcome. Divide the nuisance factors into controllable and uncontrollable.

6. **Choose the significance level and desired power.** Consider the issues of Type I (rejecting a true null hypothesis, e.g. concluding that a gene is differentially regulated when the gene is in fact unchanged) and Type II errors (not rejecting a false null hypothesis, e.g. not detecting a true differentially regulated gene). Recall from Chapter 5 that there is always a compromise between false positives and false negatives. For instance, having stringent requirements for when a gene is differentially regulated (low alpha) means that many truly differentially regulated genes may not be detected (low power).

7. **Design your experiment.** Block the controllable nuisance variables and randomize the others. Calculate the number of replicates at every level from the power requirements.

8. **Perform the experiment and collect the data.** Record every detail and check the quality at every step. Any error in experimental procedure will destroy the experimental plan and validation! This author has seen a situation in which a PhD student performed about 100 hybridizations as part of a randomized block design, without ever checking the scanned images. At the end of the experiment, it turned out that the large majority of arrays provided no signal whatsoever on one of the channels. The lack of mRNA prevented repeating the arrays and the experiment was seriously compromised.

9. **Perform the data analysis.** The data analysis is still a challenge but at least the necessary conditions are met. Tools useful at this stage are discussed in Chapters 10, 11 and 13.

10. **Extract the biological meaning from the results of the data analysis**. Statistical methods are not the untouchable proof – it is the biological meaning of your experiments that validates the work! Translate the lists of differentially regulated genes into biological knowledge by mapping differentially regulated genes to the biological processes involved, affected pathways, etc. This step is discussed in detail in Chapter 14.

$$\text{Factor}$$

1	2	$\cdots$	i	$\cdots$	a
X_{11}	X_{21}	$\cdots$	X_{i1}	$\cdots$	X_{a1}
X_{12}	X_{22}	$\cdots$	X_{i2}	$\cdots$	X_{a2}
$\vdots$	$\vdots$	$\vdots$	$\vdots$	$\vdots$	$\vdots$
X_{1n}	X_{2n}	$\cdots$	X_{in}	$\cdots$	X_{an}

FIGURE 8.4:   The data layout for a fixed effect design with one factor.

8.6 A short synthesis of statistical experiment designs

This section will review in a very concise manner several important experiment designs. In this section, a source of variability under our control will be considered a factor. Factors can include: source of mRNA sample, treatment applied to various patients of experiment animals, cell culture, etc. Each factor will have several levels, i.e. possible values. For instance, a factor such as the dye will have two values (or levels): cy3 and cy5. A factor such as a drug concentration will have as many levels as required by the study undertaken. A factor such as mRNA source will have as many levels as there are mRNA samples, etc.

8.6.1 The fixed effect design

In this design, there is only one factor and data are collected repeatedly, at the various levels of the factor. In this design, the data can be laid out as shown in Fig. 8.4. This is very similar to the data layout used in the discussion of the model I ANOVA in Chapter 7, only that the conditions have been now been replaced by the more general factor levels. In this example, there are a different levels for the factor under study and there are n different measurements for each level. Because the number of observations within each treatment level is the same, we say this is a **balanced design**. If the number of observations within each treatment were different, the design would be **unbalanced**. A balanced design has two advantages: i) the model is less sensitive to departures from the equal variance assumption and ii) has a better power with respect to an unbalanced design.

This design, either balanced or unbalanced, has the general model:

$$X_{ij} = \mu + \tau_i + \epsilon_{ij} \tag{8.2}$$

where μ is the overall mean, τ_i is the effect of the factor level (treatment) i and ϵ_{ij} is the term corresponding to the random noise or unexplained variability. In this equation i takes values from 1 to a and corresponds to the various levels of the factor

Block	Factor 1	2	$\cdots$	i	$\cdots$	a
1	X_{11}	X_{21}	$\cdots$	X_{i1}	$\cdots$	X_{a1}
2	X_{12}	X_{22}	$\cdots$	X_{i2}	$\cdots$	X_{a2}
$\vdots$	$\vdots$	$\vdots$	$\vdots$	$\vdots$	$\vdots$	$\vdots$
b	X_{1b}	X_{2b}	$\cdots$	X_{ib}	$\cdots$	X_{ab}

FIGURE 8.5: The data matrix for a randomized complete block design. Every treatment is measured in each block; the distribution of the treatments to specific blocks is random.

considered; j takes values from 1 to n and corresponds to the various measurements for each factor level. This experiment design can be analyzed as discussed in Section 7.2.1.

8.6.2 Randomized block design

Recall that the data are influenced by: i) the factors studied and ii) nuisance factors. In turn, nuisance factors can be a) controllable and b) uncontrollable. A fully randomized design would assign treatments to experimental units (e.g. hybridizations) in a completely random manner. A **block design** considers the individual groups of measurements that are expected to be more homogeneous than the others. Such groups are called blocks, and various treatments are assigned randomly to such blocks. If all treatments are present on every block, the design is a **randomized complete block design**. If some blocks do not include some treatments, the design is an **incomplete block design**. The layout of the data matrix for a randomized complete block design is illustrated in Fig. 8.5.

The model for the block design is:

$$X_{ij} = \mu + \tau_i + \beta_j + \epsilon_{ij} \tag{8.3}$$

where μ is the overall mean, τ_i is the effect of treatment i, β_j is the effect of the block j and ϵ_{ij} is the effect of the random noise. The index i takes values from 1 to the number of factor levels a whereas the index b takes values from 1 to the number of blocks b. This experiment design can be analyzed as discussed in Section 7.3.1.

8.6.3 Balanced incomplete block design

Sometimes it is not possible to run all treatment combinations in each block as the randomized complete block design requires. For instance, if more than two treatments are compared using a two-channel cDNA microarray (e.g. using cy3-cy5), an array, which is a block, will only be able to provide information about two samples at any given time. A design that does not include all treatment combinations on every

	Nuisance factor 2			
Nuisance factor 1	1	2	3	4
1	A	B	C	D
2	B	C	D	A
3	C	D	A	B
4	D	A	B	C

FIGURE 8.6: A 4x4 Latin square design. Each treatment A, B, C and D is measured once for each combination of the nuisance factors.

block is an **incomplete block design**. A **balanced incomplete block design** makes sure that any pair of treatments occurs together the same number of times as any other pair.

The data layout of the incomplete block design is the same as the one for the randomized complete block design shown in Fig. 7.3.1 with the only difference that the matrix will have missing elements since not all treatment combinations are available on each block. The statistical model for the incomplete block design is the same as for the complete block design (Eq. 8.3).

8.6.4 Latin square design

The randomized block design above takes into consideration one factor studied and one nuisance factor. However, many times there is more than one nuisance factor. For instance, if two such factors exist, one would need to measure the value of each level of the treatment for each combination of the two nuisance factors. Let us assume that the factor to be studied has 4 levels denoted by the 4 Latin letters: A, B, C and D. Furthermore, let us assume that there are two nuisance factors, each of them having 4 levels. A Latin square design for such an experiment can be illustrated as in Fig. 8.6. The Latin square has the property that each row and each column contain each treatment exactly once. If numbers are used instead of the Latin letters the sum of the elements of every row and every column would be the same. The matrix for a Latin square design can be obtained easily by starting with a random first row containing each symbol once. Each subsequent row can be obtained from the one above by shifting the elements by one position. An alternative approach would apply the same procedure to the columns.

The statistical model for a Latin square design is:

$$X_{ijk} = \mu + \alpha_i + \tau_j + \beta_k + \epsilon_{ijk} \tag{8.4}$$

where μ is the overall mean, α_i is the effect of the i-th row (the i-th level of the first nuisance factor), τ_j is the effect of the j-th treatment, β_k is the effect of the k-th column (or k-th level of nuisance factor 2) and ϵ_{ijk} is the random noise. Note that this model is a strictly additive model i.e. it does not take into consideration potential interactions between the factors considered.

B	A							
	1		2				a	
1	X_{111}		X_{211}		$\cdots$		X_{a11}	
		$\ddots$		$\ddots$				$\ddots$
		X_{11n}		X_{21n}				X_{a1n}
	$\vdots$		$\vdots$				$\vdots$	
2	X_{121}		X_{221}				X_{a21}	
		$\ddots$		$\ddots$		$\ddots$		$\ddots$
		X_{12n}		X_{22n}				X_{a2n}
	$\vdots$		$\vdots$				$\vdots$	
b	X_{1b1}		X_{2b1}				X_{ab1}	
		$\ddots$		$\ddots$		$\cdots$		$\ddots$
		X_{1bn}		X_{2bn}				X_{abn}

FIGURE 8.7: The data layout for a factorial design with two factors. Factor A has a levels, factor B has b levels and there are n replicates for each combination of factor levels. This design requires $a \cdot b \cdot n$ measurements.

8.6.5 Factorial design

The factorial design is an experiment design that takes into consideration all possible combinations of the levels considered. Furthermore, a factorial design allows us to analyze the interactions between factors. An experiment with two factors A and B, with factor A having a levels and factor B having b levels, will require $a \cdot b$ measurements. If there are n replicates for each combination, the experiment will require a total of $a \cdot b \cdot n$ measurements. The data layout for a factorial design with two factors is shown in Fig. 8.7.

The statistical model for a factorial design with two factors is:

$$X_{ijk} = \mu + \tau_i + \beta_j + (\tau\beta)_{ij} + \epsilon_{ijk} \tag{8.5}$$

where μ is the overall mean, τ_i and β_j are the main effects of the two factors, $(\tau\beta)_{ij}$ is the interaction between the two factors and ϵ_{ijk} is the random noise.

The equations Eq. 8.5 and Eq. 8.3 might seem similar inasmuch they both use only two factors β and τ. However, in the model 8.3, β is a nuisance factor whereas in Eq. 8.5 it is a factor under study. We can control the levels for a factor we study while we can only block the level for a nuisance factor, i.e. make sure that all measurements in a block are affected by the same level of the nuisance factor. Furthermore, the model in Eq. 8.5 also takes into consideration the interaction between the two factors, interaction which is not considered in Eq. 8.3.

The factorial model can be generalized to a situation in which there are a levels for factor A, b levels for factor B, c levels for factor C, etc. In general, if the complete

Source of Variation	Sum of Squares	Degrees of Freedom	Mean Squares	Expected Mean Square	F_0
A	SS_A	$a-1$	MS_A	$\sigma + \frac{bcn\sum\tau_i^2}{a-1}$	$F_0 = \frac{MS_A}{MS_E}$
B	SS_B	$b-1$	MS_B	$\sigma^2 + \frac{acn\sum\beta_j^2}{b-1}$	$F_0 = \frac{MS_B}{MS_E}$
C	SS_C	$c-1$	MS_C	$\sigma^2 + \frac{abn\sum\gamma_k^2}{c-1}$	$F_0 = \frac{MS_C}{MS_E}$
AB	SS_{AB}	$(a-1)(b-1)$	MS_{AB}	$\sigma^2 + \frac{cn\sum\sum(\tau\beta)_{ij}^2}{(a-1)(b-1)}$	$F_0 = \frac{MS_{AB}}{MS_E}$
AC	SS_{AC}	$(a-1)(c-1)$	MS_{AC}	$\sigma^2 + \frac{bn\sum\sum(\tau\gamma)_{ik}^2}{(a-1)(c-1)}$	$F_0 = \frac{MS_{AC}}{MS_E}$
BC	SS_{BC}	$(b-1)(c-1)$	MS_{BC}	$\sigma^2 + \frac{an\sum\sum(\beta\gamma)_{jk}^2}{(b-1)(c-1)}$	$F_0 = \frac{MS_{BC}}{MS_E}$
ABC	SS_{ABC}	$(a-1)(b-1)(c-1)$	MS_{ABC}	$\sigma^2 + \frac{n\sum\sum\sum(\tau\beta\gamma)_{ijk}^2}{(a-1)(b-1)(c-1)}$	$F_0 = \frac{MS_{ABC}}{MS_E}$
Error	SS_E	$abc(n-1)$	MS_E	σ^2	
Total	SS_T	$abcn-1$			

FIGURE 8.8: The ANOVA table for the general factorial design with 3 factors.

experiment is to be replicated n times, there will be $a \cdot b \cdot c \cdots n$ measurements. Note that it is necessary to have at least two replicates ($n \geq 2$) of the complete experiment in order to determine a sum of squares for the error if all possible interactions are included in the model.

As discussed in Chapter 7, the test statistic for each main effect or interaction can be calculated by dividing the corresponding mean square for the given effect or interaction by the mean square error. The number of degrees of freedom for any main effect is the number of levels of the factor minus one. The number of degrees of freedom for any interaction is the product of the degrees of freedom of the component. For instance, the three factor ANOVA model for a factorial design is:

$$X_{ijkl} = \mu + \tau_i + \beta_j + \gamma_k + (\tau\beta)_{ij} + (\tau\gamma)_{ik} + (\beta\gamma)_{jk} + (\tau\beta\gamma)_{ijk} + \epsilon_{ijkl} \quad (8.6)$$

with $i = 1, 2, \ldots, a$, $j = 1, 2, \ldots, b$, $k = 1, 2, \ldots, c$, $l = 1, 2, \ldots, n$. The degrees of freedom, mean squares, the expected mean squares and the formulae for the computation of the F test are given in Figure 8.8.

8.6.6 Confounding in the factorial design

The complete factorial design requires a very large number of experimental runs. For instance, factors that might be taken into consideration in a cDNA microarray experiment can include [285]:

1. type of fluorescent label;

2. age of the fluorescent label;

3. type of RT enzyme;

4. age of dNTPs;

5. incubation time for transcription;

6. the use of total RNA or poly(A) RNA.

If each of these factors is considered at two levels, a complete factorial experiment would require $2 \times 2 \times 2 \times 2 \times 2 \times 2 = 64$ runs.

In most microarray application, it would be impossible to perform a complete replicate of a factorial design in one block. In consequence, we will not be able to obtain information about all factors and all interactions. However, not all factors and interactions are equally important. **Confounding** is a technique that allows the designer to group the various runs of an experiment in blocks, where the block size is smaller than the number of treatment combinations in one replicate [214]. Because of this, the effect of certain factors or factor interactions will be indistinguishable from the effect of the blocks. However, if this is done by design, we can make sure that the confounded factors are factors in which we are not interested. Note that the variability due to these confounded factors is still subtracted from the total variability. Thus, the confounded factors are not lost from under control and the confounding does not necessarily decrease the reliability of our conclusions about the factors of interest. In principle, any factor or any interaction can be confounded. This is done by grouping together all the measurements corresponding to the chosen factor and assigning them to a block. For instance, if a variety (mRNA sample) is always assigned to a dye, the two effects A_i and D_j will be confounded. Confounding the main variety effect with the main dye effect may not be necessarily bad. Recall that in most cases, we are particularly interested in the variety-gene interaction $(VG)_{kg}$, which is to say we are interested in how specific genes change from one mRNA sample to another. As another example of purposeful confounding design, if a variety appears always in the same array-dye combination, the variety will be confounded with the array-dye interaction (see for instance Eq. 8.10).

8.7 Some microarray specific experiment designs

This section will use the notation and methodology developed in Chapter 7 to discuss and compare a few experiment designs and the associated data analysis methods.

	Mouse 1		Mouse 2			
	5		3			Mouse effect
	cy3	cy5	cy3	cy5		
	3	2	3	2		Dye effect
Array 1	62	61	60	59	4	
Array 2	64	63	62	61	6	Array effect
	50					Overall mean effect

TABLE 8.1: An example of a superposition of effects in a linear model. There are 4 main effects: the mouse effect, dye effect, array effect and overall mean effect. The mouse effect adds 5 to all values measured on Mouse 1, and 3 to all values measured on Mouse 2. The dye effect adds 3 to all values measured with cy3 and 2 to all values measured with cy5. The array effect adds 4 to array 1 and 6 to array 2. Finally, the overall mean effect is 50. The value measured in each condition is the result of the addition of the corresponding effects.

8.7.1 The Jackson Lab approach

Let us consider some real-world examples of experiment design in the analysis of microarray data. Kathleen Kerr and Gary Churchill were the first to recognize the suitability of the ANOVA approach for studying microarray data. They also did pioneering work in data analysis and associated experimental design. One of their first papers on this topic proposed the following model [183]:

$$\log(y_{ijkg}) = \mu + A_i + D_j + V_k + G_g + (AG)_{ig} + (VG)_{kg} + \epsilon_{ijkg} \qquad (8.7)$$

The model assumes that these effects can be summed in a linear way. Several researchers showed that this is a reasonable assumption in logarithmic scale. This means the y_{ijkg} are the *logs of* the intensities read from the microarray. The particular base of the logarithm is not very important but base 2 is convenient because it makes the interpretation easier.

In order to illustrate how such an additive model works, let us consider the data in Table 8.1. In this example, there are 4 main effects: the mouse effect, dye effect, array effect and overall mean effect. The mouse effect adds 5 to all values measured on Mouse 1, and 3 to all values measured on Mouse 2. The dye effect adds 3 to all values measured with cy3 and 2 to all values measured with cy5. The array effect adds 4 to array 1 and 6 to array 2. Finally, the overall mean effect is 50. The value measured in each condition is the result of the addition of the corresponding effects. For instance, in Table 8.1, the value measured for Mouse 1, on array 1, on the cy5 channel is:

$$50 + 5 + 2 + 4 = 61$$

In this situation, there are no interactions between factors. This means that if two factors are applied simultaneously the result would simply be the sum of the effects of the individual factors. If the result of the two factors acting together were larger than the sum of the individual effects, we would have a **synergy**. If the result of the two factors acting together were smaller than the sum of the individual effects, we would have an **interference**. Also, in this case, there is no noise since the values measured are exactly the values obtained by adding the individual effects. The noise is sometimes referred to as the unexplained variability and is the difference between the actual value and the sum of the factors considered.

In light of this example, let us revisit Eq. 8.7:

$$\log(y_{ijkg}) = \mu + A_i + D_j + V_k + G_g + (AG)_{ig} + (VG)_{kg} + \epsilon_{ijkg} \qquad (8.8)$$

In this model, there are five terms describing main effects and two terms describing interactions. The term μ corresponds to the overall mean effect. This is a quantity that is present in all measurements, independent on what array, channel or gene the value was measured on. The term A_i represents the effect of array i. This quantity accounts for a change at the array level such as a longer hybridization time. Such an array effect would affect all genes on the given array. The two dyes used in cDNA microarray experiments have different chemical properties, which may be reflected in their differential incorporation. One dye may be consistently brighter than the other one. The term D_j represents the effect of the dye j, such as a higher overall efficiency of one of the dyes. Also, some genes have been observed to have consistently higher or lower expressions than others. Additionally some sequences show differential labelling and hybridization efficiencies. The term G_g represents the contribution of such gene specific effects. Finally, the last main effect, V_k, is the effect of variety k. In this context a variety is an mRNA sample. Since Fisher's pioneering work on ANOVA compared crop varieties, the term variety has remained in the terminology denoting different sub-species to be compared.

Besides the main factors above, the model in Eq. 8.8 also considers certain factor interactions. The variety-gene interactions $(VG)_{kg}$ reflect differences in expression for particular genes under the effect of a given variety. This is the effect that we need in order to answer questions such as: "what are the genes differentially regulated in condition A versus condition B." The last interaction term considered is the $(AG)_{ig}$ interaction. This term accounts for the interaction between arrays and genes. When several arrays are printed, the spots for a gene on the different arrays will vary from each other due to differences in cDNA available for hybridization, printing differences or other causes. Therefore, the location of a gene on an array will have its effect on the final outcome. This is the so-called "spot" effect represented by the interactions term $(AG)_{ig}$.

Finally, ϵ_{ijkg} is the effect of the random noise. This is assumed to be normally distributed and have a zero mean. As explained in Chapter 7, once the analysis is done, the effects above should be used to calculate the residuals which are estimates of effects of the ϵ_{ijkg} term. The inspection of these residuals will validate or invalidate the data analysis and implicitly the experiment design. If the distribution of the residuals shows non-random features, the model used did not capture all sources of systematic variability. In some cases, the data are such that an alternative model may be used to re-analyze the data. However, many times, a non-random distribution of the residuals can indicate an inadequate experiment design.

The model in Eq. 8.7 does not take into account the interactions between the dyes and genes. If one desires to consider this interaction, the model can be augmented by adding the (DG) term [178]:

$$\log\left(y_{ijkg}\right) = \mu + A_i + D_j + V_k + G_g + (AG)_{ig} + (VG)_{kg} + (DG)_{jg} + \epsilon_{ijkg} \quad (8.9)$$

Yet another model can be constructed by adding a term for the interaction between arrays and dyes [182]:

$$\log\left(y_{ijkg}\right) = \mu + A_i + D_j + G_g + (AD)_{ij} + (AG)_{ig} + (VG)_{kg} + (DG)_{jg} + \epsilon_{ijkg}$$
$$(8.10)$$

A careful analysis of this model shows that it is complete inasmuch as all combinations of arrays, dyes and varieties are directly or indirectly accounted for. The missing term corresponding to the variety (V_k) is indirectly accounted for by the (AD) term since a variety appears only once for every array-dye combination. This design will not be able to distinguish between the variability due to the variety alone and the variability due to the array-dye interaction but this is not important since these are normalization issues that are not interesting. The important fact is that the variability due to all these factors is subtracted from the overall variability allowing us to test more accurately the hypotheses related to the $(VG)_{kg}$ term which corresponds to the differential regulation between varieties (mRNA samples). More detailed comparisons between the various models above can be found in papers such as [179] and [180].

8.7.2   Ratios and flip-dye experiments

A common and simple problem is to compare two conditions A and B looking for the genes that are expressed differently. The simplest experiment design could compare these varieties directly. If a two channel technology is used (e.g. cDNA using cy3 and cy5 labels), the two samples A and B can be compared using a single array. A common approach is to calculate the ratio of the values measured on the two channels. This is illustrated in Fig. 8.9. Most genes are expected to be expressed at similar levels and therefore most ratios will be around the value 1. The thinking is that the differentially regulated genes will be the ones with unusually large or unusually small ratios. Therefore, two thresholds can be used to select the tails of the ratio distribution and hence select the differentially regulated genes.

Let us consider this experiment in more detail. We can write the expression corresponding to the measurements on the two channels cy3 and cy5 according to the model Eq. 8.7. The values for cy3 are collected from array 1, channel 1, variety 1 and gene g and can be written as:

$$y_{111g} = \mu + A_1 + D_1 + V_1 + G_g + (VG)_{1g} + (DG)_{1g} + (AG)_{1g} + \epsilon_{111g} \quad (8.11)$$

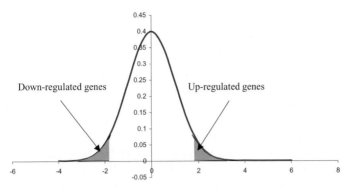

FIGURE 8.9: Two varieties can be compared directly using a cy3-cy5 cDNA microarray. A ratio can be calculated between the values measured on the two channels for each gene and represented graphically as a histogram (frequency vs. ratio value). Most genes are expected to be expressed at similar levels and therefore most ratios will be around the value 1 (0 if log values are used). Two thresholds can be used to select the tails of the ratio distribution and hence select the differentially regulated genes.

The values for cy5 are collected from the same array1, but channel 2 and variety 2. These values can be written as:

$$y_{122g} = \mu + A_1 + D_2 + V_2 + G_g + (VG)_{2g} + (DG)_{2g} + (AG)_{1g} + \epsilon_{222g} \quad (8.12)$$

Let us recall that these are log values. Taking the ratios corresponds to subtracting the two expression above since:

$$\log \frac{a}{b} = \log a - \log b \quad (8.13)$$

Subtracting equations 8.11 and 8.12, we obtain:

$$y_{111g} - y_{122g} = (D_1 - D_2) + (V_1 - V_2) + (DG)_{1g} - (DG)_{2g} + (VG)_{1g} - (VG)_{2g} + \epsilon_g \quad (8.14)$$

Let us consider the remaining terms in the above equation. The term $(D_1 - D_2) + (V_1 - V_2)$ is the average-log-ratio bias term. Normalization can eliminate the above term, since this is not gene related. The term $(VG)_{1g} - (VG)_{2g}$ is the interaction between the variety and the genes, i.e. the gene regulation due to the treatment. This is the effect of interest. However, the result also contains another term, $(DG)_{1g} - (DG)_{2g}$, which is the dye-gene interaction or the gene specific dye effect. In this experiment design, we cannot separate the effect of the uninteresting dye-gene interaction from the effect of the interesting variety-gene interaction. This is an example of *confounded effects* very similar to the example in which the two varieties of corn were planted on two different fields and the field effect could not be distinguished from the variety effect (see Fig. 8.1).

In order to be able to separate these two effects, more data are necessary. However, simply repeating the hybridization with another array will not help. The two effects would still remain undistinguishable from each other. In order to be able to separate them, we must flip the dyes. This experiment design is sometimes called the dye swap experiment or the flip fluor experiment.

This experiment design will provide two measurements for every variety with the property that each variety is measured on each dye exactly once. This experiment design provides data allowing us to calculate the sum of squares corresponding to the gene-dye interaction and therefore subtract it from the overall variability in order to estimate more precisely the gene-variety interaction.

It should be noted that the VG effects are orthogonal to all other effects, which means that other factorial effects will not bias estimates of VG effects. If the experiment is designed properly, the precision of the VG estimates is not affected by other sources of variation. In the dye swap experimental design VG effects are orthogonal to gene-specific dye effects DG.

8.7.3 Reference design vs. loop design

The pioneers of the microarray technology started by using microarrays in what Gary Churchill calls a **reference design** [66]. In this approach, a number of conditions or time points in a time series $c_1, c_2, \ldots, c_n$ are pairwise compared to a reference r.

Researchers using the reference design would use one dye to label the reference variety, and the other dye to label the varieties of interest. This would have the advantage that there are only $n + 1$ labelling reactions, one for each variety (mRNA sample) and one for the reference. Thus, n two-channel experiments would compare each condition to the reference as follows: experiment 1 compares c_1 labelled with cy3 to r labelled with cy5, experiment 2 compares $c_2/cy3$ to $r/cy5$, ... , experiment n compares $c_n/cy3$ to $r/cy5$. This is illustrated in Fig. 8.10. The same design is represented in Fig. 8.11 using the Kerr-Churchill notation. In this notation varieties, or mRNA samples, are represented by squares. An arrow linking two squares represents an array (a hybridization). A direction can be chosen arbitrarily (e.g. from red to green in Fig. 8.11) to represent the two channels used (cy3/cy5).

Kerr and Churchill made the very interesting observation that this design collects the most information exactly on the least interesting variety, namely the reference, since this reference appears on every array. Thus, the reference is measured n times while each interesting variety is measured only once.

Another and more serious criticism of the reference design is the fact that the dye effects are completely confounded with variety effects. Thus, the variability introduced by the dyes cannot be subtracted from the overall variability and erroneous conclusions may be drawn. For instance, if something happens in the second labelling reaction that reduces the amount of dye incorporated in the target by 10%, all values measured for variety 2 will be 10% lower. No matter what analysis method

	A_1	A_2	A_3
Red	R	R	R
Green	V_1	V_2	V_3

FIGURE 8.10: A classical reference design. In this design, each condition is compared with the reference. R and G denote the red and green channels, respectively. A variety is a condition such as treated/not treated or a time point in a time series analysis such as sporulation. In this design the reference is measured n times while each of the n varieties is measured only once.

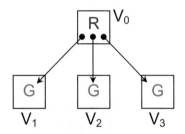

FIGURE 8.11: A classical reference design in the Kerr-Churchill notation. R and G denote the red and green channels, respectively. Arrays are represented by arrows. For instance, the arrow from V_0 to V_1 represents the array in which variety V_0 was labelled with red and variety V_1 with green.

is used, there will be no way to distinguish between a gene that is 10% lower due to the dye and a gene that is 10% lower due to the label. In order to provide data allowing this separation between dye effects and array effects, one needs to swap the dyes. This would require $2n$ arrays and would measure each variety twice and the reference $2n$ times.

One might try to optimize the information collected by measuring the reference fewer times in favor of more measurements for the various varieties while still swapping dyes. This is achieved by the **loop design** illustrated in Fig. 8.12. This design uses the same number of arrays as the reference design but collects twice as much data on the varieties of interest. The same design is shown in Fig. 8.13 in the Kerr-Churchill notation. In a loop design, experiment 1 would compare r labelled with cy3 vs. c_1 labelled with cy5, experiment 2 would compare c_1 labelled with cy3 vs. c_2 labelled with cy5, experiment 3 would compare c_2 labelled with cy3 vs. c_3 labelled with cy5, etc. until the last experiment which closes the loop by comparing c_n labelled with cy3 vs. r labelled with cy5. In this way, the loop experiment design maximizes the amount of information that can be extracted from the data with a given number of arrays.

Although the loop design is an elegant theoretical solution, problems might appear in the practical use. Each sample must be labelled with both Cy5 and Cy3 dyes, which doubles the number of labelling reactions and thus increases the time and cost of the experiment. Furthermore, if many varieties are involved and the loop becomes very

	A_1	A_2	A_3
Red	V_1	V_2	V_3
Green	V_2	V_3	V_1

FIGURE 8.12: A loop design. This design compares each condition with every other condition. Furthermore, each condition is measured once on every channel (dye).

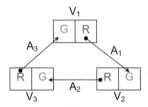

FIGURE 8.13: A loop design in the Kerr-Churchill notation. In the classical reference design, each condition is compared with the reference. In the loop design, conditions are compared to each other, flipping dyes at every comparison. R and G denote the red and green channels, respectively. For instance, the arrow $A1$ represents the array in which variety $V1$ was labelled with red and variety $V2$ with green. In this design, each variety is measured twice. This is also a balanced design since all combinations variety-dye appear together the same number of times.

	A_1	A_2	A_3	A_5	A_6	A_7
Red	V_1	V_2	V_3	V_4	V_6	V_7
Green	V_2	V_3	V_4	V_5	V_7	V_1

FIGURE 8.14: A design involving many varieties creates a large loop. In this example, comparing V_3 and V_6 can only be done indirectly by comparing $V_3 \rightarrow V_4 \rightarrow V_5 \rightarrow V_6$ or $V_3 \rightarrow V_2 \rightarrow V_1 \rightarrow V_7 \rightarrow V_6$.

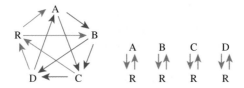

FIGURE 8.15: A comparison between the full loop design with reference and a flip-dye reference design. Each letter represents a condition (such as treatments with various drugs). An arrow represents a two-channel array on which the source condition in labelled with cy3 and the destination condition is labelled with cy5.

large, the data will not provide direct comparisons between all pairs. For instance, in Fig. 8.14, comparing varieties V_3 and V_6 can only be done indirectly by comparing $V_3 \rightarrow V_4 \rightarrow V_5 \rightarrow V_6$ or $V_3 \rightarrow V_2 \rightarrow V_1 \rightarrow V_7 \rightarrow V_6$. Either path would involve 4 comparisons with 3 other arrays. In order to alleviate this problem, the loop design can be improved by adding direct comparisons between varieties that are not neighbors on the loop. Furthermore, the loop design in Fig. 8.12 does not provide any information about the reference. Although the focus is on studying the varieties, sometimes it is useful to compare the reference directly with the conditions studied. If nothing else, the inclusion of the reference provides an internal control and the basis for the alignment of the data gathered with the rest of the data from the literature. A loop design in which the control is included as another variety is called a **loop with reference**. Fig. 8.15 shows a comparison between a full loop with reference and a flip-dye reference. The loop design with reference requires 10 arrays as opposed to the 8 arrays required by the classical reference with flip-dye design.

These and many other issues related to the statistical design of cDNA microarray experiments are discussed in detail in the work of Kerr and Churchill [180, 182, 183].

8.8 Summary

This chapter presented the main concepts related to experiment design. Replication, randomization and blocking have been discussed as the main tools used in experiment design. Some guidelines for experimental design were outlined. The chapter also included a discussion of some classical experiment designs such as: the fixed effect design, the randomized block design, the balanced incomplete block design, the Latin square design and the factorial design. Confounding was discussed in the context of increasing the efficiency of the factorial design by reducing the number of experimental runs. Finally, the chapter discussed in detail several of the ANOVA models proposed by Kerr and Churchill for cDNA microarrays, as well as a related ratio-based experiment design. The reference design, loop design and loop-with-reference design were also discussed. More details about microarray specific experiment design issues can be found in the work of Kerr and Churchill [183, 182, 183]. A very complete treatment of the experiment design topic can be found in [214].

Chapter 9

Multiple comparisons

Quantitative accumulations lead to qualitative changes.

—Karl Marx

9.1 Introduction

The problem of multiple comparisons is probably the most challenging topic for the typical life scientist. We will introduce this problem through an example that illustrates the dangers that lurk behind multiple comparisons. We will then approach the problem from a statistical perspective and prove that multiple comparisons need to be treated in a special way. Once this is clear, the chapter will present a few classical solutions to the problem as well as the best choices from a microarray data analysis perspective.

9.2 The problem of multiple comparisons

Let us consider an experiment comparing the gene expression levels in two different conditions such as healthy tissue vs. tumor. Let us consider that we have 5 tumor samples, 5 healthy tissue samples and we are following 20 genes. The data have been pre-processed and normalized. The last step in the normalization was a division by the global maximum. This made all values between zero and one. The maximum value was an internal control so the value one does not actually appear in the data. The data can be organized as shown in Table 9.1. The task is to find those genes that are differentially regulated between cancer and healthy subjects.

An examination of Table 9.1 shows that each gene has a number of measurements for both cancer and healthy subjects. A simple approach would be to consider each gene independently and perform a test for means involving two samples, as discussed in Chap. 6. As it was discussed there, this test for means should be preceded by a test for variance. For simplicity, we will assume that the variance of the cancer population is equal to the variance of the healthy population. Using this assumption,

| Gene | Tumor | | | | | Controls | | | | |
	T1	T2	T3	T4	T5	C1	C2	C3	C4	C5
g 1	0.340	0.232	0.760	0.610	0.224	0.238	0.075	0.624	0.978	0.198
g 2	0.155	0.724	0.163	0.100	0.143	0.257	0.833	0.062	0.578	0.796
g 3	0.659	0.273	0.003	0.202	0.332	0.752	0.010	0.585	0.694	0.201
g 4	0.034	0.918	0.749	0.748	0.643	0.807	0.760	0.520	0.930	0.638
g 5	0.887	0.532	0.091	0.254	0.487	0.380	0.075	0.936	0.730	0.362
g 6	0.630	0.177	0.352	0.638	0.555	0.151	0.765	0.619	0.833	0.593
g 7	0.676	0.243	0.673	0.289	0.066	0.494	0.553	0.277	0.159	0.962
g 8	0.374	0.334	0.619	0.095	0.287	0.831	0.952	0.077	0.802	0.601
g 9	0.591	0.771	0.407	0.473	0.647	0.937	0.804	0.881	0.394	0.524
g 10	0.200	0.761	0.681	0.960	0.734	0.005	0.439	0.355	0.745	0.852
g 11	0.342	0.220	0.025	0.149	0.045	0.500	0.222	0.091	0.899	0.828
g 12	0.794	0.122	0.279	0.311	0.046	0.507	0.714	0.963	0.941	0.879
g 13	0.455	0.083	0.409	0.756	0.268	0.868	0.442	0.098	0.619	0.194
g 14	0.239	0.913	0.990	0.754	0.558	0.971	0.444	0.253	0.674	0.948
g 15	0.332	0.569	0.065	0.956	0.543	0.510	0.842	0.851	0.800	0.307
g 16	0.473	0.817	0.076	0.239	0.054	0.154	0.432	0.582	0.396	0.525
g 17	0.282	0.296	0.448	0.801	0.871	0.672	0.532	0.278	0.575	0.774
g 18	0.778	0.212	0.170	0.624	0.790	0.372	0.468	0.611	0.255	0.017
g 19	0.670	0.397	0.767	0.159	0.909	0.798	0.258	0.080	0.904	0.930
g 20	0.594	0.517	0.078	0.336	0.802	0.077	0.964	0.059	0.751	0.207

TABLE 9.1: Expression data from two groups of subjects: cancer patients and healthy controls. The data are already normalized.

Gene	Tumor					Controls					p-value
	T1	T2	T3	T4	T5	C1	C2	C3	C4	C5	
g 1	0.340	0.232	0.760	0.610	0.224	0.238	0.075	0.624	0.978	0.198	0.959
g 2	0.155	0.724	0.163	0.100	0.143	0.257	0.833	0.062	0.578	0.796	0.230
g 3	0.659	0.273	0.003	0.202	0.332	0.752	0.010	0.585	0.694	0.201	0.419
g 4	0.034	0.918	0.749	0.748	0.643	0.807	0.760	0.520	0.930	0.638	0.522
g 5	0.887	0.532	0.091	0.254	0.487	0.380	0.075	0.936	0.730	0.362	0.825
g 6	0.630	0.177	0.352	0.638	0.555	0.151	0.765	0.619	0.833	0.593	0.438
g 7	0.676	0.243	0.673	0.289	0.066	0.494	0.553	0.277	0.159	0.962	0.604
g 8	0.374	0.334	0.619	0.095	0.287	0.831	0.952	0.077	0.802	0.601	0.115
g 9	0.591	0.771	0.407	0.473	0.647	0.937	0.804	0.881	0.394	0.524	0.323
g 10	0.200	0.761	0.681	0.960	0.734	0.005	0.439	0.355	0.745	0.852	0.366
g 11	0.342	0.220	0.025	0.149	0.045	0.500	0.222	0.091	0.899	0.828	0.072
g 12	0.794	0.122	0.279	0.311	0.046	0.507	0.714	0.963	0.941	0.879	0.014
g 13	0.455	0.083	0.409	0.756	0.268	0.868	0.442	0.098	0.619	0.194	0.787
g 14	0.239	0.913	0.990	0.754	0.558	0.971	0.444	0.253	0.674	0.948	0.870
g 15	0.332	0.569	0.065	0.956	0.543	0.510	0.842	0.851	0.800	0.307	0.382
g 16	0.473	0.817	0.076	0.239	0.054	0.154	0.432	0.582	0.396	0.525	0.607
g 17	0.282	0.296	0.448	0.801	0.871	0.672	0.532	0.278	0.575	0.774	0.864
g 18	0.778	0.212	0.170	0.624	0.790	0.372	0.468	0.611	0.255	0.017	0.342
g 19	0.670	0.397	0.767	0.159	0.909	0.798	0.258	0.080	0.904	0.930	0.953
g 20	0.594	0.517	0.078	0.336	0.802	0.077	0.964	0.059	0.751	0.207	0.815

TABLE 9.2: Expression data from two groups of subjects: cancer patients and healthy controls. The last column shows the p-values of a t-test done gene by gene assuming the populations are normal with equal variance. The null hypothesis that the cancer and healthy measurements are coming from the same distribution can be rejected for gene 11 and 12 at 10% significance level. In other words, genes 11 and 12 differ significantly between cancer and healthy.

we can calculate the value of the t-statistic and the associated p-value. Recall that the p-value is the probability of rejecting a true null hypothesis or the probability associated with a false positive (a gene that is declared to be differentially regulated although it is not). We do not have any a priori expectations so we will use a two-tail test. We choose to work at a 10% significance level.

Now that we have clearly formulated our hypotheses and assumptions, we can use a spreadsheet to quickly generate these p-values.[1] The data together with the computed p-values are shown in Table 9.2. An examination of the last column shows that the null hypothesis that the cancer and healthy measurements are coming from the same distribution can be rejected for gene 11 and 12 at our chosen 10% significance level.

---

[1] The appropriate Excel function is "ttest(array1, array2, 2, 2)". The third parameter uses the value 2 for a homoscedastic test (equal variance) and the fourth parameter indicates we are performing a 2-tail test.

In other words, genes 11 and 12 differ significantly between cancer and healthy. Also gene 8 is really close to the threshold with a p-value of 0.115 so we might want to take a closer look at it, as well. These values came from 10 different subjects, 10 different mRNA preparations, 10 different hybridizations, etc. Let us have a close look at gene 11:

	Tumor					Controls			
Gene	T1	T2	T3	T4	T5	C1	C2	C3	C4	C5
g 11	0.342	0.220	0.025	0.149	0.045	0.500	0.222	0.091	0.899	0.828

The mean of the cancer values is 0.156 and the mean of the healthy values is 0.508.[2] Taking into consideration that the values are all normalized to the (0,1) interval, this is a considerable difference. Furthermore, even the t-test tells us that the change is significant for this gene, even at a significance level of 5%. There seems to be a lot of statistical evidence that this is a gene which is down-regulated in cancer. Right? Wrong! In fact, there is little if any evidence that this gene is indeed differentially regulated. The gene may or may not be so but this conclusion does not follow from these data. The problem here is that multiple comparisons have been done in parallel. The significance level of 5% is the probability of a type I error that we are prepared to accept. The definition of the significance level tells us that it is likely that we will make one mistake (a false positive) every 20 times we apply the test. Well, it turns out that there are exactly 20 genes in the table. By calculating the p-value from the t-test for each gene, we have, in fact, applied the test 20 times. Sure enough, one gene appeared to have a p-value lower than the threshold of 0.05.

In fact, the data in Table 9.1 were obtained from a random number generator. This was just a small scale demonstration of what happens when multiple comparisons are performed without any special care. Applying the t-test to a list of 10,000 genes (a number comparable with the number of genes on the Affymetrix HG95Av2 array) will produce approximatively 500 genes that appear to be regulated even if they are in fact random. Of course, real genes interact in complex ways and their values are not random. However, the example does show it is perfectly possible for tens or hundreds of genes to appear as being significantly regulated even if they are not. We recommend the reader to perform a similar experiment on a larger scale as follows.

Example 9.1

Using a spreadsheet, generate a matrix of 10,000 x 20 random numbers.[3] Copy these data and use the command "Paste Special" to paste them, as values only[4] to a different sheet. Now consider the first 10 columns as cancer data and the remaining

---

[2]Note that there is no special link between any cancer patient and any control subject. Thus, a regular t-test was used instead of a paired one. A paired t-test would have produced a p-value of 0.006. A paired t-test would have been appropriate for instance if the data had been collected before and after treatment for the same subjects.

[3]The function rand() generates a random number between 0 and 1 in Excel.

[4]If the content of the cells is not pasted as values only, the numbers will change at every step.

10 columns as control data. Use the t-test to calculate a p-value for every gene in a separate column and sort the whole block of data by the p-values. The top 500 "genes" should have p-values lower than 0.05. ▯

An even more convincing example can be made with one's own real data. If you have data involving two groups of samples, merely change the labels of the sample or, equivalently, move them randomly from one group to another. A subsequent t-test will always provide approximately 5% "regulated" genes for any labelling and hopefully many more when the real labels are used.

Now that we have established that this approach will produce a list of "differentially regulated" genes no matter what the data look like, let us try to discuss the reasons for this very unpleasant outcome. The typical objection brought by the young statistical mind is that the gene should not care whether it is tested in the presence of the other 9999 genes or by itself. Thus, this one gene was measured 5 times, using 5 different mRNA samples, 5 different mRNA preparations and 5 different hybridizations for the cancer. The same was also done for the controls. This gene has exhibited a behavior *consistently different* between the two conditions. The means of the two conditions are considerably different and even when the variance is taken into consideration by calculating the t-test, the two means appear to be significantly different, as well. It seems that everything is in place and the conclusion cannot be denied. And yet, there is a problem. The problem is related to the fact that this gene *has been selected* from a very, very large number of similar genes. Had this experiment been performed on this one gene alone, the same data would have been very convincing because, it is very, very unlikely that a single gene exhibits such consistently different behavior between two conditions in 10 hybridizations. However, the microarray experiment involved perhaps 10,000 such genes. In such a very large number of genes, the probability that was negligible for any one gene suddenly becomes large enough for things to happen. With so many genes, it will be likely that a few genes are affected by random effects that will make them appear to be consistently lower in a condition and consistently higher in the other condition. "Yes, the young statistician can object, but the gene didn't know that it was hybridizing in the presence of another 9999 genes and, therefore, it could not have used this information in order to alter its values." This is entirely true. This individual gene did not know about the other genes. However, *we did!* This gene was selected based on the very fact that its p-value was low. In performing this selection, we have used the fact that there were many genes. If there had been fewer, we probably would not have been able to find such an outrageous false positive that was affected so consistently by the random factors to make it appear significant.

At this point, the balance starts to tip and our young statistician starts to believe that the gene under scrutiny is, indeed, nothing more than a false positive. However, the balance goes all the way and now our young statistician has adopted the opposite point of view: "If this gene cannot be trusted in spite of its consistent behavior across samples, then no individual gene coming from a large microarray can. After all, a gene can hardly be more consistent than having low values for all patients and higher values for all controls." That is not the case. In fact, any individual gene can be trusted as long as it is chosen *before* the experiment and not based on the p-values coming from the experiment. "Hold on!" my young opponent jumps. "What if I had chosen precisely the same gene as before? You know, the one that had a low p-value after selecting it from the many other genes?" Well, this would have been a Type I error and we would have been wrong. However, the probability of a gene exhibiting such consistent behavior due exclusively to random chance is extremely small. In fact, this is similar to a lottery with extremely low odds of winning, only that the outcome is reversed. In the lottery, the rare event is winning. In our experiment, the rare event is being wrong. Picking up a gene from the beginning and then being so unlucky that precisely this particular gene is the one in many thousands that is perturbed by the random factors in a mischievous way is similar to being so lucky to win the lottery. However, it is not difficult at all to pick the unlucky gene[5] *after* the experiment in the same way it is not difficult at all to pick the winning lottery number after the draw. No weight should be given to such a gene, the same way no prize is associated to picking the winning lottery number after the draw.

9.3　A more precise argument

Let us study this phenomenon using a more rigorous point of view. The significance level α was defined as the acceptable probability of a Type I error. This corresponds to a situation in which the null hypothesis is rejected when it is in fact true. The genes that are called differentially regulated when in fact they are not will be **false positives**.

Let us now think in the terms of hypothesis testing. When the t statistic for a gene is more extreme than the threshold t_α, we will call this gene differentially regulated. However, the gene may be so just due to random effects. This will happen with probability α. If this happens and we call this gene differentially regulated, we will be making an erroneous decision. Therefore, the probability of making a mistake of this kind is exactly α. If we do not make a mistake, we will be drawing the correct conclusion for that given gene. This will happen with probability:

$$\text{Prob(correct)} \;=\; 1 - p$$

Now we have to take into consideration the fact that there are many such genes. Let us consider there are R such genes. For each of them we will follow the same reasoning. However, at the end, we would like to draw the correct conclusion from all of them. This means, we have to have the correct conclusion for the first gene AND for the second gene AND ... AND for the last gene. We have seen that the probability of such an event is the multiplication of the probabilities corresponding

---

[5] It is unlucky because it has a low p-value just by chance.

Number of genes	significance level used for individual genes			
	0.01	0.05	0.1	0.15
10	0.095617925	0.401263061	0.65132156	0.803125596
20	0.182093062	0.641514078	0.878423345	0.961240469
50	0.394993933	0.923055025	0.994846225	0.999704235
100	0.633967659	0.994079471	0.999973439	0.999999913
500	0.993429517	1	1	1
1000	0.999956829	1	1	1
5000	1	1	1	1
10000	1	1	1	1

TABLE 9.3: The probability of making a Type I error (at least one false positive) in a multiple comparison situation. An array with as few as 20 genes has a probability of 87.84% of having at least one false positive if the gene level test is performed at a gene significance value of 0.1. For an array with 100 genes, the same probability becomes 99.99%.

to the individual events (see Chap. 4). Therefore, the probability of drawing the correct conclusion from all experiments is:

$$\text{Prob(globally correct)} \quad = \quad (1-p) \cdot (1-p) \cdots (1-p) = (1-p)^R$$

We can now calculate the probability of being wrong somewhere. This would be 1 minus the probability of being correct in all experiments:

$$\text{Prob(wrong somewhere)} \quad = \quad 1 - \text{Prob(globally correct)} \quad = \quad 1 - (1-p)^R$$

In this situation, being wrong means drawing the wrong conclusion for at least one gene. This is in fact the α value for the whole experiment. Table 9.3 shows the values of this probability for various significance levels and various sizes of the array. An array with as few as 20 genes has a probability of 87.84% of having at least one false positive if the gene level test is performed at a gene significance value of 0.1. For an array with 100 genes, the same probability becomes 99.99%. Although this is worrisome, the table does not paint the whole picture. After all, having a false positive from time to time may be deemed to be acceptable. Microarrays cannot be trusted completely anyway.[6] One might argue that any gene found as differentially regulated using microarrays should be confirmed with alternative assays such as quantitative real time polymerase chain reactions (Q-RT-PCR) and further biological experiments. The question then becomes how many such false positives are expected for a given array size and gene level significance? Table 9.4 shows these numbers. For instance a small array with 5,000 genes on which the gene level analysis is performed at 0.01 significance level is expected to produce about 500 false positives mixed up with whatever true positives are there in the given condition. For

---

[6]Firstly, individual gene hybridizations are inherently unreliable [193]. Secondly, they only reflect the phenomena at the mRNA level and completely ignore the translation and post-translational modifications.

Number of genes	gene significance level			
	0.01	0.05	0.1	0.15
10	< 1	< 1	1	1.5
20	< 1	1	2	3
50	< 1	2.5	5	7.5
100	1	5	10	15
500	5	25	50	75
1000	10	50	100	150
5000	50	250	500	750
10000	100	500	1000	1500

TABLE 9.4: The expected number of false positives for a given gene significance level and size of the array if no correction for multiple comparison is performed.

such numbers, performing alternative assays in order to sort out the true positives from the false positives is not an option anymore.

An experiment involving multiple comparisons is a good example of a situation in which small quantitative changes accumulate until a qualitative change occurs. In the multiple comparison, a hypothesis testing approach that was perfectly valid for a single test (e.g. any one gene chosen before the experiment is performed) or a small number of such tests, becomes inadequate for analyzing data coming from large arrays. Our task is to control the global or experiment level significance level. This is the probability of having a Type I error anywhere. This probability is also known as the **family-wise error rate** (FWER).

9.4 Corrections for multiple comparisons

9.4.1 The Šidák correction

After discussing the problem both from an intuitive perspective and from a statistical one, let us try to focus on how we can address the problem. The issue here is that we would like to control the overall probability of making a Type I error. This probability is equal to the probability of making at least one such mistake, calculated above:

$$\text{Prob(wrong somewhere)} = 1 - (1 - p)^R$$

This can be re-written as:

$$\alpha_e = 1 - (1 - \alpha_c)^R \tag{9.1}$$

where α_e is the probability of a Type I error at the experiment level and α_c is the probability of a Type I error at the gene level (single comparison). The task is to calculate the α level that we need to use for individual genes (α_c) in order to ensure that the global, or experiment level Type I error is less or equal to α_e which was

Genes	Šidák	Bonferroni
1	0.05	0.05
10	0.005116197	0.005
20	0.002561379	0.0025
100	0.000512801	0.0005
1000	0.000051292	0.00005
5000	0.0000102586	0.00001
10000	0.00000512932	0.000005
20000	0.00000256466	0.0000025

TABLE 9.5: The significance levels that need to be used at individual gene level in order to ensure an overall significance level of 0.05. Both Šidák and Bonferroni corrections require that tests at gene level be performed with extremely high significance which is unfeasible in gene expression experiments.

chosen. Using simple algebraic manipulations, we can extract α_c from the equation above:

$$\alpha_c = 1 - \sqrt[R]{1 - \alpha_e} \qquad (9.2)$$

This is the so-called **Šidák correction** for multiple comparisons [272].

9.4.2 The Bonferroni correction

Bonferroni [41, 42] noted that for small p, Eq. 9.1 can be approximated by taking only the first two terms of the binomial expansion of $(1 - p)^R$:

$$\alpha_e = 1 - (1 - \alpha_c)^R = 1 - (1 - R \cdot \alpha_c + \cdots) \approx R \cdot \alpha_c \qquad (9.3)$$

Using this approximation, we can calculate the experiment level α_c value as:

$$\alpha_e = \alpha_c \cdot R \Rightarrow \alpha_c = \frac{\alpha_e}{R} \qquad (9.4)$$

This is the **Bonferroni correction** for multiple comparisons. This is a very simple formula but it is only an approximation of the exact value given by Eq. 9.2. Bonferroni starts to depart from the exact values even for as few as 20 genes (see Table 9.5). However, this is the least of our problems. Unfortunately, both Bonferroni and Šidák corrections are unsuitable for gene expression analysis because for large number of genes R, the required significance at the gene level becomes very small, very quickly. Table 9.5 shows the significance levels that need to be used at the individual gene level in order to ensure an overall significance level of 0.05. For arrays involving more than 1000 genes, the technology is simply not able to provide values precise enough such that the genes will appear significant at those levels. At such stringent significance levels, the hypothesis testing approach will not be able to reject the null hypothesis for many genes. It is said that Bonferroni and Šidák are conservative methods in the sense that if a gene is significant after either Bonferroni or Šidák adjustments,[7] then the gene is truly different between the groups. However, if a gene is not significant according to these adjustments, then it may still be truly different. In other words, Bonferroni and Šidák are sufficient but not necessary conditions.

9.4.3 Holm's step-wise correction

A family of methods that allow less conservative adjustments of the p-values is the Holm step-down group of methods [146, 148, 149, 244]. These methods order the genes in increasing order of their p-value and make successive smaller adjustments. Let us consider we have a set of R genes. Each gene is measured in two groups, e.g. patients and controls. For a given gene, g_i, the null hypothesis is that the mean of the values of gene i measured in controls is the same as the mean of the values measured in patients i.e. $H_i : \mu_{ic} = \mu_{ip}$. For each gene, we will use an independent test statistic Y_i (e.g. a t-test between the patient's group and the control's group) to generate a p_i value. The p_i value will be the probability of the corresponding test statistic to have the observed value just by chance, i.e. when the null hypothesis is true. Holm's step-wise correction proceeds as follows:

Holm's step-wise correction procedure:

1. Choose the experiment level significance level α_e.

2. Order the genes in the increasing order of individual p-values:

Genes	g_{i_1}	g_{i_2}	$\cdots$	g_{i_k}	$\cdots$	g_{i_R}
Increasing p-values	p_1	p_2	$\cdots$	p_k	$\cdots$	p_R

3. Compare the p-values of each gene with a threshold that depends on the position of the gene in the list of ordered values. The thresholds are as follows: $\frac{\alpha_e}{R}$ for the first gene, $\frac{\alpha_e}{R-1}$ for the second gene, etc.

Genes	g_{i_1}	g_{i_2}	$\cdots$	g_{i_k}	$\cdots$	g_{i_R}
p-values	p_1	p_2	$\cdots$	p_k	$\cdots$	p_R
Test	$p1 < \frac{\alpha_e}{R}$	$p_2 < \frac{\alpha_e}{R-1}$	$\cdots$	$p_k < \frac{\alpha_e}{R-k+1}$	$\cdots$	$p_R < \frac{\alpha_e}{1}$

4. Let k be the largest i for which $p_i < \frac{\alpha_e}{R-i+1}$. Reject the null hypotheses H_i for $i = 1, 2, \ldots, k$. These genes are indeed different between the two groups at chosen α_e significance level.

Now it should be clear why Holm's procedure is called step-wise. Unlike Šidák and Bonferroni where the corrected threshold was unique for all genes and calculated in a single step procedure, the Holm's thresholds are different for every gene and they depend on their order in the ordered list of uncorrected p-values.

---

[7]Between the two of them, Bonferroni is slightly more conservative than Šidák (see Table 9.5).

9.4.4 The false discovery rate (FDR)

Bonferroni, Šidák and Holm's step-down adjustment are statistical procedures that assume the variables are independent. However, the genes of an organism are actually known to be involved in complex dependencies and regulatory mechanisms [83, 84]. The False Discovery Rate (FDR) correction procedure was initially proven for independent variables [30] but was recently extended to allow for some dependencies [31].

The FRD procedure adjusts the p-values in a manner similar to Holm's. The genes are ordered in increasing order of the p-values provided by the individual independent tests. However, the threshold for the i-th p-value will be: $p_i < \frac{i}{n} \frac{\alpha_e}{p_0}$ where α_e is the chosen experiment level significance level and p_0 is the proportion of the null hypotheses H_i that are actually true. Since this proportion is not known (if we knew which of the genes are not regulated we wouldn't need to do this), the p_0 can be conservatively estimated as being 1. This assumes that all null hypotheses are actually true and there are no differentially regulated genes. The conservative aspect means that if the null hypothesis can be rejected for any particular gene in these circumstances, then the null hypothesis will still be rejected if some of the null hypotheses are actually false and $p_0 < 1$.

False Discovery Rate (FDR) correction procedure:

1. Choose the experiment level significance level α_e.

2. Order the genes in the increasing order of individual p-values:

Genes	g_{i_1}	g_{i_2}	$\cdots$	g_{i_k}	$\cdots$	g_{i_R}
Increasing p-values	p_1	p_2	$\cdots$	p_k	$\cdots$	p_R

3. Compare the p-values of each gene with a threshold that depends on the position of the gene in the list of ordered values. The thresholds are as follows: $\frac{1}{R}\alpha_e$ for the first gene, $\frac{2}{R}\alpha_e$ for the second gene, etc.

Genes	g_{i_1}	g_{i_2}	$\cdots$	g_{i_k}	$\cdots$	g_{i_R}
p-values	p_1	p_2	$\cdots$	p_k	$\cdots$	p_R
Test	$p_1 < \frac{1}{R}\alpha_e$	$p_2 < \frac{2}{R}\alpha_e$	$\cdots$	$p_k < \frac{k}{R}\alpha_e$	$\cdots$	$p_R < \alpha_e$

4. Reject the null hypothesis for those genes that have a p-value lower than their corresponding threshold. These genes are indeed different between the two groups at chosen α_e significance level.

   Let k be the largest i for which $p_i < \frac{i}{R}\alpha_e$. Reject the null hypotheses H_i for $i = 1, 2, \ldots, k$. These genes are indeed different between the two groups at chosen α_e significance level.

9.4.5 Permutation correction

The Westfall and Young (W-Y) step-down correction [283] is a more general method that adjusts the p-value while taking into consideration the possible correlations.

Let us consider the following data set.

| Gene | Tumor | | | | | Controls | | | | | t |
	T1	T2	T3	T4	T5	C1	C2	C3	C4	C5	
g 1	0.340	0.232	0.760	0.610	0.224	0.238	0.075	0.624	0.978	0.198	t_1
g 2	0.155	0.724	0.163	0.100	0.143	0.257	0.833	0.062	0.578	0.796	t_2
g 3	0.659	0.273	0.003	0.202	0.332	0.752	0.010	0.585	0.694	0.201	t_3
g 4	0.034	0.918	0.749	0.748	0.643	0.807	0.760	0.520	0.930	0.638	t_4
g 5	0.887	0.532	0.091	0.254	0.487	0.380	0.075	0.936	0.730	0.362	t_4
⋮	⋮	⋮	⋮	⋮	⋮	⋮	⋮	⋮	⋮	⋮	⋮

This procedure starts by changing the measurements randomly between the patient's and control's groups. Alternatively, the same result can be achieved by randomly assigning the "patient" and "control" labels to the various measurements. A first such permutation may be:

Gene	C5	C3	T3	C1	T5	T2	C2	T1	T4	C4	t
g 1	0.340	0.232	0.760	0.610	0.224	0.238	0.075	0.624	0.978	0.198	t_{11}
g 2	0.155	0.724	0.163	0.100	0.143	0.257	0.833	0.062	0.578	0.796	t_{12}
g 3	0.659	0.273	0.003	0.202	0.332	0.752	0.010	0.585	0.694	0.201	t_{13}
g 4	0.034	0.918	0.749	0.748	0.643	0.807	0.760	0.520	0.930	0.638	t_{14}
g 5	0.887	0.532	0.091	0.254	0.487	0.380	0.075	0.936	0.730	0.362	t_{15}
⋮	⋮	⋮	⋮	⋮	⋮	⋮	⋮	⋮	⋮	⋮	⋮

New p-values are calculated using the chosen gene level test (e.g. t-test) for this permutation and the values are corrected for multiple experiments using Holm's step-down method discussed above. Then, a new permutation is done and new p-values resulting from this permutation are calculated:

Gene	C3	C5	C1	T5	T3	T4	C2	T1	T2	C4	t
g 1	0.340	0.232	0.760	0.610	0.224	0.238	0.075	0.624	0.978	0.198	t_{21}
g 2	0.155	0.724	0.163	0.100	0.143	0.257	0.833	0.062	0.578	0.796	t_{22}
g 3	0.659	0.273	0.003	0.202	0.332	0.752	0.010	0.585	0.694	0.201	t_{23}
g 4	0.034	0.918	0.749	0.748	0.643	0.807	0.760	0.520	0.930	0.638	t_{24}
g 5	0.887	0.532	0.091	0.254	0.487	0.380	0.075	0.936	0.730	0.362	t_{25}
⋮	⋮	⋮	⋮	⋮	⋮	⋮	⋮	⋮	⋮	⋮	⋮

This whole process (random labelling + testing) is repeated thousands or tens of thousands of times. Finally, the p-value for a gene i will be the proportion of times the value of t calculated for the real labels t_i is less or equal to the value of t calculated for a random permutation:

$$\text{p-value for gene } i: \quad \frac{\text{number of permutations for which } u_j^{(b)} \geq t_i}{\text{total number of permutations}}$$

where $u_j^{(b)}$ are the values corrected as in Holm's step-down method for permutation b. More details and an example of applying this method to microarray data can be found in [98].

The main important advantage of the W-Y approach is that it fully takes into consideration all dependencies between genes. This is extremely important for tightly correlated genes such as those being involved in the same pathways. Disadvantages include the fact that it is an empirical process lacking the elegance of a more theoretical approach. Also, the label permutation process is extremely computationally intensive and, therefore, inherently slow.

This method is a refinement of a more general approach known as bootstrapping [108, 181, 283]. The method samples with replacement the pool of observations to create new data sets and calculates p-values for all tests. For each data set, the minimum p-value on the resampled data sets is compared with the p-value on the original test. The adjusted p-value will be the proportion of resampled data where the minimum pseudo-p-value is less than or equal to an actual p-value. Bootstrap used with sampling without replacement is known as the permutation method [54, 142].

9.4.6 Significance analysis of microarrays (SAM)

Tusher et al. have reported that the step-down adjustment method of Westfall and Young was still too stringent for their microarray data [268]. In response to this, they have developed another method called **significance analysis of microarrays** (SAM). SAM assigns a score to each gene taking into consideration the relative change of each gene expression level with respect to the standard deviation of repeated measurements. The basic statistic used is similar to that of the t-test used in Chap. 6, section 6.3.2.1. The basic idea of a t-test is to calculate a difference between means divided by an estimate of the standard deviation. In other words, the purpose is to express the difference between means in units of standard deviations. How exactly the estimate of the standard deviation is calculated depends on whether the two populations are known (or assumed) to have the same variance, etc. SAM calculates the following statistic which is very similar to the t-statistic discussed in Chap. 6:

$$d_i = \frac{\overline{x}_{i1} - \overline{x}_{i2}}{s_i + s_0} \tag{9.5}$$

This is, again, a difference between means over a standard deviation. The second term of the denominator, s_0, is a "fudge" term. Its purpose is to prevent the computed statistic d_i from becoming too large when the estimated variance s_i is close to zero. The estimated variance is calculated based on the specific problem. In many cases, the assumption of equal variance is reasonable and the pooled variance is calculated as for the t-test involving two samples, equal variance in section 6.3.2.1:

$$s_i = \sqrt{\frac{(n_1 - 1) \cdot s_1^2 + (n_2 - 1) \cdot s_2^2}{n_1 + n_2 - 2} \left(\frac{1}{n_1} + \frac{1}{n_2} \right)} \tag{9.6}$$

Note that this variance is exactly the same as the one used in Eq. 6.39. If the assumptions of the experiment change, the test statistic above can be modified accordingly. However, the general approach remains the same.

Note that SAM calculates a gene by gene variance which will allow for the selection of the appropriate genes independently of their expression levels. However, the issue of multiple comparisons still remains since many thousands of genes are analyzed at the same time. SAM uses the same permutation idea to estimate the percentage of genes identified just by chance (the false discovery rate). Many permutations of the labels are done. For each permutation i, the value of the test statistic t_i is calculated. Each such t_i is actually an observation when the null hypothesis is true (the labels are actually random). These values are used to construct an empirical distribution for t_i values. In practice, the algorithm is as follows:

False discovery rate in SAM:

1. Fix a threshold for differentially expressed genes

2. Count how many genes are reported as differentially expressed in each permutation (false positives)

3. Calculate the median number of false positives across all permutations

4. Calculate FDR as the number of false positives divided by the number of genes in the original data.

SAM is available (freely for non-profit use) as an Excel macro that can be downloaded from: http://www-stat.stanford.edu/ tibs/SAM/index.html.

9.4.7 On permutations based methods

A word of caution should be added for all methods using permutations. In order for these methods to work, a large number of random permutations is necessary. However, the total number of distinct permutations is limited by the number of samples in each group. For instance, if there are only 3 patients and 3 controls, there are only

$$C_6^3 = \binom{6}{3} = \frac{6 \cdot 5 \cdot 4}{3 \cdot 2 \cdot 1} = 20$$

distinct permutations, which is completely insufficient in order to construct a good re-sampling distribution. If the number of patients and controls is increased to 6 each, the number of distinct permutation becomes:

$$C_{12}^6 = \binom{12}{6} = \frac{12 \cdot 11 \cdot 10 \cdot 9 \cdot 8 \cdot 7}{6 \cdot 5 \cdot 4 \cdot 3 \cdot 2 \cdot 1} = 924$$

which is more acceptable. In general, the number of permutations should be at least 1,000 or so. One order of magnitude more permutations ($\approx 10,000$) will make the results much more trustworthy.

9.5 Summary

This chapter discussed the general issue of correcting for multiple comparisons. An example and an informal discussion introduced the problem. The Šidák and Bonferroni corrections were presented as single step methods. Both methods are conservative and, therefore, they can be used only when few multiple comparisons are used. The two methods are not appropriate for microarray data when many thousands of genes are compared since, in these cases, the adjusted p-values are too small and the methods yield many false negatives. Holm's correction method is less conservative but assumes the variables are independent. In many cases, this assumption does not hold for microarray experiments involving interacting genes. The false discovery rate (FDR) is able to cope with some degree of interaction and is computationally efficient. Bootstrapping based methods such as permutation (Westfall and Young) and SAM are very powerful inasmuch they take into account any dependencies and correlations between variables but can be very computationally intensive. Significance Analysis of Microarrays is another method recently proposed for selecting differentially regulated genes while also correcting for multiple experiments.

Chapter 10

Analysis and visualization tools

Here is the answer that I will give to President Roosevelt. . . . Give us the tools and we will finish the job.

—Sir Winston Churchill, British Prime Minister, Radio Broadcast, 9 Feb. 1941

10.1   Introduction

This chapter will present briefly a few basic analysis and visualization tools used in gene expression analysis. All tools with the exception of gene pies are equally effective in the analysis of both Affymetrix and cDNA data. Gene pies are most useful for the visualizing of two-channel cDNA array data.

10.2   Box plots

A **box plot** is a plot that represents graphically several descriptive statistics[1] of a given data sample. The box plot usually has a box including a central line and two tails. The central line in the box shows the position of the median (the value located halfway between the largest and smallest data value). The upper and lower boundaries of the box show the location of the **upper quartile** (UQ) and **lower quartile** (LQ), respectively. The upper and lower quartiles are the 75th and 25th percentiles, respectively. Thus, the box will represent the interval that contains the central 50% of the data. The interval between the upper and the lower quartiles is called the **interquartile distance** (IQD). The length of the tails is usually $1.5 \cdot IQD$. Data points that fall beyond $UG + 1.5 \cdot IQD$ or $LQ - 1.5 \cdot IQD$ are considered outliers. Fig. 10.1 shows a box plot in the GeneSight package from BioDiscovery. The sliders on the left hand side allow the user to change the criteria for defining the outliers by adjusting the definition of box. The upper slider changes the upper boundary of the box and the lower slider changes the lower boundary of the box.

---

[1] See Chapter 4 for a discussion of several descriptive statistics including mean, median, percentiles, etc.

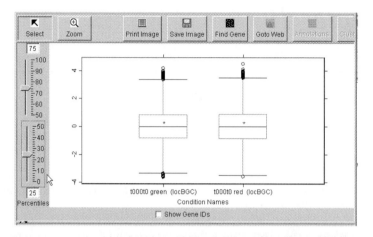

FIGURE 10.1:   A box plot in GeneSight. The central line in the box shows the position of the median (the value located halfway between the largest and smallest data value). For these settings, the upper and lower boundaries of the box show the location of the upper quartile (UQ) and lower quartile (LQ), respectively. The upper and lower quartiles are the 75-th and 25-th percentiles, respectively. The data outside the ends of the tails are outliers. The sliders on the left hand side allow the user to change the definition of the box.

10.3   Gene pies

Gene pies are nice visualization tools most useful for cDNA data obtained from two color experiments. This type of data is characterized by two types of information: absolute intensity and the ratio between the cy3 and cy5 intensities. In general, the meaning of the ratio between the channels depends essentially on the absolute intensity levels on the two channels, cy3 and cy5. If the intensities of the two channels are close to the background intensity, the ratio is completely meaningless since the ratio of two numbers close to zero can be almost anything. Furthermore, from a biological point of view, the gene is not particularly interesting since it seems to be shut off in both mRNA samples. If the intensity is high on one channel and close to the background level on the other channel, the ratio will be either close to zero or a very large number depending on which intensity is the denominator. For instance, if the values are 0.1 and 100, the ratio can be either $100/0.1 = 1000$ or $0.1/100 = 0.001$. In either case, this is a biologically significant change: the gene was expressed in a sample and not expressed in the other sample. Even though this is an interesting gene, the specific value of the ratio is not trustworthy since the value close to the background level is most probably just noise. Finally, the ratio is most informative if the intensities are well over background for both cy3 and cy5 channels. For instance,

Intensities cy3	cy5	Magnitude of cy3/cy5 ratio	Biological meaning	Interesting gene
low	low	meaningless; can be anything	the gene is not expressed in either condition	no
low	high	meaningless; close to zero	gene expressed on cy5, not expressed on cy3	yes
high	low	meaningless; very large	gene expressed on cy3, not expressed on cy5	yes
high	high	meaningful; can be anything	gene expressed on both channels, useful ratio	depends on the ratio

TABLE 10.1: The meaning of the ratio cy3/cy5 depends in an essential way on the absolute intensities cy3 and cy5. Furthermore, a gene may be interesting from a biological point of view even if it has a meaningless ratio.

if the values are 100 and 200, the ratio is either 0.5 or 2 and is highly informative either way. These observations are summarized in Table 10.1

A gene pie conveys information about both ratios and intensities. The maximum intensity is encoded as the diameter of the pie chart while the ratio is represented by the relative proportion of the two colors within any pie chart. From Table 10.1 it follows that the genes that may be biologically interesting are those with at least one large individual channel intensity. Since the gene pies encode the maximum intensity as the diameter of the pie, the net effect is that the genes that might be biologically relevant are made more conspicuous independently on their ratio. An example of a gene pie plot is given in Fig. 10.2.

10.4 Scatter plots

The **scatter plot** is probably the simplest tool that can be used to analyze DNA expression levels. In a scatter plot, each axis corresponds to an experiment and each expression level corresponding to an individual gene is represented as a point. Scatter plots are very useful to convey information about 2 dimensional data and they have been used in a number of research papers [226, 229, 260, 265]. If a gene G has an expression level of e_1 in the first experiment and that of e_2 in the second experiment, the point representing G will be plotted at coordinates (e_1, e_2) in the scatter plot. Fig. 10.3 presents a scatter plot.

In such a plot, genes with similar expression levels will appear somewhere on the first diagonal (the line $y = x$) of the coordinate system. A gene that has an expression level that is very different between the two experiments will appear far from the diagonal. Therefore, it is easy to identify such genes very quickly. For instance in

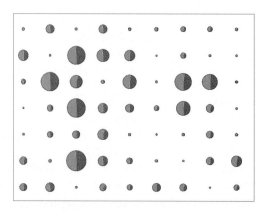

FIGURE 10.2:   A gene pie plot. Gene pies convey information about both ratios and intensities. The maximum intensity is encoded in the diameter of the pie chart while the ratio is represented by the relative proportion of the two colors within any pie chart. The net effect is that the genes that might be biologically relevant are made more conspicuous independently on their ratio (which may not always be meaningful). Figure obtained with GeneSight 3.5.2.

Fig. 10.3, gene A has higher values in the experiment represented on the horizontal axis (below the diagonal $y = x$) and gene B has a higher value in the experiment represented on the vertical axis (above the diagonal). The further away the point is from the diagonal, the more significant is the variation in expression levels from one experiment to another. In principle, the scatter plot could be used as a tool to identify differentially regulated genes.

Simple as it may be, the scatter plot does allow us to observe certain important features of the data. Fig. 12.4 presents a scatter plot of the logs[2] of the background corrected values in a typical cy5/cy3 experiment. The cy3 was used to label the control sample and is plotted on the horizontal axis; the cy5 was used to label the experiment sample and is plotted on the vertical axis. The data appear as a comma (or banana) shaped blob. Note that in this plot, most genes appear to be down-regulated in the experiment vs. control since most of the genes are plotted below the $x = y$ diagonal. However, in the huge majority of experiments involving living organisms, most genes are expected to be expressed at roughly the same expression levels. If too many genes change too much at any given time, it is likely that the functioning of the organism would be disrupted so badly that it would probably die. In other words, we expect most genes to be placed somewhere around the diagonal. The fact that we do not see this indicates that the data on one channel are consistently lower than the data on the other channel. This shows the need for some data preprocessing and normalization. Furthermore, the specific comma shape of the data is not an accident. This

---

[2]Chapter 12 will explain in detail the role of the logarithmic function.

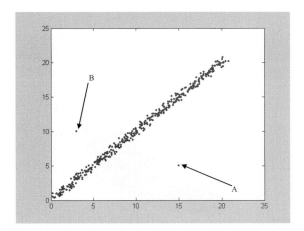

FIGURE 10.3: Expression levels in two experiments visualized as a scatter plot (synthetic data). The control is plotted on the horizontal axis while the experiment is plotted on the vertical axis. Points above the diagonal $y = x$ represent genes with expression levels higher in the experiment (e.g. gene B) whereas points below the diagonal represent genes with expression levels higher in control (gene A).

specific shape indicates that the dyes introduce a non-linear effect distorting the data. Again, this should be corrected through an appropriate normalization procedure as will be discussed in Chapter 12.

Another type of plot commonly used for two-channel cDNA array data is the ratio vs. intensity plot. In this plot, the horizontal axis represents a quantity directly proportional with the intensity. This quantity may be the intensity of one of the channels, or a sum of the intensities on the two channels $\log(cy3) + \log(cy5)$. Note that if log values are used, it follows from the properties of the logarithmic function that the sum of the log intensities is equal to the log of the product of the intensities:

$$\log(cy3) + \log(cy5) = \log(cy3 \cdot cy5) \tag{10.1}$$

The vertical axis represents the ratio of the two channels $\log(\frac{cy5}{cy3})$. Once again, one can note the typical non-linearity introduced by the dyes (the comma or banana shape). Note that according to the same properties of the logarithmic function:

$$\log\left(\frac{cy5}{cy3}\right) = \log(cy5) - \log(cy3) \tag{10.2}$$

Therefore, if most of the genes have equal expression values in both control and experiment, the expression above will be zero. In other words, we expect the points to be grouped around the horizontal line $y = 0$. The fact that most of the points are off this line shows that there is a systematic tendency for the values measured on one of the two channels to be higher than the values measured on the other channel.

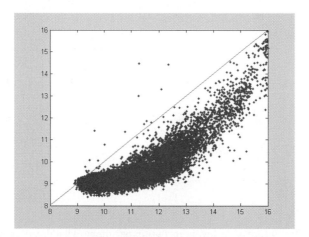

FIGURE 10.4:   Typical graph of the cy5 (vertical axis) vs. cy3 (horizontal axis) scatter plot of the raw data obtained from a cDNA experiment (real data). The cy3 was used to label the control sample; the cy5 was used to label the experiment sample. The straight line represents the first diagonal cy3=cy5.

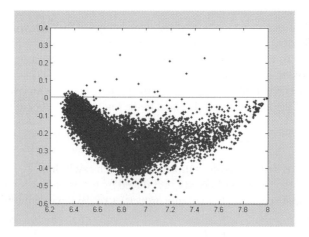

FIGURE 10.5:   In this ratio vs. intensity plot, the horizontal axis is used to plot the log of the product of the intensities (which is the sum of the log intensities) of the two channels. The vertical axis plots the ratio between the channels. The typical raw data set will exhibit the same banana shape visible in the cy3 vs. cy5 plot.

Many times, the data will exhibit a high variance at low intensities and a lower variance at high intensities. This is due to the fact that there is an error inherent to the measurement of the fluorescent intensity. In general, this error is higher for low intensities. Even if the error were constant, at high intensities the error will represent a small part of the signal read while at low intensities, the same absolute value of the error will represent a much larger proportion of the signal. In consequence, the same absolute value will affect more genes expressed at low levels.

For example, let us consider two genes g_1 and g_2, for which the intensity read on the control array is 1,000 and 30,000, respectively.[3] Let us assume that in a treated animal, both genes go up 1.5 times to 1,500 and 45,000, respectively. Let us also assume that the error in reading the intensity values is ± 100. The following table shows how the individual intensities will vary.

Untreated sample		Treated sample	
actual	read	value	read
1000	900-1100	1500	1400-1600
30000	29900 - 30100	45000	44900 - 45100

However, if we calculate ratios between such intensity, the variance will increase (see also Chapter 4). The lowest possible ratios will be obtained by combining the lowest possible value of the numerator with the highest possible value of the denominator. Similarly, the highest possible ratio will be obtained by combining the highest value of the numerator with the lowest value of the denominator. The following table shows what happens to the ratios as well as the log ratios.

values read		range for ratios	range for log ratios
$\frac{1400}{1100}$	$\frac{1600}{900}$	$1.272 - 1.777$	0.104 - 0.249
$\frac{44900}{30100}$	$\frac{45100}{29900}$	$1.491 - 1.508$	0.173 - 0.178

Note that the same absolute error in reading the intensity leads to wildly different ranges for the log ratios of the two genes. The ratio of the low intensity gene can more than double: the highest possible ratio is 0.249 while the lowest possible ratio is 0.101. However, the ratio of the high intensity gene remains within a very reasonable range. This means that at the low end of the intensity range, the data points will appear to be more scattered than at the high end. This phenomenon is the cause of the funnel shape of a scatter plot of the ratios vs. intensities. An example of this is shown in Fig. 10.6.

10.4.1 Scatter plot limitations

The main disadvantage of scatter plots is the fact that they can only be applied to data with a very small number of components since they can only be plotted in two or

---

[3]Recall that the image file is a 16-bit tiff file that can capture values from 0 to 65,535.

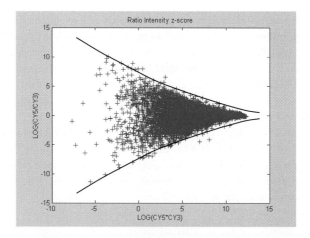

FIGURE 10.6: Typically, the data will exhibit high variance at low intensities and lower variance at higher intensities. This will produce a funnel shape. This funnel shape is more accentuated in the ratio plots such as this one. Note that here the data have been normalized and most of the points lie near the $y = 0$ reference.

three dimensions. In practice, this means that one can only analyze data corresponding to two or three different slides at the same time. Data involving hundreds or thousands of individuals or time points in a time series cannot be analyzed with scatter plots. Dimensionality reduction techniques such as Principal Component Analysis (PCA) are usually used to extend the usefulness of scatter plots.

10.4.2 Scatter plot summary

1. The scatter plot is a two or three dimensional plot in which a vector is plotted as a point having the coordinates equal to the components of the vector.

2. Scatter plots are highly intuitive and easy to understand. Limitations include the reduced number of dimensions in which data can be plotted.

3. In a scatter plot of two expression experiments including many genes or whole genomes, in which each experiment is represented on an axis, most data points are expected to lie around the line $y = x$ (or $z = y = x$). The assumption here is that most genes will not change. Points far from the diagonal are differentially expressed in the experiments plotted. The assumption may not hold if the set of genes has been preselected in some relevant way.

4. In a scatter plot of two expression experiments including many genes or whole genomes in which the ratio is plotted against an intensity, most data points are expected to lie around the line $y = 0$.

5. A consistent departure from the reference lines above may indicate a systematic trend in the data and the need for a normalization procedure.

6. Many scatter plots of microarray data will exhibit a funnel shape with the wider variance toward low intensities.

10.5 Histograms

A **histogram** is a graph that shows the frequency distribution of the values in a given data set. The horizontal axis of a histogram spans the entire range of values encountered in the data set. The vertical axis shows the frequency of each value. Usually, the histogram is represented as a bar graph. If this is the case, a bar of height y is drawn at position x to represent the fact that the value x appears y times in the data set. A histogram may also be drawn as a graph $y = f(x)$ with the same meaning: a point plotted at coordinates (x, y) means that the value x appears y times in the given data set. Fig. 10.7 shows an example of a histogram of the normalized log ratios of the background subtracted cy3/cy5 ratios[4] from the publicly available yeast cell cycle data set [1, 66]. This data set includes the expression of 6118 genes of the YSC328 yeast strain collected over 7 times points. The sporulation was induced by transferring the cells into a nitrogen-deficient medium. Data was collected at 0.5, 2, 5, 6, 7, 9 and 11.5 hours. The histogram included only the data corresponding to the first time point at 0.5 hours.

In general, histograms provide information about the shape of the distribution that generated the data. The histogram may be used as an empirical probability density function (pdf) since the frequency of a certain value will be directly proportional to the probability density for that value.

Let us recall that most expression data are real numbers. Real numbers may have a large number of decimal places depending on the computer used to process the data. Currently, most computers store floating point numbers using 64 or 128 bits which gives the ability to represent a staggering number of decimal places. This makes it likely that almost all values extracted from the image will be distinct even if they are very close to each other. For instance, the numbers 2.147, 2.148 and 2.151 are different even though they are all close approximations of 2.15. Furthermore, even in log scale, it is almost sure that the least significant digits are completely irrelevant to the gene expression which can only be effectively measured with a much lower accuracy. If the frequency of each distinct real value is computed, it is likely that the histogram will be a very wide graph with as many bars as data points and all bars having the height equal to 1 since each real data point is likely to be unique. In order to avoid this and make the histogram more meaningful, one must define some

---

[4]A full discussion of the pre-processing and normalization techniques can be found in Chap. 12.

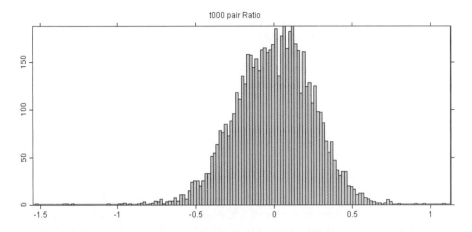

FIGURE 10.7: A histogram is a graph showing the frequency of various values in the data set. The horizontal axis spans the range of values while the vertical axis shows the frequency. A bar of height y is drawn at position x to represent the fact that the value x appears y times in the data set. Figure obtained with GeneSight 3.5.2.

bins. A bin is an interval that is used as an entity in counting the frequencies. The histogram will contain a bar for each bin and the height of the bar will be equal to the number of values falling in the interval represented by the bin. In order to illustrate the importance of the binning and the phenomena related to it let us consider the following example.

Example 10.1
Let us consider the following data set:

   0.24   0.90   0.33   0.97   0.16   0.64   0.24   0.44   0.58   0.74   0.48   0.32   0.18
   0.48   0.14   0.08   0.48   0.56   0.93   0.04   0.10   0.24   0.37   0.20   0.33   0.11

As we have seen, if we calculate the frequency of each value, we notice that all values are unique. A histogram with no binning will contain all values on the x axes and a number of bars equal to the number of values, all bars having height equal to 1. This is not very informative. In order to obtain more information about the data, we divide the range of values (0.04, 0.97) into a number of equal bins. Fig. 10.8 shows from the top down the histograms obtained with 100, 40, 20 and 10 bins, respectively. Note that the histogram obtained for 100 bins is very similar to the one that would be produced with no binning: almost all bar heights are equal to 1. As the number of bins decreases, the intervals corresponding to each bin increase and the number of values falling into each such interval starts to increase and also vary from bin to bin.

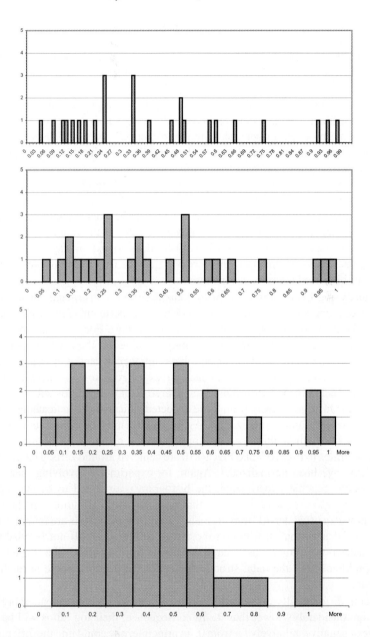

FIGURE 10.8: The effect of the bin size on the shape of the histogram. From top to bottom, the data are exactly the same but the bin sizes are 0.01, 0.025, 0.05 and 0.1. If bins are very narrow, few values fall into each bin; as bins get wider, more values fall into each bin and the shape of the histogram changes. Figure obtained with GeneSight 3.5.2.

Should we continue the process by reducing the number of bins further, the heights of the bars will continue to increase. Fig. 10.9 shows what happens when histograms are constructed for the same data with bin sizes of 0.24, 0.33 and 0.5, respectively. Interestingly, these 3 graphs appear to show a clear tendency of the data to have a higher frequency at lower values. In reality, this data set was drawn from a uniform distribution across the interval (0,1). This is a good example of a **binning artifact**. A binning artifact is an apparent property of the data that is in fact due exclusively to the binning process. In general, binning artifacts can be easily detected because they disappear as the number of bins is changed. This is why it is very important to plot several histograms across a large range of bin numbers and sizes before trying to extract any data features. For instance, the histogram in Fig. 10.7 exhibits a gap right at the peak of the histogram. One would be well advised to construct several other histograms with different bin numbers and sizes before concluding that the data really have such a feature at that point. Indeed, another histogram constructed from the same data using only 88 bins (see Fig. 10.10) does not exhibit any gap at all. Most software packages implementing this tool will give the user the ability to change the number of bins used to create the histogram.

When analyzing expression data, histograms are often constructed for the log ratios of expression values measured in two different experiments (either two different Affymetrix arrays or the two channels of a cDNA array). For such data, the peak of the histogram is expected to be in the vicinity of value 0 if the experiment has been performed on the whole genome or on a large number of randomly selected genes. This is because most genes in an organism are expected to remain unchanged which means that most ratios will be around 1 and therefore most log ratios will be around 0. A shift of the histogram away from these expected values might indicate the need for data preprocessing and normalization. Again, caution must be exercised in assessing such shifts since small shifts can also appear as the result of the binning process. Both histograms in Fig. 10.7 and Fig. 10.10 are centered around zero. This is because these data have been normalized.[5] Again, for experiments involving many genes or for whole genome experiments, the histogram is expected to be symmetrical, with a shape more or less similar to that of a normal distribution. Although the distribution can often depart from normality inasmuch as the tails are fatter or thinner than those of a normal distribution most such differences will not be visible at a visual inspection. In most cases, large departures from a normal-like shape indicate either problems with the data, strong artifacts or the need for some normalization procedures.

In general, differentially expressed genes will have ratios either considerably larger than 1 or considerably lower than 1. In consequence, their log ratios will be either positive or negative, relatively far from 0. In principle, one could find the differentially regulated genes in a given data set by calculating the histogram of the ratios or log-ratios and subsequently selecting the tails of the histogram beyond a certain threshold. For instance, if the values are normalized first and then the log is taken

---

[5]Various normalization procedures are discussed in Chapter 12.

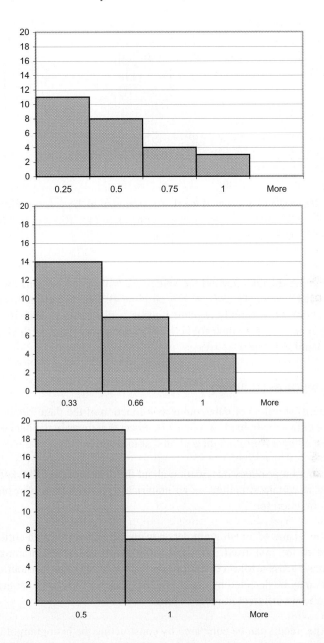

FIGURE 10.9:   An artifact of the binning process. From top to bottom, the same data are shown with bin sizes of 0.25, 0.33 and 0.5, respectively. Note that all 3 histograms suggest a clear tendency for higher frequencies at lower values. The data are in fact uniformly distributed and this tendency is entirely an artifact of the binning process and the relatively small sample size (26 data points).

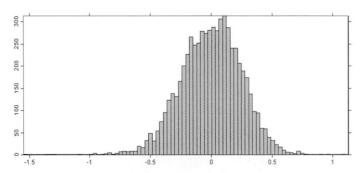

FIGURE 10.10:   The histogram of the first experiment in the yeast sporulation data constructed using 88 bins. The data are the same as in Fig. 10.7 but this histogram does not have the artifactual gap near the peak.

in base 2, selecting the tails beyond the values ± 2 will select the genes with a fold change of at least 4 ($\log_2 \frac{ex_1}{ex_2} > 2 \Rightarrow \frac{ex_1}{ex_2} > 2^2 = 4$). The use of the histogram as a tool for selecting differentially regulated genes is illustrated in Fig. 10.11. Note that selecting genes based on their fold change **is not** a very good method as will be discussed in detail in Chap. 13.

10.5.1   Histograms summary

- Histograms are plots of data counts as a function of the data values. Usually they are drawn using bars. A value x that occurs y times in the data set will be represented by a bar of height y plotted at location x.

- The histogram provides information about the distribution of the data and can be used in certain situations as an empirical approximation of a probability density function (pdf).

- The exact shape of a histogram depends on the number of data collected and the size of the bins used. A small sample size and/or the binning process may create gross artifacts distorting the nature of the data distribution. Such artifacts may be detected by comparing the shapes of several histograms constructed using different number of bins.

- Two experiments can be compared by constructing the histogram of the ratios of the corresponding values. If the experiments involve a large number of genes and the data are suitably pre-processed and normalized, the histogram of the ratios is expected to be centered on either zero (if logs are used) or 1 (no logs are used) and be approximatively symmetrical.

- Differentially regulated genes will be found in the tails of such a histogram.

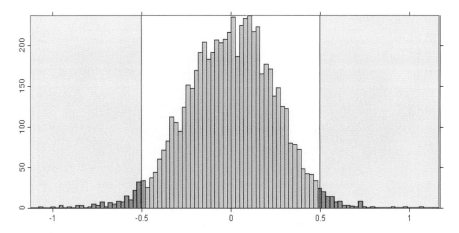

FIGURE 10.11: The histogram of the normalized log ratios can be used to select the genes that have a minimum desired fold change. In this histogram, the tails beyond ± 0.5 contain the genes with a fold change of more than 1.4 ($\log_2 x = 0.5 \Rightarrow x = 2^{0.5} = 1.41$).

10.6 Time series

A time series is a plot in which the expression values of genes are plotted against the time points when the values were measured. Fig. 10.12 shows an example of the time series of the normalized log ratios of the background subtracted R/G ratios[6] from the publicly available yeast cell cycle data set [1, 66]. The horizontal axis is the time. The vertical axis represents the measured expression values. In most cases the time axis will plot the time points at equal intervals. Since the time collection points are usually chosen in a manner that is suitable to the biological process under study and are not equidistant, this graphical representation will introduce a distortion. For instance, the data in the sporulation data set were collected at t = 0, 0.5, 2, 5, 6, 7, 9 and 11.5 hours after transfer to the sporulation medium. This means that the first segment on the plot represents 0.5 hours while the last segment represents 2.5 hours. A hypothetical gene that would increase at a constant rate will be represented by line segments with two different slopes in the two segments. An illustrative example is gene YLR107W shown in Fig. 10.13. This gene seems to exhibit a small increase between the first two time points and a large increase between the last two time points. This seems to suggest that the gene is much more active in the last interval than it was during the first interval. In fact, the rate of change is higher during the

---

[6]A full discussion of the pre-processing and normalization techniques can be found in Chap. 12.

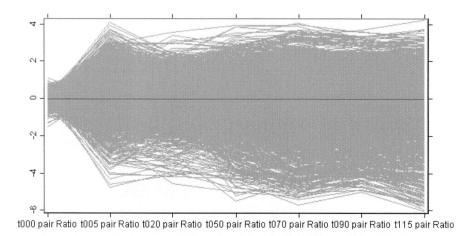

t000 pair Ratio t005 pair Ratio t020 pair Ratio t050 pair Ratio t070 pair Ratio t090 pair Ratio t115 pair Ratio

FIGURE 10.12:   A time series plots the expression values of the genes as a function of the time points when the measurements were made. The horizontal axis is the time although in most cases the time will not be represented to scale. The vertical axis represents the measured expression values, in this case the normalized log ratios of the two channels, measured at t = 0, 0.5, 2, 5, 6, 7, 9, and 11.5 hours.

first interval. During the first interval, the gene changed from 0 to approximately 0.5 over 0.5 hours, i.e. change at a rate of approximately 1 log/hour. During the last interval, the gene changed from 0.5 to 2.5 over 2.5 hours, i.e. changed at a rate of approximately 0.8 log/hour. This example illustrates why mere data visualization may sometimes be misleading.

10.7   Principal component analysis (PCA)

One very common difficulty in many problems is the large number of dimensions. In gene expression experiments each gene and each experiment may represent one dimension. For instance, a set of 10 experiments involving 20,000 genes may be conceptualized as 10,000 data points (genes) in a space with 10 dimensions (experiments) or 10 points (experiments) in a space with 20,000 dimensions (genes). Both situations are well beyond the capabilities of current visualization tools and frankly, probably well beyond of the visualization capabilities of our brains.

A natural approach is to try to reduce the number of dimensions and, thus, the complexity of the problem, by eliminating those dimensions that are not "important." Of course, the problem now shifts to defining what an important dimension is. A common statistical approach is to pay attention to those dimensions that account for a

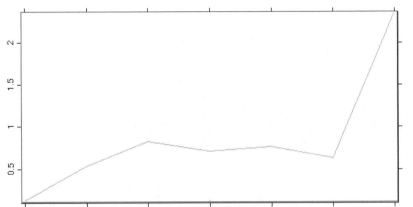

t000 pair Ratio t005 pair Ratio t020 pair Ratio t050 pair Ratio t070 pair Ratio t090 pair Ratio t115 pair Ratio

FIGURE 10.13: An example of a distortion introduced by a non-uniform time scale. The yeast gene YLR107W seems to exhibit a small increase between the first two time points and a large increase between the last two time points of the sporulation time course. This seems to suggest that the gene is more active in the last time segment. In fact, the rate of change is higher in the first interval. The distortion is due to the fact that the first segment on the time scale represents 0.5 hours while the last segment represents 2.5 hours.

large variance in the data and to ignore the dimensions in which the data do not vary much. This is the approach used by Principal Component Analysis (PCA).

PCA works by calculating a new system of coordinates. The directions of the coordinate system calculated by PCA are the **eigenvectors of the covariance matrix of the patterns** (see Chapter 4 for a definition of the covariance matrix). An **eigenvector** of a matrix A is defined as a vector $\mathbf{z}$ such as:

$$A\mathbf{z} = \lambda\mathbf{z} \tag{10.3}$$

where λ is a scalar called **eigenvalue**. Each eigenvector has its own eigenvalue although it is possible that the eigenvalues of different eigenvectors have the same numerical value. For instance, the matrix:

$$A = \begin{bmatrix} -1 & 1 \\ 0 & -2 \end{bmatrix} \tag{10.4}$$

has the eigenvalues $\lambda_1 = -1$ and $\lambda_2 = -2$ and the eigenvectors $z_1 = \begin{bmatrix} 1 \\ 0 \end{bmatrix}$ and $z_2 = \begin{bmatrix} 1 \\ -1 \end{bmatrix}$.

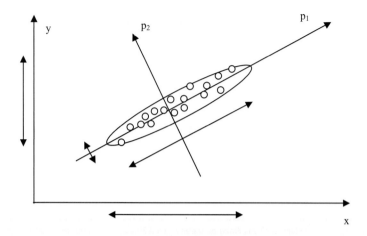

FIGURE 10.14: In spite of the fact that each point has two coordinates, this data set is essentially one dimensional: most of the variance is along the first eigenvector p_1 with the variance along the second direction p_2 being probably due to the noise. The PCA will find a new coordinate system such as the first coordinate is the direction on which the data have maximum variance (the first eigenvector), the second coordinate is perpendicular on the first and captures the second largest variance, etc.

It can be verified that:

$$A\mathbf{z_1} = \begin{bmatrix} -1 & 1 \\ 0 & -2 \end{bmatrix} \begin{bmatrix} 1 \\ 0 \end{bmatrix} = (-1) \cdot \begin{bmatrix} 1 \\ 0 \end{bmatrix} = \lambda_1 \mathbf{z_1} \qquad (10.5)$$

and

$$A\mathbf{z_2} = \begin{bmatrix} -1 & 1 \\ 0 & -2 \end{bmatrix} \begin{bmatrix} 1 \\ -1 \end{bmatrix} = (-2) \cdot \begin{bmatrix} 1 \\ -1 \end{bmatrix} = \lambda_2 \mathbf{z_2} \qquad (10.6)$$

In intuitive terms, the covariance matrix **captures the shape of the set of data points**. If one imagines an n-dimensional hyper-ellipsoid including the data, the eigenvectors of the covariance matrix, or the directions found by the PCA, will be the directions of the main axes of the ellipse. This is illustrated in Fig. 10.14.

The essential aspect of the PCA is related to the fact that the absolute value of the eigenvalues are directly proportional to the dimension of the multidimensional ellipse in the direction of the corresponding eigenvector. Since the eigenvectors are obtained from the covariance matrix that captures the shape occupied by the points in the original n-dimensional space, it follows that the absolute values of the eigenvalues will tell us how the data are distributed along these directions. In particular, **the eigenvalue with the largest absolute value will indicate that the data have the largest variance along its eigenvector** (which is a particular direction in space).

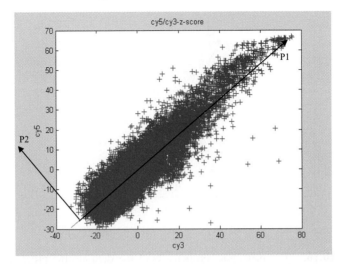

FIGURE 10.15:   The PCA can also be used to separate the within-experiment variation from the inter-experiment variation. The new axis P1 will be aligned with the
direction of the within-experiment variation. The within-experiment variation stems
from the fact that genes are expressed at all levels in either sample. The new axis
P2 will be perpendicular on P1 and will capture the inter-experiment variation. The
inter-experiment variation is the interesting variations if we want to find those genes
that differ between the two samples.

Second largest eigenvalue will indicate the direction of the second largest variance,
etc. It turns out that many times, only few directions manage to capture most of the
variability in the data. For instance, Fig. 10.14 shows a data set that lies essentially
along a single direction although each particular data point will have two coordinates. This is a set of data points in a two dimensional space and each data point is
described by two coordinates. However, most of the variability in the data lies along
a one-dimensional space that is described by the first principal component. PCA will
be able to discover the relevant directions as well as indicate the amount of variance
that each new axis captures. In this example, the first principal component is sufficient
to capture most of the variance present in the data and the second principal component
may be discarded.

Note that the direction of the highest variance may not always be the most useful. For
instance, in Fig. 10.15, PCA is used to distinguish between the within-experiment
variation and the inter-experiment variation. Fig. 10.15 shows the new axes found by
the PCA applied to a typical cy3-cy5 data set. Note that the longer axis, $P1$, corresponds to the within-experiment variation. In either sample, genes will be expressed
at various levels and they will appear at various positions along this axis: genes
expressed at low levels will be closer to the origin while genes expressed at high
levels will be further up the axis. This variation is not particularly interesting since we

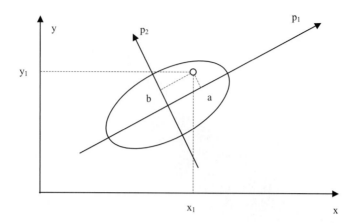

FIGURE 10.16: The principal components of the data are the projections of the data on the new coordinate system. In this figure, the data point shown has the coordinates (x_1, x_2) in the original coordinate system xOy. The ellipse represents the area in which most of the data points lie. The first eigenvector, p_1, will show the direction of the largest variance. The second eigenvector will be perpendicular on the first one. The principal components of the data point (x_1, x_2) will be (a, b) with a being the coordinate along the first eigenvector and b the coordinate along the second eigenvector.

know *a priori* that genes will be expressed at all levels. However, the other axis, $P2$, will be aligned with the direction of the inter-experiment variation. The genes that are differentially regulated will appear far from the origin of this axis while genes that are expressed at the same level in both mRNA samples will be projected close to the origin of this axis. Note that this happens independently of the level of expression of the given gene. Therefore, in principle, the second principal component would be sufficient to look for genes expressed differentially between the two experiments.

The nice aspect of the PCA is that the computation involved is always the same, independently of whether we are interested in the directions of the large variances or low variances. If the input data is n dimensional, i.e. each input point is described by n coordinates, the PCA will provide a complete set of n directions together with the associated eigenvalues and will leave the choice of what to use up to us. Most packages implementing PCA can provide an explicit list of the eigenvectors and associated eigenvalues and allow the user to select a small subset on which to project the data.

The projections of the data points in the new coordinate system found by the PCA are called **the principal components of the data** (see Fig. 10.16). These can be easily obtained by multiplying each data point (which is an n-dimensional vector) by the matrix of the eigenvectors.

Depending on the problem, we will select a small number of directions (e.g. 2 or

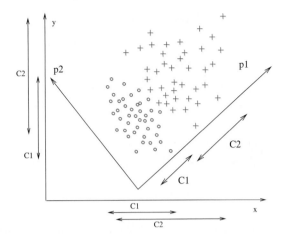

FIGURE 10.17:   Principal Component Analysis (PCA) – If one of the two original axes x or y is eliminated, the classes cannot be separated. The co-ordinate system found by PCA ($p1$, $p2$) allows the elimination of $p2$ while preserving the ability to separate the given classes. It is said that PCA performs a dimensionality reduction.

3) and look at the projection of the data in the coordinate system formed with only those directions. By projecting the n dimensional input data into a space with only 2 or 3 coordinates, one achieves a **dimensionality reduction**.

However, one should note that the dimensionality reduction obtained through PCA may not always be a true reduction of the number of variables we need to consider. For instance, in the example shown in Fig. 10.16, the PCA found the axis p_1 that captures most of the variance in the data. Under these circumstances, one could discard the second principal component p_2 and describe the data only through its first component. However, each component is a linear combination of the original variables. For instance, the direction of the first component p_1 can be described by an equation such as $a \cdot x + b \cdot y + c = 0$ where x and y are the original variables. In order to conclude that any of the original variables may be discarded, one needs to look at the coefficients of the equations describing the chosen PCA directions. If a particular variable (such as y) has coefficients (such as b) that are close to zero in all chosen PCA directions then one could indeed conclude that this original variable is not important under the given circumstances. Since the original coordinates correspond to either genes or experiments, this is a way of selecting those genes or experiments that are truly important for the phenomenon under study.

An example of using PCA is shown in Fig. 10.17. The data, which include patterns from two classes (red crosses and blue circles), are given in the original coordinate system with axes x_1 and x_2. If the data are projected on each of the two axes, the clusters corresponding to the two classes overlap and the classes cannot be separated using any single dimension. Therefore, a dimensionality reduction is not possible in

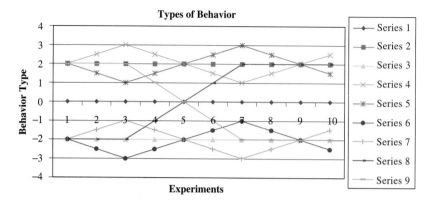

FIGURE 10.18: Typical gene profiles. The prototypes include genes that are constant at -2 (Series 3), 0 (Series 1) and 2 (Series 2), genes that decrease over the 10 experiments (Series 9), genes that increase (Series 8) and genes that oscillate in various ways (Series 4, 5, 6 and 7).

the original coordinate system. In order to separate the two classes, one needs both x_1 and x_2. However, the PCA approach can analyze the data and extract from it a new coordinate system with axes p_1 and p_2. The direction of the new axes will be the direction in which the data have the largest and second largest variance. If the two clusters are projected on the axes of the new coordinate systems one can notice that the situation is now different. The projections of the two classes on p_2 overlap completely. However, the projections of the two classes on p_1 yield two clusters that can be easily separated. In these conditions, one can discard the second coordinate p_2. In effect, we have achieved a dimensionality reduction from a space with two dimensions to a space with just one dimension while retaining the ability to separate the two classes.

PCA also provides good results in those cases in which the data have certain properties that become visible in the principal component coordinate system while they may not be so in the original coordinate system. In order to illustrate this we will use a sample data set kindly provided by John Quackenbush,[7] as well as the Multiple Experiment Viewer (MEV) software from TIGR.

Example 10.2

This synthetic data set includes 610 genes measured in 10 experiments. The data set includes several prototype genes with a predetermined behavior across the set

---

[7]We modified the data set by adding 150 completely random genes in order to make it more realistic but the credit of creating such a wonderful example remains entirely John's.

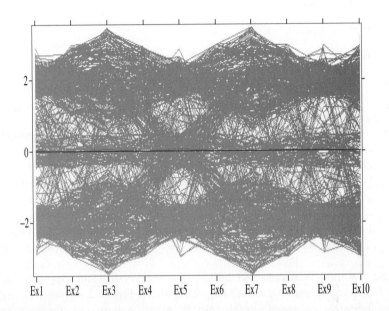

FIGURE 10.19: The data used for the PCA example. These genes were constructed by adding random noise to each prototype in Fig. 10.18 as well as by adding 150 completely random genes. Each prototype was the base for constructing 50 genes of similar behavior. The set shown here includes a total of 610 genes. Plot created with GeneSight 3.5.0.

of experiments. The set includes genes that exhibit one of the following types of behavior across the 10 experiments:

- Are relatively constant:

  - around zero

  - at some positive value

  - at some negative value

- Increase steadily

- Decrease steadily

- Oscillate

  - up then down then up around a positive, zero or negative value

  - down then up then down around a positive, zero or negative value

In total, there are 9 such typical gene profiles shown in Fig. 10.18. A number of other genes are created by adding random noise to each of these profiles. Fifty genes are created by adding random noise to each profile for a total of 450 prototype-based genes. Furthermore, an additional 150 genes are created with totally random values. The graphs of these genes across the set of 10 experiments are shown in Fig. 10.19. This graph looks fairly typical in the sense that there is no particular behavior that can be distinguished for any of the genes. Furthermore, this graph does not show that there are some typical profiles, nor that there are several genes behaving in a similar way.

We can now apply PCA, construct the new coordinate system and select from the new axes the ones corresponding to the 3 largest variances. Fig. 10.20 shows the 610 genes in this 3 dimensional space. In this case, PCA has been used to effectively reduce the dimensionality of the experiment space from 10 to 3. This PCA space reveals a lot of information about the data. We can see from the plot that there are 9 distinct types of gene behavior represented by the small spherical clusters at the extremities of the axes and the small cluster centered in the origin. It can be seen that for each such behavior, there are a number of genes that behave in that particular way. Finally, one can also conclude that there are some genes that behave in a completely random way since they are distributed uniformly in a large area of the space. The 9 clusters in Fig. 10.20 can be distinguished more easily by comparing it with Fig. 10.21 from which the completely random genes have been removed.

$\square$

In the previous example, PCA was applied to reduce the dimensionality of the experiment space. The result was a space in which the axes were the eigenvectors of the correlation matrix of the experiments or "principal experiments." In this space, each data point is a gene. Such a plot provides information about the structure contained in the behavior of the genes. As shown in Example 10.2, this can be used to see

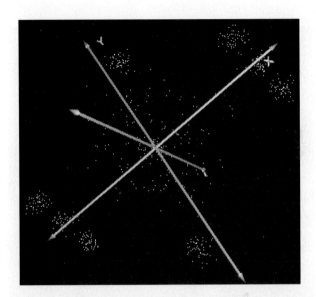

FIGURE 10.20: Using PCA for visualization. The figure shows the 610 genes from Fig. 10.19 plotted in the space of the first 3 principal components of the 10 experiments (principal experiments). In this plot, each dot represents a gene. The coordinates of each dot are equal to the first three principal components of the respective gene. The plot shows clearly that there are 9 distinct types of gene behavior (the tight spherical clusters at the extremities of the axes and the tight cluster centered in the origin) as well as some random genes (the more dispersed collection of points around the origin). Figure obtained with Multiple Experiment Viewer.

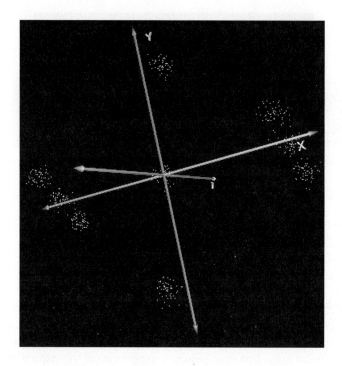

FIGURE 10.21:   The PCA plot of the 460 non-random genes. The plot shows clearly 9 clusters corresponding to each of the profiles present in the data. Figure obtained with Multiple Experiment Viewer.

whether there are genes that behave in similar ways, whether there are distinct patterns of gene behavior, etc. However, PCA can also be applied to the gene expression vectors. In this case, PCA will calculate the eigenvectors of the correlation matrix of the genes or "principal genes." In this space, each point is an experiment. Such a plot provides information about the structure contained in the various experiments: whether there are similar experiments, whether various experiments tend to have a certain profile, etc.

PCA has been shown to be extremely effective in many practical problems including gene expression data [104, 144, 225] and is currently available in a number of software tools.

10.7.1 PCA limitations

In spite of its usefulness, PCA has also drastic limitations. Most such limitations are related to the fact that PCA only takes into consideration the variance of the data which is a first-order statistical characteristic of the data. Furthermore, the eigenvectors of the correlation matrix are perpendicular on each other which means that any axes found by the PCA will also be perpendicular on each other. In effect, this means that the transformation provided by the change of coordinate system from the original system to the principal components system is a rotation perhaps followed by a scaling proportional to the eigenvalues of the same covariance matrix. There are cases in which such a system is not suitable. For instance, in Fig. 10.22, the coordinate system (p_1, p_2) found by PCA is not useful at all since both classes have the same variance on each of the p_1 and p_2 axes.

Furthermore, the direction of the new axes found by PCA is determined exclusively based on the variance of the data. The class of each data point, i.e. whether the point is a blue circle or red cross, is not taken into consideration by the PCA algorithm. Therefore, PCA may not always be as useful as in the example shown in Fig. 10.17. This point is illustrated in Fig. 10.23 which shows a data set in which the coordinates of the points are the same as in Fig. 10.17 but their classes are different. The class is represented by the color/shape of the individual data points with red crosses representing cancer patients and blue circles representing healthy controls. Since the numerical values are the same, PCA will calculate the same coordinate system based on the variance of the data and disregarding their class. In turn, this will produce the same principal components as in Fig. 10.17. However, in this case, this principal components coordinate system is not particularly useful.

10.7.2 PCA summary

Principal Component Analysis is a technique that uses first order statistical properties of the data in order to construct a new coordinate system. The direction of the new axes will be the eigenvectors of the correlation matrix of the data. The PCA has the following properties:

1. The directions are chosen in decreasing order of the amount of data variance

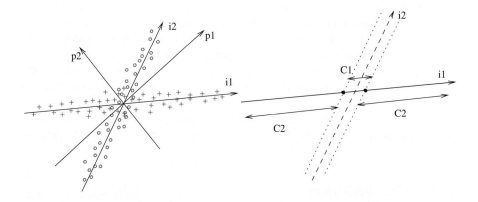

FIGURE 10.22:   The difference between principal component analysis (PCA) and independent component analysis (ICA). The coordinate system $\{p_1, p_2\}$ found by PCA is less useful than the $\{i_1, i_2\}$ found by ICA. The classes cannot be separated using any one axis from the coordinate system found by PCA. The right panel shows how the classes can be separated with minimal misclassification using just the projections on i_1.

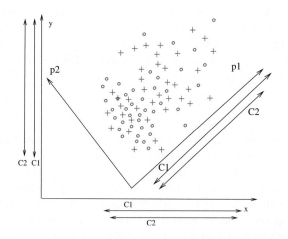

FIGURE 10.23:   An example in which the Principal Component Analysis (PCA) is not very useful. The coordinate system found by PCA ($p1$, $p2$) is based exclusively on the spatial distribution of the data points, i.e. on their position, and disregards completely their type (blue circles or red crosses). This figure illustrates a problem in which the data are placed in the same positions as in Fig. 10.17 but have different labels (e.g. red circles are cancer and blue crosses are healthy); in this case, the coordinate system found by the PCA algorithm is not helpful in separating the two classes.

they explain.

2. The directions are perpendicular on each other.

3. PCA can be performed on either genes or experiments. PCA performed on genes will produce a graph in which each point is an experiment. PCA performed on experiments will produce a graph in which each point is a gene.

4. PCA can be used to achieve dimensionality reduction by choosing either high variance axes or low variance axes depending on the goal of the analysis. In most cases, one selects the directions that explain most of the variance in the data (high variance axes).

5. The dimensionality reduction achieved through PCA can be useful in visualization and classification.

10.8 Independent component analysis (ICA)

Independent Component Analysis (ICA) [26] is a technique that is able to consider higher order statistical dependencies like skew and kurtosis. In Fig. 10.22, i_1 and i_2 are the two axes of the coordinate system found by ICA. The blue circles class has a very large variance along i_1 and a very low variance along i_2, whereas the red crosses class has a very large variance along i_2 and a very low variance along i_1. Therefore, the two classes may be separated reasonably well using just one of the two dimensions.

ICA has been used very successfully in different variations of the blind source separation problem. A particular example of the blind source separation is the "cocktail party problem". In this problem, the n original signals are n people speaking in the same room. The available information is a set of recordings coming from n microphones placed somewhere in the room. Each microphone captures a mixing of the n voices speaking at the same time. Thus, each individual recording is a jumbled superposition of all voices in the room. The task is to use the n recordings to recover the speech of each individual person. This is possible because the individual sources (speakers) are statistically independent. This problem has been tackled very successfully by the ICA approach. Currently, there are software programs that are able to process such recordings and produce signals corresponding to individual speakers [26].

In the blind source separation problem, there are a number of n sources $s_1, s_2, \ldots, s_n$ that are mixed by an unknown mixing matrix A. Nothing is known about the sources or about the mixing process. All that we observe is a set of n mixed signals that form a vector x:

$$s = [s_1 s_2 \ldots s_n] \tag{10.7}$$

$$\begin{bmatrix} x_1 \\ x_2 \\ \vdots \\ x_n \end{bmatrix} = As = A \begin{bmatrix} s_1 \\ s_2 \\ \vdots \\ s_n \end{bmatrix} \tag{10.8}$$

The task is to recover the original sources by finding a square matrix W which is a permutation and rescaling of the inverse of the unknown matrix A:

$$\begin{bmatrix} \overline{s_1} \\ \overline{s_2} \\ \vdots \\ \overline{s_n} \end{bmatrix} = W \begin{bmatrix} x_1 \\ x_2 \\ \vdots \\ x_n \end{bmatrix} \tag{10.9}$$

The same approach can be used in the DNA microarray data analysis. Here, the n genes expressed on a particular array can be seen as the n signals to be separated. One array corresponds to one microphone in the cocktail party problem. A series of experiments done over the range of a given variable (e.g. time for sporulation data, dosage for drug tests, etc.) will provide the signals analogous to the microphone recordings in the cocktail party problem. Given a large number of genes, arrays and time points (as usually available with microarray data) the task is to unveil "functional signals" of each individual gene, i.e. to be able to follow the evolution in time of each individual gene.

10.9 Summary

This chapter presented a number of general purpose tools useful in the analysis and visualization of microarray data. For each tool, the discussion included a detailed explanation of the tool as well as a discussion of its advantages, limitations, as well as several potential problems that can occur in its practical use.

Box plots are informative graphical tools that present in a concise manner several statistics of the data: mean, top and bottom quartiles (or other chosen percentiles) and the interquartile distance. Box plots afford a convenient way of identifying the outliers in a given data set.

Gene pies are tools particularly useful in the analysis of two-channel cDNA data. Gene pies represent each gene as a pie chart divided into two-colored regions. The ratio of the two channels corresponds to the ratio of the areas of the two regions. A useful feature is the ability to code the maximum absolute intensity on one of the two channels as the diameter of the pie chart. This allows an easy identification of those ratios that correspond to genes expressed at higher levels in at least one of the two channels. Such genes are usually biologically meaningful.

The scatter plot is a two or three dimensional plot in which a vector is plotted as a point having the coordinates equal to the components of the vector. Limitations include the reduced number of dimensions in which data can be plotted. In a scatter plot of two expression experiments including many genes or whole genomes, in which each experiment is represented on an axis, most data points are expected to lie around the line $y = x$ (or $z = y = x$). The assumption here is that most genes will not change. Points far from the diagonal are differentially expressed in the experiments plotted. The assumption may not hold if the set of genes has been preselected in some relevant way. In a scatter plot of two expression experiments including many genes or whole genomes in which the ratio is plotted against an intensity, most data points are expected to lie around the line $y = 0$. A consistent departure from the reference lines above may indicate a systematic trend in the data and the need for a normalization procedure. Many scatter plots of microarray data will exhibit a funnel shape with the wider variance toward low intensities.

Histograms are plots of data counts as a function of the data values. Usually they are drawn using bars. A value x that occurs y times in the data set will be represented by a bar of height y plotted at location x. The histogram provides information about the distribution of the data and can be used in certain situations as an empirical approximation of a probability density function (pdf). The exact shape of a histogram depends on the number of data collected and the size of the bins used. A small sample size and/or the binning process may create gross artifacts distorting the nature of the data distribution. Such artifacts may be detected by comparing the shapes of several histograms constructed using different number of bins. Two experiments can be compared by constructing the histogram of the ratios of the corresponding values. If the experiments involve a large number of genes and is suitably pre-processed and normalized, the histogram is expected to be centered on either zero (if logs are used) or 1 (no logs are used) and be approximatively symmetrical. Differentially regulated genes will be found in the tails of a histogram of log ratios.

Time series are plots in which expression values are plotted against the time they were measured at. A non-uniform time scale is often used because of the nature of the biological experiments. Such as scale may produce very misleading results.

Principal Component Analysis is a technique that uses first order statistical properties of the data in order to construct a new coordinate system. The direction of the new axes will be the eigenvectors of the correlation matrix of the data. The directions are chosen in decreasing order of the amount of data variance they explain and they are perpendicular on each other. PCA can be used to achieve dimensionality reduction by choosing either high variance axes or low variance axes depending on the goal of the analysis. In most cases, one selects the directions that explain most of the variance in the data (high variance axes).

Independent Component Analysis is a technique that takes into consideration higher order statistical properties of the data. Furthermore, the directions of the new axes found by ICA are not necessarily perpendicular on each other. ICA has been used successfully to solve the blind source separation problem and may be used in a similar way in genomics to separate the signals from individual genes.

Chapter 11

Cluster analysis

Birds of a feather flock together.

—Unknown

11.1 Introduction

Cluster analysis is currently the most frequently used multivariate technique to analyze gene sequence expression data. Clustering is appropriate when there is no a priori knowledge about the data. In such circumstances, the only possible approach is to study the similarity between different samples or experiments. In a machine learning framework, such an analysis process is known as unsupervised learning since there is no known desired answer for any particular gene or experiment.

Clustering has become so popular in this field that most authors presenting results obtained with microarrays feel the need to include some type of clustering diagram in their papers [2, 29, 50, 68, 104, 106, 120, 139, 141, 221, 262, 267, 269, 275, 284, 301, 304]. In fact, the popularity of the clustering techniques is so great that sometimes clustering is mistakenly taken as a very fuzzy and all-inclusive ultimate goal of microarray data analysis. This author has been approached by several accomplished life scientists seeking help in order to "do their clustering." Subsequent probing revealed that "doing the clustering" could mean anything from selecting a subset of differentially regulated genes to identifying gene interactions to building classifiers based on gene expression data. In fact, clustering is **the process of grouping together similar entities**. Clustering can be done on any data: genes, samples, time points in a time series, etc. The particular type of input makes no difference to the clustering algorithm. The algorithm will treat all inputs as a set of n numbers or **an n-dimensional vector**.

If one is to group together things that are similar, one should start by defining the meaning of similarity. In other words, we need a very precise **measure of similarity**. Such a measure of similarity is called a **distance** or a **metric**. A distance is a formula that takes two points in the input space of the problem and calculates a positive number that contains information about how close the two points are to each other. The input space of the problem is an n-dimensional space so the two points can be for instance two genes measured across n experiments or two experiments, each

263

represented by the expression values of n genes. There are many different ways in which such a measure of similarity can be calculated. The final result of the clustering depends in a very essential way on the exact formula used. In the following we will discuss a number of distances used in the analysis of gene expression data.

11.2   Distance metric

A **distance metric** d is a function that takes as arguments two points x and y in an n-dimensional space $\mathbb{R}^n$ and has the following properties:

1. **Symmetry**. The distance should be symmetric, i.e.:

$$d(x, y) = d(y, x) \qquad (11.1)$$

   This means that the distance from x to y should be the same as the distance from y to x.

2. **Positivity**. The distance between any two points should be a real number greater than or equal to zero:

$$d(x, y) \geq 0 \qquad (11.2)$$

   for any x and y. The equality is true if and only if $x = y$, i.e. $d(x, x) = 0$.

3. **Triangle inequality**. The distance between two points x and y should be shorter than or equal to the sum of the distances from x to a third point z and from z to y:

$$d(x, y) \leq d(x, z) + d(z, y) \qquad (11.3)$$

   This property reflects the fact that the distance between two points should be measured along the shortest route (see Fig. 11.1).

It turns out that many different distances can be defined. The only properties shared by all distances are the three properties above. Other properties may intuitively appear to be associated with distances but they may only hold true for certain ways of defining the distance. Furthermore, certain familiar concepts such as that of a circle are strongly influenced by the implicit distance used to define it. Distances are discussed extensively in the literature in the context of clustering and classification [96, 168].

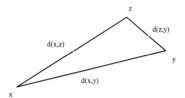

FIGURE 11.1: Triangle inequality: the distance between two points x and y should be shorter than or equal to the sum of the distances from x to a third point z and from z to y. The equality holds true only when z is on the line between x and y.

11.2.1 Euclidean distance

The Euclidean distance between two n-dimensional vectors $\mathbf{x} = (x_1, x_2, \ldots, x_n)$ and $\mathbf{y} = (y_1, y_2, \ldots, y_n)$ is:

$$d_E(\mathbf{x}, \mathbf{y}) = \sqrt{(x_1 - y_1)^2 + (x_2 - y_2)^2 + \cdots + (x_n - y_n)^2} = \sqrt{\sum_{i=1}^{n}(x_i - y_i)^2}$$

(11.4)

This is the usual distance that we use for most practical purposes. Its numerical value comes from the Pythagorean Theorem (see Fig. 11.2). This distance will have all properties extrapolated from our common sense concept of distance. For instance, if both points are translated in space, the distance between them will remain the same, a circle will be the familiar round and symmetrical shape, etc.

As an example, let us calculate the distance from the origin $O(0,0)$ to the point $A(3, 4)$ using the Euclidean metric:

$$d_E(O, A) = \sqrt{3^2 + 4^2} = \sqrt{25} = 5$$

(11.5)

Let us now assume that the coordinates of point A are measured incorrectly due to some experimental error. Let us assume that the measured point is $A'(4, 4)$. The distance between the origin and the measured A' is:

$$d_E(O, A) = \sqrt{4^2 + 4^2} = \sqrt{32} = 5.65$$

(11.6)

which represents a change of $\frac{5.65}{5} = 1.13$. A change of one unit in one of the coordinates determined a change of 13% with respect to the true distance.

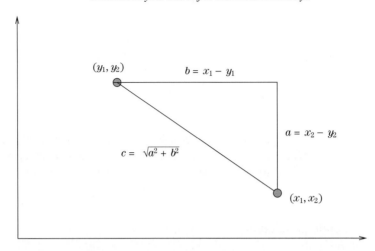

FIGURE 11.2: The Euclidean distance is computed in accordance to the Pythagorean Theorem.

11.2.2 Manhattan distance

The Manhattan distance between two n-dimensional vectors $\mathbf{x} = (x_1, x_2, \ldots, x_n)$ and $\mathbf{y} = (y_1, y_2, \ldots, y_n)$ is:

$$d_M(\mathbf{x}, \mathbf{y}) = |x_1 - y_1| + |x_2 - y_2| + \cdots + |x_n - y_n| = \sum_{i=1}^{n} |x_i - y_i| \qquad (11.7)$$

where $|x_i - y_i|$ represents the absolute value of the difference between x_i and y_i.
The Manhattan distance, or city-block distance, is named after the well known New York borough because it represents the distance that one needs to travel in an environment in which one can move only along directions parallel to the x and y axes (no diagonal movements). This is similar to the Manhattan borough where most streets are straight and cross each other at right angles. Note that in the city-block world, as expected, the distance between two points depends only on the points themselves and not on the path followed to travel between the points (see Fig. 11.3).
As an illustration of the fact that the definition of the distance changes dramatically the properties of any object or concept that uses distances even in an implicit way, Fig. 11.4 shows a comparison between a circle in a space using a Manhattan distance and the usual circle in a Euclidean space.
Let us calculate the same distance OA using the Manhattan distance:

$$d_M(O, A) = 3 + 4 = 7 \qquad (11.8)$$

The first observation is that the city-block distance provided a larger absolute value. If one of the measurements is incorrect and A is measured as $A'(4, 4)$, the distance between the origin and the measured A' is:

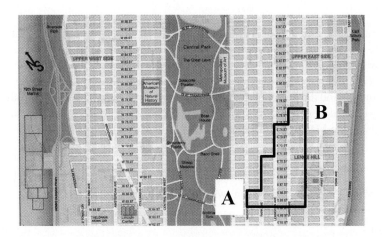

FIGURE 11.3: The Manhattan (or city-block) distance is the distance that one needs to travel in an environment in which one can move only along directions parallel to the x and y axes (no diagonal movements). The Manhattan distance is independent on the path travelled between the two points.

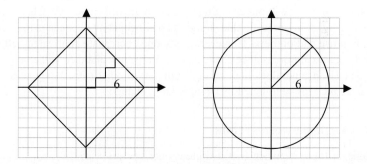

FIGURE 11.4: The distance confers essential properties to the space and objects therein. For instance, a circle is the locus of all points situated at a constant distance from a fixed point called the center. The right panel shows a circle of radius 6 in a Euclidean metric space. The left panel shows the same circle of radius 6 in a Manhattan metric space.

$$d_M(O, A) = 4 + 4 = 8 \tag{11.9}$$

which represents a change of $\frac{8}{7} = 1.14$. A change of one unit in one of the coordinates leads to a change of 14% with respect to the true measurement. We can see that, in comparison to the Euclidean distance, the city-block distance tends to yield a larger numerical value for the same relative position of the points. If the two given points are an outlier and the center of a given cluster, we can say that the city-block distance slightly emphasizes the outliers of a data set: an outlier will appear a bit further away when using the Manhattan distance.

11.2.3   Chebychev distance

The Chebychev distance between two n-dimensional vectors $\mathbf{x} = (x_1, x_2, \ldots, x_n)$ and $\mathbf{y} = (y_1, y_2, \ldots, y_n)$ is:

$$d_{\max}(\mathbf{x}, \mathbf{y}) = \max_i |x_i - y_i| \tag{11.10}$$

The Chebychev distance simply picks the largest difference between any two corresponding coordinates. For instance if the vectors $\mathbf{x} = (x_1, x_2, \ldots, x_n)$ and $\mathbf{y} = (y_1, y_2, \ldots, y_n)$ are two genes measured in n experiments each, the Chebychev distance will pick the one experiment in which these two genes are most different and will consider that value the distance between the genes. The Chebychev distance is to be used when the goal is to reflect any big difference between any corresponding coordinates. The Chebychev distance behaves inconsistently with respect to outliers since it only looks at one dimension. If any or all other coordinates are changed due to measurement error without changing the maximum difference, the Chebychev distance will remain the same. In this situation the Chebychev distance is resilient with respect to noise and outliers. However, if any one coordinate is affected sufficiently such that the maximum distance changes, the Chebychev distance will change. Thus, this distance is in general resilient to small amounts of noise even if they affect several coordinates but will be affected by a single large change.

11.2.4   Angle between vectors

The angle distance between two n-dimensional vectors $\mathbf{x} = (x_1, x_2, \ldots, x_n)$ and $\mathbf{y} = (y_1, y_2, \ldots, y_n)$ is:

$$d_\alpha(\mathbf{x}, \mathbf{y}) = \cos(\theta) = \frac{\mathbf{x} \cdot \mathbf{y}}{\|\mathbf{x}\| \, \|\mathbf{y}\|} \tag{11.11}$$

where $\mathbf{x} \cdot \mathbf{y}$ is the dot product of the two vectors:

$$\mathbf{x} \cdot \mathbf{y} = x_1 y_1 + x_2 y_2 + \cdots + x_n y_n = \sum_{i=1}^{n} x_i y_i \tag{11.12}$$

and $\|\cdot\|$ is the norm, or length, of a vector:

$$\|\mathbf{x}\| = \sqrt{x_1^2 + x_2^2 + \cdots + x_n^2} = \sqrt{\sum_{i=1}^{n} x_i^2} \qquad (11.13)$$

Note that if this distance is used, if a point A is moved anywhere on the line that goes through its original position and the origin, in a new location A', the distance $d(OA')$ will be the same as $d(OA)$. In particular if a point is shifted by scaling all its coordinates by the same factors, the angle distance will not change.

11.2.5 Correlation distance

The Pearson correlation distance between two n-dimensional vectors $\mathbf{x} = (x_1, x_2, \ldots, x_n)$ and $\mathbf{y} = (y_1, y_2, \ldots, y_n)$ is:

$$d_R(\mathbf{x}, \mathbf{y}) = 1 - r_{xy} \qquad (11.14)$$

where r_{ik} is the Pearson correlation coefficient of the vectors $\mathbf{x}$ and $\mathbf{y}$:

$$r_{xy} = \frac{s_{xy}}{\sqrt{s_x}\sqrt{s_y}} = \frac{\sum_{i=1}^{n}(x_i - \bar{x})(y_i - \bar{y})}{\sqrt{\sum_{i=1}^{n}(x_i - \bar{x})^2}\sqrt{\sum_{i=1}^{n}(y_i - \bar{y})^2}} \qquad (11.15)$$

Note that since the Pearson correlation coefficient r_{xy} varies only between -1 and 1, the distance $1 - r_{xy}$ will take values between 0 and 2.

The Pearson correlation focuses on whether the coordinates of the two points change in the same way (e.g. corresponding coordinate increase or decrease at the same time). The magnitude of the coordinates is less important since the denominator will be proportional to the magnitudes of the vectors. If the vector is, for instance, a set of measurements of given genes in a particular experiment and two such experiments are compared, the Pearson distance will be high if the genes vary in a similar way in the two experiments even if the magnitude of the change differs greatly.

A related issue is the problem of outliers. If a gene is measured incorrectly in one of the experiments, its coordinate along that particular dimension can be very different. This can produce a low overall correlation. In order to address this, the **jackknife correlation** calculates the correlation n times, each time leaving one dimension out and calculating the correlation only on the remaining $n - 1$. This will produce n different correlation values for any two vectors (e.g. genes). The distance between the two vectors is taken to be the minimum correlation distance between the n different values. If one of the measurements is wrong, the correlations will be low every time the incorrect measurement is taken into consideration. However, since the jackknife discards a different dimension each time, the incorrect measurement will be eventually discarded. For that particular computation, the correlation between the given genes will be much higher and therefore their distance $(1 - r)$ will be lower. This is the value that will represent the jackknife distance between the given genes. This distance can be written as:

$$d_J(\mathbf{x}, \mathbf{y}) = \min\left\{ d_R^1(\mathbf{x}, \mathbf{y}), d_R^2(\mathbf{x}, \mathbf{y}), \ldots, d_R^n(\mathbf{x}, \mathbf{y}) \right\} \tag{11.16}$$

where $d_R^k(\mathbf{x}, \mathbf{y})$ is the correlation distance between $\mathbf{x}$ and $\mathbf{y}$ calculated disregarding the k-th component of the vectors $\mathbf{x}$ and $\mathbf{y}$.

Sometimes the jackknife correlation is too radical since it takes the least value of these correlation coefficients as the measure of the similarity. Furthermore, the method works by discarding data which is not always a good idea. Finally, the jackknife correlation is only robust to a single outlier. For more outliers, a more general definition of jackknife correlation is needed which makes the method more computationally intensive and more dangerous (for n outliers, n data points will be ignored at every step).

11.2.6 Squared Euclidean distance

The squared Euclidean distance between two n-dimensional vectors $\mathbf{x} = (x_1, x_2, \ldots, x_n)$ and $\mathbf{y} = (y_1, y_2, \ldots, y_n)$ is:

$$d_{E^2}(\mathbf{x}, \mathbf{y}) = (x_1 - y_1)^2 + (x_2 - y_2)^2 + \cdots + (x_n - y_n)^2 = \sum_{i=1}^{n} (x_i - y_i)^2 \tag{11.17}$$

When compared to the Euclidean distance, the squared Euclidean distance tends to give more weights to the outliers due to the lack of the square root. The squared Euclidean distance from the origin $O(0, 0)$ to the point $A(3, 4)$ is:

$$d_{E^2}(O, A) = 3^2 + 4^2 = 25 \tag{11.18}$$

The distance between the origin and the measured A' is:

$$d_{E^2}(O, A) = 4^2 + 4^2 = 32 \tag{11.19}$$

which represents a change of $\frac{32}{25} = 1.28$. A change of one unit in one of the coordinates leads to a change of 28% with respect to the true measurement. When we compare this with the change of 13% for Euclidean and 14% for Manhattan we see why an outlier will be overemphasized by the squared Euclidean distance.

11.2.7 Standardized Euclidean distance

All distances discussed so far give exactly the same importance to all dimensions. The idea behind standardized Euclidean is that not all directions are necessarily the same. For instance, a data set may be known to have a non-homogeneous distribution that is characterized by a larger variance across the y axis and a smaller variance across the x axis. In these conditions, a point that is a certain distance away from the center along x may be very unusual, and therefore interesting, while a point that is the same distance away from the center along the y axis may be well within the usual distribution of the patterns (see Fig. 11.5). In this situation, the border

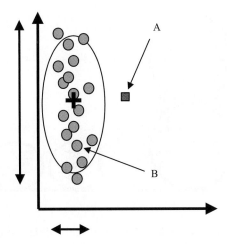

FIGURE 11.5: Data with unequal coordinate variances. A point A that is a certain distance away from the center along x may be very unusual, and therefore interesting, while a point B that is the same distance away from the center along the y axis may be well within the usual distribution of the patterns. The ellipse represents a contour of equal probability: the points outside the ellipse are unlikely to come from the same distribution as the points inside.

marking a region of constant probability will be an ellipse instead of a circle. The standardized Euclidean distance takes this into consideration by dividing with the standard deviation of each dimension. The standardized Euclidean distance between two n-dimensional vectors $\mathbf{x} = (x_1, x_2, \ldots, x_n)$ and $\mathbf{y} = (y_1, y_2, \ldots, y_n)$ is:

$$d_{SE}(\mathbf{x}, \mathbf{y}) = \sqrt{\frac{1}{s_1^2}(x_1 - y_1)^2 + \cdots + \frac{1}{s_n^2}(x_n - y_n)^2} = \sqrt{\sum_{i=1}^{n} \frac{1}{s_i^2}(x_i - y_i)^2}$$

$$(11.20)$$

Because of this, the standardized Euclidean distance may provide better results in certain situations. Fig. 11.6 shows an example in which the data are distributed as two long and thin clusters. In this case, the usual Euclidean distance would not be able to form the correct clusters while the standardized Euclidean has no difficulty in doing so. In Fig. 11.6, the points B and A belong to the same cluster. However, the Euclidean distance between B and A is larger than the Euclidean distance between B and C. Therefore, a clustering algorithm using the usual Euclidean distance will group B and C in the same cluster as shown in the left panel. A clustering algorithm using a standardized Euclidean distance will effectively equalize the variances on each axis and produce the clusters in the right panel which reflect better the structure of the data.

The standardization can be done using the sample variance, as in Eq. 11.20, or by

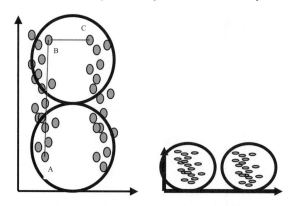

FIGURE 11.6: A data set that has a larger variance along the y axis. Even though the points B and A belong to the same cluster, they are further apart than B and C. A clustering algorithm using the usual Euclidean distance will group B and C in the same cluster as shown in the left panel. A clustering algorithm using a standardized Euclidean distance will effectively equalize the variances on each axis and produce the clusters in the right panel which reflect better the structure of the data. Image courtesy of BioDiscovery Inc.

using the range:

$$d_{SE}(\mathbf{x}, \mathbf{y}) = \sqrt{\frac{1}{R_1^2}(x_1 - y_1)^2 + \cdots + \frac{1}{R_n^2}(x_n - y_n)^2} = \sqrt{\sum_{i=1}^{n} \frac{1}{R_i^2}(x_i - y_i)^2}$$

(11.21)

where R_i is the range of the data along dimension i.

11.2.8 Mahalanobis distance

The standardized Euclidean distance used the idea of weighting each dimension by a quantity inversely proportional to the amount of variability along that dimension. This is equivalent to distorting the space by shrinking it along the axes of large variance and expanding it along the axes of low variance. This can be generalized. One might want to distort the space in an arbitrary way, not necessarily along the axes. This is achieved by the Mahalanobis distance.

The Mahalanobis distance between two n-dimensional vectors $\mathbf{x} = (x_1, x_2, \ldots, x_n)$ and $\mathbf{y} = (y_1, y_2, \ldots, y_n)$ is:

$$d_{Ml}(\mathbf{x}, \mathbf{y}) = \sqrt{(\mathbf{x} - \mathbf{y})^T S^{-1}(\mathbf{x} - \mathbf{y})}$$

(11.22)

where S is any $n \times n$ positive definite matrix and $(\mathbf{x} - \mathbf{y})^T$ is the transposition of $\mathbf{x} - \mathbf{y}$.

The role of the matrix S is to distort the space as desired. Usually, this matrix is the covariance matrix of the data set. This achieves the same purpose as the standardized Euclidean only that the directions of distortion can be arbitrary, as best suited to the data. This is in contrast to the distortion introduced by the standardized Euclidean which was limited to scaling along the axes. If the space warping matrix S is taken to be the identity matrix, the Mahalanobis distance reduces to the classical Euclidean distance.

$$d_{Ml}(\mathbf{x}, \mathbf{y}) = \sqrt{(\mathbf{x} - \mathbf{y})(\mathbf{x} - \mathbf{y})^T} = \sqrt{\sum_{i=1}^{n}(x_i - y_i)^2} \qquad (11.23)$$

11.2.9 Minkowski distance

The Minkowski distance is a generalization of the Euclidean and Manhattan distance. The Minkowski distance between two n-dimensional vectors $\mathbf{x} = (x_1, x_2, \dots, x_n)$ and $\mathbf{y} = (y_1, y_2, \dots, y_n)$ is:

$$d_{Mk}(\mathbf{x}, \mathbf{y}) = \{|x_1 - y_1|^m + |x_2 - y_2|^m + \cdots + |x_n - y_n|^m\}^{\frac{1}{m}} =$$

$$= \left\{\sum_{i=1}^{n} |x_i - y_i|^m\right\}^{\frac{1}{m}} \qquad (11.24)$$

Recalling that $x^{\frac{1}{m}} = \sqrt[m]{x}$, we note that for $m = 1$ the Minkowski distance reduces to Manhattan, i.e. a simple sum of absolute differences. For $m = 2$, the Minkowski distance reduces to Euclidean distance.

11.2.10 When to use what distance

With so many distances, a natural question is when to use what? This section will discuss briefly the issues and criteria that can be taken into consideration when choosing the distance to be used in clustering.

Sometimes, different types of variables need to be mixed together. In order to do this, any of the distances above can be modified by applying a weighting scheme. For instance mixing clinical data with gene expression values can be done by assigning different weights to each type of variable in a way that is compatible with the purpose of the study.

In many cases, it is necessary to normalize and/or pre-process the data.[1] One possible step in the normalization procedure is to standardize genes or arrays. This may be necessary or desirable in order to compare the amount of variation of two different genes or arrays from their respective central locations. Standardizing genes can be done by applying a z-transform, i.e. subtracting the mean and dividing by the

---

[1] See Chapter 12 for a more complete discussion of normalization issues.

Distance	Variable standardization	Observation standardization
Euclidean	different	different
Standardized Euclidean	same	different
Correlation	different	different
Mahalanobis	different	different

TABLE 11.1: The effect of standardization upon several distances.

standard deviation (see Eqs. 4.63 and 4.64 in Chap. 4). For a gene g and an array i, standardizing the gene means adjusting the values as follows:

$$x_{gi} = \frac{x_{gi} - \overline{x}_{g.}}{s_{g.}} \tag{11.25}$$

where $\overline{x}_{g.}$ is the mean of the gene g over all arrays and $s_{g.}$ is the standard error of the gene g over the same set of measurements. The values thus modified will have a mean of zero and a variance of one across the arrays.

Standardizing the arrays means adjusting the values as follows:

$$x_{gi} = \frac{x_{gi} - \overline{x}_{.i}}{s_{.i}} \tag{11.26}$$

where $\overline{x}_{.i}$ is the mean of the array and $s_{.i}$ is the standard error of the array across all genes. Similar standardization can be performed using median instead of mean as a more robust estimator of central tendency (see Chap. 4) and the median absolute deviation as an estimator of the amount of variability.

In some sense, gene standardization makes all genes similar. A gene that is affected only by the inherent measurement noise will be indistinguishable from a gene that varies 10 fold from one experiment to another. Although there are situations in which this is useful, gene standardization may not necessarily be a wise thing to do every time. Standardizing the arrays, on the other hand, is applicable in a larger set of circumstances. However, standardizing the arrays is rather simplistic if used as the only normalization procedure. More normalization issues will be discussed in Chap. 12. Table 11.1 shows how various distances behave with respect to gene or experiment standardization.

There is an important interaction between the choice of the distance and the type of values being compared. It is commonly believed that the Affymetrix technology measures the absolute abundance of mRNA and hence is an absolute measure of the expression level of a gene. On the other hand, cDNA arrays are used most commonly with two channels, one of which is the reference.[2] Furthermore, the spots are often characterized by a ratio or log-ratio of the values measured on the two channels. For these reasons, the cDNA technology is often thought to measure a relative expression of the expression level in a condition with respect to a reference. In reality, the difference is only superficial since one could, in principle, pair each Affymetrix

---

[2]We have seen in Chap. 8 that this may not be the best practice.

Distance	Clustering genes	Clustering samples
Euclidean	different	same
Manhattan	different	same
Correlation	same	different

TABLE 11.2: The effect of using absolute or relative expression values with several distances. If the distances are the same, the clustering will also be the same.

array exploring a condition with another Affymetrix array hybridized with a control and then take the ratio of the corresponding genes. This would provide relative measurements with respect to that condition. Alternatively, one could measure different conditions on the two channels of the same cDNA microarray and analyze the values without calculating ratios with respect to the reference (see for instance the ANOVA loop model discussed in Chap. 8). This approach would extract "absolute" expression levels from cDNA arrays. In conclusion, whether we are using absolute or relative expression values has little to do with the technology itself. However, as noted before, there is a strong link between the type of values measured (absolute or relative) and the distances. Specifically, the question is whether the distance between two given genes or two given experiments depends on whether the values are relative or absolute for a given distance. The answers are summarized in Table 11.2 for the most commonly used distances.

11.2.11 A comparison of various distances

1. Euclidean distance – the usual distance as we know it from our environment. It will be used as a reference when summarizing the other distances.

2. Squared Euclidean – tends to emphasize the distances. Same data clustered with squared Euclidean might appear more sparse and less compact.

3. Angle between vectors – takes into consideration only the angle, not the magnitude. For instance, a gene g_1 measured in two experiments, $g_1 = (1, 1)$, and a gene $g_2 = (100, 100)$ will have the distance (angle):

$$cos(\theta) = \frac{\mathbf{x} \cdot \mathbf{y}}{\|\mathbf{x}\| \, \|\mathbf{y}\|} = \frac{[100 \ 100] \begin{bmatrix} 1 \\ 1 \end{bmatrix}}{\sqrt{100^2 + 100^2} \cdot \sqrt{1^2 + 1^2}} = \frac{100 + 100}{100 \cdot \sqrt{2} \cdot \sqrt{2}} = 1 \tag{11.27}$$

Therefore, the angle between these two vectors is zero. Clustering with this distance will place these two genes in the same cluster although their absolute expression levels are very different.

4. Correlation distance – will look for similar variation as opposed to similar numerical values. Let us consider a set of 5 experiments and a gene g_1 that has an expression of 1, 2, 3, 4 and 5 in the 5 experiments, respectively. This gene can be represented as

$$g_1 = (1, 2, 3, 4, 5)$$

Let us also consider the genes

$$g_2 = (100, 200, 300, 400, 500)$$

and

$$g_3 = (5, 4, 3, 2, 1)$$

The correlation distance will place g_1 in the same cluster with g_2 and in a different cluster from g_3 because $(1, 2, 3, 4, 5)$ and $(100, 200, 300, 400, 500)$ have a high correlation $(d(g_1, g_2) = 1 - r = 1 - 1 = 0)$ whereas $(1, 2, 3, 4, 5)$ and $(5, 4, 3, 2, 1)$ are anti-correlated $(d(g_1, g_3) = 1 - (-1) = 2)$. However, the Euclidean distance will place g_1 in the same cluster with g_3 and in a different cluster from g_2 because $d_E(g1, g2) = 734.20$ while $d_E(g1, g3) = 6.32$.

5. Standardized Euclidean – eliminates the variance information. All directions will be equally important. If genes are standardized, genes with a small range of variation (e.g. affected only by noise) will appear the same as genes with a large range of variation (e.g. changing several orders of magnitude).

6. Manhattan – the set of genes or experiments being equally distant from a reference does not match the similar set constructed with Euclidean distance (see Fig 11.4).

7. Jackknife – robust with respect to one or few erroneous measurements.

8. Chebychev – focuses on the most important difference: $(1, 2, 3, 4)$ and $(2, 3, 4, 5)$ have distance 2 in Euclidean and 1 in Chebychev. $(1, 2, 3, 4)$ and $(1, 2, 3, 6)$ have distance $\sqrt{2}$ in Euclidean and 2 in Chebychev.

9. Mahalanobis – can warp the space in any convenient way. Usually, the space is warped using the correlation matrix of the data.

10. Minkowski – a generalization of Euclidean and Manhattan.

11.3   Clustering algorithms

Before we start to discuss specific clustering approaches and algorithms, it is useful to clarify the terminology as well as make a few general observations. Any clustering algorithm can be used to group genes or experiments or any set of homogeneous entities described by a set of numbers usually arranged as a vector. We shall refer to such entities as patterns or instances. Similar patterns grouped together by the algorithm form **clusters**. A set of clusters including all genes or experiments considered form a **clustering, cluster tree** or **dendrogram**.

There are a few general observations that can be made about clustering. Contrary to the popular belief, clustering is not a goal in itself and, by itself, is seldom convincing. **Anything can be clustered**. In order to prove this, let us consider the well known leukemia data set [127]. This data set contains measurements corresponding to acute lymphoblastic leukemia (ALL) and acute myeloid leukemia (AML) samples from bone marrow and peripheral blood. There are 11 AML samples and 27 ALL samples for a total of 38 samples. These samples were analyzed with Affymetrix 6800 arrays containing approximately 6,800 genes. The genes that appeared to be different between the two data sets can be found using a t-test. We ordered the genes in the increasing order of their p-values and we selected the top 35 genes. These genes are expected to be most different between the two groups. The top panel in Fig. 11.7 shows a clustering of these top 35 genes.

In order to illustrate the fact that anything can be clustered, we generated a number of fake expression values using a random number generator. These numbers were organized in a matrix with 6,800 rows and 38 columns. Each row represents a fake gene and each column represents a fake experiment. We divided the experiments in two classes, labelled ALL and AML. Respecting the structure of the original data, we picked 11 experiments for the AML group and 27 experiments for the ALL group. We then selected the genes that appeared to be different in both data sets using a t-test. We ordered the genes by p-values and picked the top 35 genes. These genes were then clustered. The bottom panel in Fig. 11.7 shows the results for the random data set. In both panels, one can observe two distinct groups of genes: i) genes with lower expression values in the ALL group (green) and higher values in the AML (red) and ii) genes with lower values in AML and higher values in ALL.

The astute reader will notice that the two dendrograms are not *exactly* the same: the one constructed from the random data has a more grainy appearance whereas the dendrogram of the real data is smoother. This is in fact due to the completely random data. However, a similar experiment can be done by using real expression data coming from some other experiment or even using the same leukemia data set in which the experiment labels have been assigned randomly between the ALL and AML groups. Once this random assignment has been done, a t-test can be used to select the genes that are different between the two groups. A clustering obtained from the top 35 genes will be equally smooth and real-looking as the clustering obtained using the correct labels.

In conclusion, the first major observation is that given enough genes, the **genes will always cluster**. Given the large number of genes in most organisms, there is no surprise and therefore no scientific value in the fact that there are genes that behave in a similar way. The scientific value should always come from what can be said about the genes that fall in the same cluster and what can be done with said genes. In the leukemia data set, the interesting result was that a classifier constructed based on these genes could distinguish correctly between ALL and AML on a different data set used for validation.[3] In other data sets, the interesting result may be related to the

---

[3]It turns out that the two types of samples were collected from two different sources in a biased way:

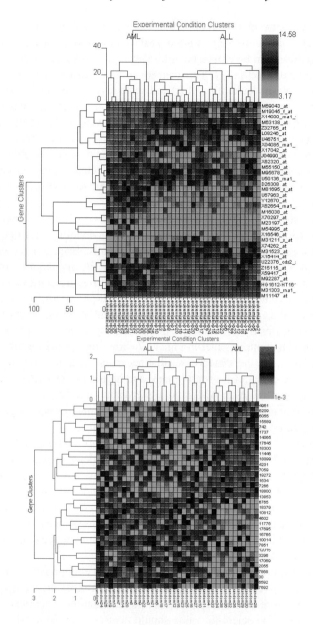

FIGURE 11.7: Anything can be clustered. The top panel shows the top 35 genes in the Leukemia data set [127]; the bottom panel shows the top 35 genes in a random data set. In both panels, one can observe two distinct groups: i) genes with lower expression values in the ALL group (green) and higher values in the AML (red) and ii) genes with lower values in AML and higher values in ALL. Image obtained with GeneSight 3.5.2.

functional analysis of the genes that are clustered together (see Chapter 14) or other biologically meaningful relationships between the members of a given cluster.

The second general observation is that, in most cases, the clustering produced by a given algorithm is **highly dependent on the distance metric used**. Changing the distance metric may affect dramatically the number and membership of the clusters as well as the relationship between them. Looking at a clustering without knowing the distance used to generate it is uninformative and can be very misleading. Let us consider for instance the 3 genes discussed in Section 11.2.11: $g_1 = (1, 2, 3, 4, 5)$, $g_2 = (100, 200, 300, 400, 500)$ and $g_3 = (5, 4, 3, 2, 1)$. The correlation distance will place g_1 in the same cluster with g_2 and in a different cluster from g_3 while the Euclidean distance will place g_1 in the same cluster with g_3 and in a different cluster from g_2. The mere observation that two genes are close, and therefore the clustering itself, can be interpreted only if the distance is clearly specified.

Another observation is that the clustering is not necessarily deterministic. This means that **the same clustering algorithm applied to the same data may produce different results**. Many clustering algorithms have an intrinsically non-deterministic component. For instance, the initialization of the clusters in both k-means and self-organizing feature maps (SOFMs) is done entirely randomly. The stochastic aspects may involve, for instance, a random choice of the initial cluster centers or a random choice of the patterns to be used as initial clusters. It follows that the membership or any particular gene to any particular cluster should be taken with a grain of salt and analyzed carefully. For instance, one should always check whether the given gene would fall into the same cluster if the same algorithm were applied again, etc.

Finally, one must note that in most clustering algorithms (e.g. k-means and hierarchical clustering) **the position of the patterns within the clusters does not reflect their relationship in the input space**. For instance, in Fig. 11.8 genes M11147_at[4] and M55150_at (in the second gene cluster from the top) are right next to each other in the dendrogram. However, their expression profiles are probably as different as they get between any two genes within that cluster. The gene M55150 has a profile more similar to that of gene U50136_rna1_at which is plotted 5 genes away.

Furthermore, the fact that two patterns belong to a given cluster does not necessarily mean that they are close to each other. In fact, a pattern belonging to a cluster A may be closer to some patterns from a different cluster B than it is to other patterns in its own cluster A. In Fig. 11.8, the experiment Exp35 (rightmost experiment in the rightmost cluster) is closer to some experiments in the AML cluster than it is to the other experiments in its own cluster. This is hardly surprising since this *is* an AML sample. Contrary to the appearances, this is neither a software bug nor an

---

most ALL samples came from one source and most AML samples came from a second source. Thus, it is not clear whether the classifier learned to distinguish between ALL/AML or between the data sources. However, no matter what the class corresponds to, the paper showed it is possible to construct a classifier based on gene expression and this in itself remains a landmark result.

[4]For convenience, we are using here the Affymetrix probe IDs as gene names. Gene identifiers can be easily converted between Affymetrix probe ID, GenBank accession IDs and UniGene cluster IDs using the Onto-Convert software available at http://vortex.cs.wayne.edu/Projects.html.

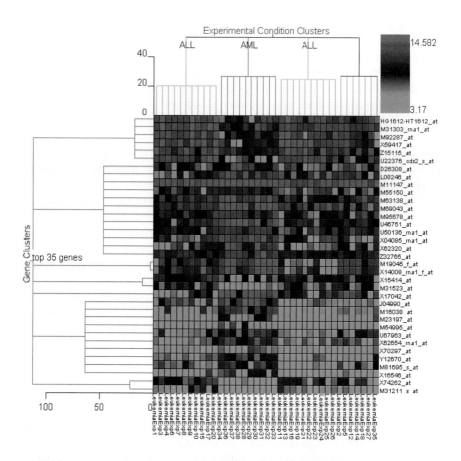

FIGURE 11.8: The way the clusters are plotted may be misleading. The data represents 35 genes that can be used to separate the ALL/AML classes in the leukemia data set [127]. M11147_at and M55150_at are next to each other in the dendrogram even though their expression profiles are not very similar. Gene M55150 has a profile more similar to that of gene U50136_rna1_at which is plotted 5 genes away. Image obtained with GeneSight 3.5.2.

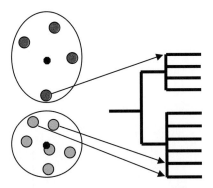

FIGURE 11.9: A 2D example in which the pattern indicated by the top arrow is closer to patterns from another cluster than it is to other patterns in its own cluster. Note that the clustering is correct inasmuch each pattern is assigned to the closest cluster center.

algorithmic mistake. Fig. 11.9 shows the same phenomenon in 2 dimensions. Note that the cluster assignment of each pattern is correct: each pattern is assigned to the closest cluster center. However, the patterns that appear to be far from each other in the clustering are in fact very close to each other. Furthermore, the pattern in the upper cluster is closer to the patterns indicated by arrows than it is to other patterns in its own cluster.

11.3.1 *k*-means clustering

The k-means algorithm is one of the simplest and fastest clustering algorithms. In consequence, it is also one of the most widely used algorithms. The k-means clustering algorithm takes the number of clusters, k, as an input parameter. This is usually chosen by the user. The program starts by randomly choosing k points as the centers of the clusters (see upper-left panel in Fig. 11.10). These points may be just random points in the input space, random points from more densely populated volumes of the input space or just randomly chosen patterns from the data itself.

Once some cluster centers have been chosen, the algorithm will take each pattern and calculate the distance from it to all cluster centers. Each pattern will be associated with the closest cluster center. A first approximate clustering is obtained after allocating each pattern to a cluster. However, since the cluster centers were chosen randomly, it is not said that this is the correct clustering. The second step starts by considering all patterns associated to one cluster center and calculating a new position for this cluster center (upper-right panel in Fig. 11.10). The coordinates of this new center are usually obtained by calculating the mean of the coordinates of the points belonging to that cluster (i.e. the center is calculated as the centroid of the

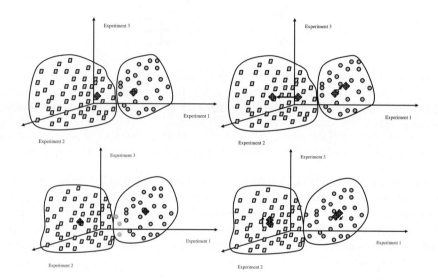

FIGURE 11.10: The *k*-means algorithm with $k = 2$. Upper-left: two cluster centers are chosen randomly and patterns are assigned to each cluster based on their distance to the cluster center. Upper-right: new centers are calculated based on the patterns belonging to each cluster. Bottom-left: patterns are re-assigned to the new clusters based on their distance to the new cluster centers. Three patterns move from the left cluster to the right cluster. Bottom-right: new cluster centers are calculated based on the patterns assigned to each cluster. The algorithm continues to try to reassign patterns but no pattern will need to be moved between clusters. Since no patterns are moved from one cluster to another, the cluster centers remain the same and the algorithm stops.

group of patterns). Since the centers have moved, the pattern membership needs to be updated by recalculating the distance from each pattern to the new cluster centers (in the bottom-left panel in Fig. 11.10 three patterns move from one cluster to the other). The algorithm continues to update the cluster centers based on the new membership and update the membership of each pattern until the cluster centers are such that no pattern moves from one cluster to another. Since no pattern has changed membership, the centers will remain the same and the algorithm can terminate (bottom right in Fig. 11.10).

11.3.1.1 Characteristics of the k-means clustering

The k-means algorithm has several important properties. First of all, the results of the algorithm, i.e. the clustering or the membership of various patterns to various clusters, *can change* between successive runs of the algorithm (see Fig. 11.11). Furthermore, if some clusters are initialized with centers far from all patterns, no patterns will fall into their sphere of attraction and they will produce empty clusters. In order to alleviate these problems, care should be taken in the initialization phase. A common practice initializes centers with k points chosen randomly from the existing patterns. This ensures that i) the starting cluster centers are in the general area populated by the given data and ii) each cluster will have at least one pattern. This is because if a pattern is initialized as a center of a cluster, it will probably remain in that cluster.

In the k-means example shown in Fig. 11.9, a different initialization might produce a different clustering in which the top cluster has only 3 patterns and the bottom cluster has 7 patterns. A natural question arises regarding the meaning of the k-means clustering results: if k-means can produce different clusters every time, what confidence can one have in the results of the clustering? This issue can be refined into a number of questions that will be briefly considered in the following.

11.3.1.2 Cluster quality assessment

Given a particular clustering, how good is a particular cluster? Can one assess the quality of a specific cluster? One way to assess the goodness of fit of a given clustering is to compare **the size of the clusters vs. the distance to the nearest cluster**. If the inter-cluster distance is much larger than the size of the clusters, the cluster is deemed to be more trustworthy (see Fig. 11.12). The ratio between the distance D to the nearest cluster center and its diameter d can be calculated for each cluster and can be used as an indication of the cluster quality.

Another possible quality indicator is the average of the **distances between the members of a cluster and the cluster center**. In the dendrogram shown in Fig. 11.14, the length of the branches of the tree are proportional to the average square distance of the members of the clusters to the cluster centroid. Thus, shorter clusters are better than taller clusters.[5] Fig. 11.14 shows a clustering and the 2D PCA plot of the same

---

[5] Note that this feature is specific to GeneSight 3.5 from BioDiscovery. Other data analysis programs may

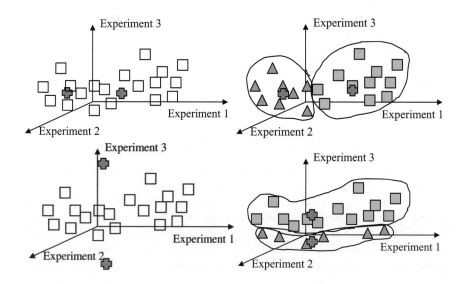

FIGURE 11.11:   The *k*-means algorithm can produce different results in different runs depending on the initialization of the cluster centers. An initialization as in the upper left panel will lead to a clustering as in the upper right panel. An initialization as in the bottom left panel will lead to a clustering as in the bottom right panel.

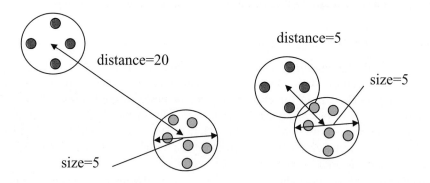

FIGURE 11.12:   Cluster quality assessment. The quality of a cluster may be assessed by the ratio between its diameter and the distance to the nearest cluster. Image courtesy of BioDiscovery Inc.

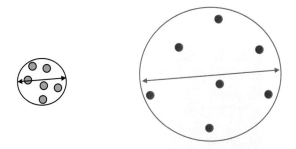

FIGURE 11.13:   Cluster quality assessment. The quality of a cluster may be assessed by calculating the average distance from the members of the cluster to the centroid or the diameter of the sphere including all members of the cluster. The cluster to the left represents a genuine similarity of the patterns while the cluster to the right was formed mainly due to the imposed number of clusters (two).

data. The height of the clusters reflects the average distance from the gene to the center of the cluster.

The **diameter of the smallest sphere** including all members of a given cluster may also be used as a cluster quality measure. Such as example is shown in Fig. 11.13. In this figure, the cluster to the left represents a genuine similarity of the patterns while the cluster to the right was formed mainly due to the imposed number of clusters (two). A potential disadvantage of this measure of cluster confidence is the fact that the diameter of the smallest sphere including all members of the cluster is determined by the furthest pattern from the cluster. In consequence, this measure is sensitive to cluster outliers. A tight cluster with only one pattern far from the center will have a large diameter.

Another interesting question is how confident can one be that a gene that fell into a cluster will fall into the same cluster if the clustering is repeated? This question can be addressed by repeating the clustering several times and following the particular gene of interest. Certain software packages offer the possibility of creating partitions based on the cluster membership. Such partitions will have all genes in a cluster coded with the same color (see Fig. 11.14). Repeating the clustering several times will reveal whether the colors remain grouped together. Those genes that are clustered together repeatedly are more likely to be genuinely similar.

Fig. 11.14 shows a clustering of the top 35 genes from the leukemia data set [127]. The left panel shows the result of a k-means clustering with k=5. The genes are colored by their cluster membership. The right panel shows the same genes in a

---

not calculate such quality information and therefore the length of the branches in many publications may have no particular significance.

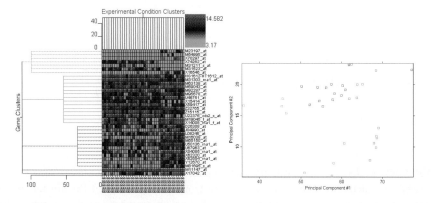

FIGURE 11.14:   A clustering of the top 35 genes from the leukemia data set [127]. The left panel shows the result of a k-means clustering with $k = 5$. The genes are colored by their cluster membership. The right panel shows the same genes in a 2D PCA plot. Note that the patterns are not grouped in any meaningful way. In spite of this, the k-means dutifully produced 5 clusters as requested. Maintaining the color coding and repeating the clustering a few times shows that the same genes are grouped differently every time. Note how the height of the clusters reflects the average distance from the gene to the center of the cluster. Image obtained with GeneSight 3.5.2.

2D PCA plot of the first two principal components. Note that the patterns are not grouped in any meaningful way. In spite of this, the k-means dutifully produced 5 clusters as requested. Maintaining the color coding and repeating the clustering a few times shows that the same genes are grouped differently every time.

The question of how confident one can be that a given pattern belongs to a given cluster can also be addressed using a **bootstrapping** approach. Bootstrapping is a general technique that allows the computation of some goodness of fit measure based on many repeats of the same experiment on slightly different data sets all constructed from the available data [103]. The bootstrapping idea was discussed in Chap. 9 as a mean of correcting for multiple experiments. In the context of cluster confidence, this idea was investigated by a number of researchers [108, 180]. In particular, one could use the replicate measurement data in combination with a bootstrap approach in order to address the issue of cluster confidence.

Fig. 11.15 illustrates this approach. The idea is to use the fact that each gene expression value is usually the result of several measurements. In cDNA arrays, the gene expression may be the mean of the background corrected mean intensities of several spots. In Affymetrix arrays, the gene expression may be the mean of the values measured on several arrays. ANOVA can be used as explained in Chap. 7 to fit a model suitable to the problem. The ANOVA will have some residuals that will correspond to the unexplained variation in the data. These residuals form a population. The bootstrapping approach proposed by Kerr and Churchill [180] draws n

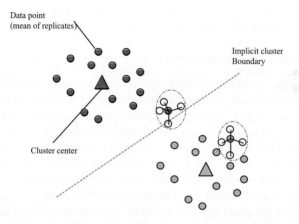

FIGURE 11.15:   The residual based cluster confidence approach [108, 180]. Each gene is represented by a mean of some individual measurements. The confidence that a given gene belongs to a cluster is inversely proportional to the number of times the gene falls into a different cluster during the bootstrapping process. The gene shown in red remains in the red cluster 3 out of 4 times which translates into a confidence of 0.75. Image courtesy of BioDiscovery Inc.

samples from this population and uses them to construct fictitious data that preserve the characteristics of the original data. In particular, the variance of the noise will be exactly the same as that of the original data. The clustering is performed on the original data. The clusters and boundaries between clusters are stored. Subsequently, the same clustering is performed n times using the data constructed from the model and the population of residuals. For each gene, the number of times that the gene moves from one cluster to another is inversely proportional with the confidence of that gene-cluster assignment. A gene that will remain in the same cluster independently of the noise will have a confidence of 1 or 100%. A gene that moves to a different cluster 50% of the time will have a confidence of 0.5. In Fig. 11.15, the number of fake measurements is $n = 4$. For 3 out of these 4 bootstrap runs, the gene close to the border between clusters remained in its cluster. However, in one of the 4 runs, the noise was such that the gene appeared to be in the adjacent cluster. This particular gene will have a confidence value of 0.75.

11.3.1.3 Number of clusters in k-means

The choice of the number of clusters is another issue that needs careful consideration. If it is known in advance that the patterns to be clustered belong to several different classes (e.g. cancer and healthy), one should cluster using the known number of classes. Thus, if there are features that clearly distinguish between the classes, the algorithm might use them to construct meaningful clusters. Note that it is not necessary to know which pattern belongs to each class but only that there are two different classes. If the analysis has an exploratory character and the number of existing classes is not known, one could repeat the clustering for several values of k and compare the results, i.e. track the genes that tend to fall in the same cluster for different values of k. This approach is heuristic in nature and its utility will vary widely depending on the particular problem studied.

11.3.1.4 Algorithm complexity

The complexity of the k-means algorithm must also be considered. It can be shown that the k-means algorithm is linear in the number of patterns, e.g. genes, N. This means that the number of computations that need to be performed can be written as $c \cdot N$ where c is a value that does not depend on N. The value c does depend on the number k of clusters chosen by the user as well as the number of iterations. However, the number of clusters is generally very small in comparison with the number of patterns. Typically, hundreds or thousands genes are clustered in a handful of clusters. Overall, one can conclude that k-means has a very low computational complexity which translates directly into a high speed.

11.3.2 Hierarchical clustering

Hierarchical clustering has been used since the very beginning of the microarray field [104, 141, 304]. Hierarchical clustering aims at the more ambitious task of providing the definitive clustering that characterizes a set of patterns in the context of a given

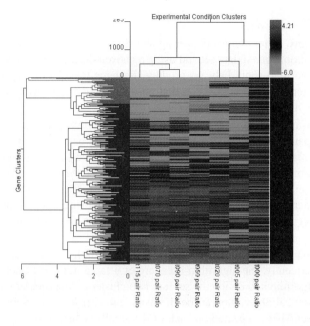

FIGURE 11.16: A hierarchical clustering of both genes and experiments in the yeast sporulation data set. [1, 66]. Image obtained with GeneSight 3.5.2.

distance metric. The result of k-means clustering is a set of k clusters. All these clusters, as well as all elements of a given cluster, are on the same level. As we have seen, no particular inferences can be made about the relationship between members of a given cluster or between clusters. In contrast, the result of a hierarchical clustering is a complete tree with individual patterns (genes or experiments) as leaves and the root as the convergence point of all branches.

The diagrams produced by the hierarchical clustering are also known as dendrograms. A dendrogram is a branching diagram representing a hierarchy of categories based on degree of similarity. Different genes and/or experiments are grouped together in clusters using the distance chosen. Different clusters are also linked together based on a cluster distance such as the average distance between all pairs of objects in the clusters. A combined dendrogram with gene clustering plotted horizontally and experiment clustering plotted vertically is presented in Fig. 11.16.

Unlike the real trees, hierarchical trees are usually drawn with the root on top and the branches developing underneath. The tree can be constructed in a **bottom-up** fashion, starting from the individual patterns and working upwards towards the root or following a **top-down** approach, starting at the root and working downwards towards the leaves. The bottom-up approach is sometimes called **agglomerative** because it works by putting smaller clusters together to form bigger clusters. Analogously, the top-down approach is sometime called **divisive** because it works by splitting large clusters into smaller ones.

Unlike k-means, a hierarchical clustering algorithm should be completely deterministic. Applied on a given data set and using a chosen distance, the same hierarchical clustering algorithm should always produce the same tree. However, different hierarchical clustering algorithms, e.g. a bottom-up approach and a top-down approach, may produce different trees.

The bottom-up method works as follows. It starts with n clusters, each consisting of a single pattern. The pattern can be either a gene or an experiment depending on what the algorithm is applied to. The algorithm calculates a table containing the distances from each cluster to every other cluster. For n points this computation will require on the order of n^2 arithmetical operations. Then, the bottom-up method repeatedly merges the two most similar clusters into a single super-cluster until the entire tree is constructed. A distance able to assess the similarity of two clusters is required.

The top-down approach starts by considering the whole set of patterns to be clustered. Subsequently, the algorithm uses any of a large number of non-hierarchical clustering algorithms to divide the set into two clusters. A particular choice of such a non-hierarchical algorithm can be the k-means with $k = 2$. Subsequently, the process is recursively repeated on each of the smaller clusters as they are obtained. The process stops when all small clusters contain a single pattern. The top-down approach tends to be faster than the bottom-up approach.

Finally, another approach to building a hierarchical clustering uses **an incremental method**. This approach can be even faster than the top-down approach. Such methods build the dendrogram by adding one pattern at a time, with minimal changes to the existing hierarchy. In order to add a new gene, the gene under consideration is compared with each cluster in the tree, starting with the root and following always the most similar branch according to the distance used. When finding a cluster containing a single gene, the algorithm adds a branch containing the gene under consideration. As mentioned, this approach can be lightning fast compared to the others. However, the weakness is that the results can depend not only on the distance metric (as any clustering) or the distance metric and some random initialization (as the top-down approach) but also on the *order* in which the points are considered.

11.3.2.1   Inter-cluster distances and algorithm complexity

The distance between clusters can be taken to be the distance between the closest neighbors (known as **single linkage** clustering), furthest neighbors (**complete linkage**), the distance between the centers of the clusters (**centroid linkage**) or the average distance of all patterns in each cluster (**average linkage**). The centroid of a group of patterns is the point that has each coordinate equal to the mean of the corresponding coordinates of the given patterns. For instance, the set of experiments: $Exp_1 = (1, 2, 3)$, $Exp_2 = (2, 3, 4)$ and $Exp_3 = (3, 4, 5)$ has the centroid in:

$$\left(\frac{1+2+3}{3}, \frac{2+3+4}{3}, \frac{3+4+5}{3} \right) \qquad (11.28)$$

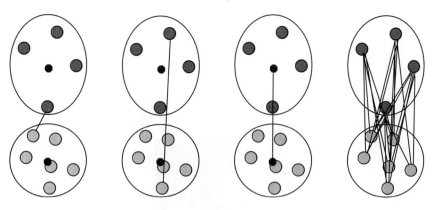

FIGURE 11.17: Linkage types in hierarchical clustering. Left to right: single linkage, complete linkage, centroid linkage and average linkage.

Clearly, the total complexity of the algorithm and therefore its speed is very much dependent on the linkage choice. Single or complete linkages require only choosing one of the distances already calculated while more elaborated linkages, such as centroid, require more computations. Such further computations are needed *every time two clusters are joined* which greatly increases the total complexity of the clustering. However, much like always, cheaper is not always better. Simple and fast methods such as single linkage tend to produce long stringy clusters, e.g. if using a Euclidian distance. More complex methods such as centroid linkage or neighbor joining [252] tend to produce clusters that reflect more accurately the structure present in the data but are extremely slow. The complexity of a bottom-up implementation can vary between n^2 and n^3 depending on the linkage chosen. In the context of gene expression, one should try to prune as much as possible the set of genes of interest before attempting to apply a bottom-up clustering with a more complex linkage.

11.3.2.2 Top-down vs. bottom-up

In general, algorithms working by division require less computation and are therefore faster. However, obtaining the results quicker may not necessarily be a reason for joy because a hierarchical clustering algorithm working by division, or top-down, may produce results inferior to the results of an algorithm working by agglomeration. This can happen because in dividing the clusters the most important splits, affecting many patterns, are performed at the beginning before accumulating enough information and two patterns inadvertently placed in different clusters by an early splitting decision will never be put together again.

The top-down clustering tends to be faster but the clusters produced tend to reflect less accurately the structure present in the data. Furthermore, theoretically, the results of a hierarchical clustering should only depend on the data and the metric chosen. However, a top-down approach will rely essentially on the qualities of the par-

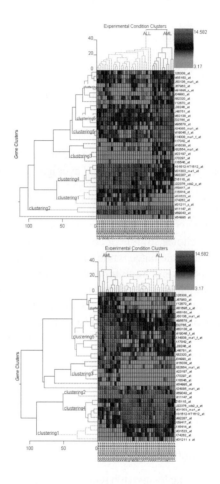

FIGURE 11.18: Two different hierarchical clustering constructed by division (top-down) using k-means to split the larger clusters. The deterministic aspect of the hierarchical clustering is lost due to the usage of the non-deterministic k-means. Both diagrams above have been obtained using the same distance (Euclidian) on the same data (35 top genes in the ALL-AML data set [127]). Image obtained with GeneSight 3.5.2.

titioning algorithm chosen. For instance, if k-means is chosen to divide clusters into sub-clusters, the overall result may be different if the algorithm is run twice with the same data. This can be due to the random initialization of the cluster centers in the k-means division. Fig. 11.18 shows two different clusterings obtained by running the same top-down hiearchical clustering algorithm twice on the top 35 genes selected from the ALL-AML data set [127].

The complexity of the top-down approach can require between $n \log n$ and n^2 computations and is therefore intrinsically faster than the bottom-up approach especially when a complex linkage is involved.

Fig. 11.19 shows the effect of the various combinations between the linkage type and the approach used. All diagrams in this figure use exactly the same data (35 top genes in the ALL-AML data set [127]) and the same distance (Euclidian). The upper left panel shows the results of the clustering using division; upper right: agglomeration with single linkage; bottom left: agglomeration with complete linkage and bottom right: agglomeration with average linkage. For this data, the centroid linkage produces the same clustering as the average linkage.

11.3.2.3 Cutting tree diagrams

A hierarchical clustering diagram may be used to divide the data into a pre-determined number of clusters. This division may be done by cutting the tree at a certain depth (distance from the root). For instance, the tree in Fig. 11.20 can be cut to generate two clusters (left panel) or five clusters (right panel). The tree can also be cut at different depths on different branches in order to reflect better the structure of the data. The lowest possible cut is at the individual pattern level. This cut will always generate as many clusters as patterns with only one pattern in each cluster. The highest possible cut is at the root level. This is equivalent to saying that there is a single cluster containing all available data.

11.3.2.4 An illustrative example

Armed with all the necessary knowledge, let us construct manually a very simple hierarchical clustering for the following example.

Example 11.1

Let us consider a gene measured in a set of 5 experiments: A, B, C, D and E. Let us consider that the values measured in the 5 experiments are: $A = 100, B = 200, C = 500, D = 900$ and $E = 1100$. We will construct the hierarchical clustering of these values using Euclidian distance, centroid linkage and an agglomerative approach.

SOLUTION The closest two values are 100 and 200. The centroid of these two values is 150. Now, we are clustering the values 150, 500, 900, 1100. The closest two values are 900 and 1100. These values are joined and the centroid is calculated. The centroid of 900 and 1100 is 1000. The remaining values to be joined are 150,

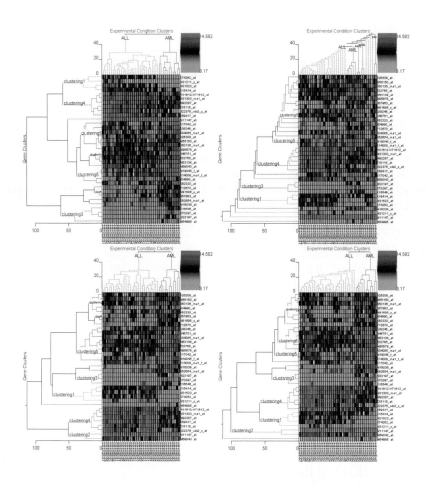

FIGURE 11.19: The effect of the various combinations between the linkage type and the approach used. All diagrams use the same data (35 top genes in the ALL-AML data set [127]) and the same distance (Euclidian). Upper left: division clustering; upper right: agglomeration with single linkage; bottom left: agglomeration with complete linkage and bottom right: agglomeration with average linkage. For this data, the centroid linkage produces the same clustering as the average linkage. Images obtained with GeneSight 3.5.2.

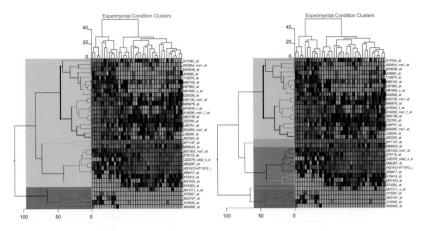

FIGURE 11.20: A complete hierarchical tree structure can be cut at various depths to obtain a different number of clusters. For instance, in the left panel, the dendrogram is cut at the depth of 1 to generate two clusters. The same dendrogram is cut at a depth of 4 to generate 5 clusters in the right panel. Hierarchical trees can also be cut at different depths on different branches in order to reflect better the structure of the data. The data used here are a subset of 35 genes from the AML-ALL data [127]. Images obtained with GeneSight 3.5.2.

500 and 1000. The closest values are 150 and 500. These values are joined together. Finally the two resulting subtrees are joined in the root of the tree. The resulting dendrogram is shown in the left panel of Fig. 11.21. ⬚

We obtained this result because we started with the experiments ordered by the measurements of the gene considered. However, when more than one gene is involved, the experiments cannot be ordered anymore. Therefore, it is important to see what happens when the values are considered in a different, arbitrary, order. The same clustering process applied to an initial arbitrary ordering will produce a different looking tree as shown in the right panel of Fig. 11.21. It is important to notice that the two trees shown in Fig. 11.21 are *exactly the same*. Indeed, tree are judged by their topology i.e. the way their branches converge. In both trees in Fig. 11.21, $A = 100$ is most similar to $B = 200$, $C = 500$ is most similar to the group (A, B) and $D = 900$ is most similar to $E = 1100$. However, the tree in the left panel shows the experiment A furthest from E while the tree in the right panel shows the experiment A closest to E. It clear from this example that **nothing can be inferred from the fact that two genes or experiments are plotted next to each other in a hierarchical dendrogram**. ⬚

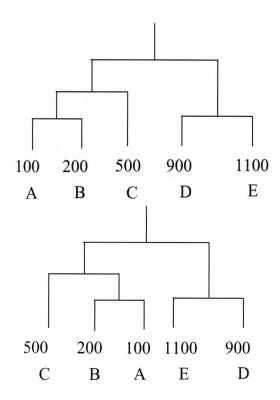

FIGURE 11.21:   Two hierarchical clusters of the expression values of a single gene measured in 5 experiments. The dendrograms are identical: both diagrams show that A is most similar to B, C is most similar to the group (A,B) and D is most similar to E. In the top panel, pattern A and E are plotted far from each other. In the bottom panel, A and E are immediate neighbors. This example shows that the proximity in a hierarchical clustering (e.g. A and E in the right panel) does not necessarily correspond to similarity.

11.3.2.5   Hierarchical clustering summary

A few conclusions can be drawn from this discussion of various hierarchical clustering methods. A first conclusion is that various hierarchical clustering implementations using the same data and the same metric can still produce different dendrograms if they use different approaches. Another important conclusion is that merely obtaining a clustering is not an issue and that the dendrogram itself is almost never the answer to the research question. A dendrogram connecting various genes in a graphically pleasant way can be obtained relatively quickly from any data set. The real problem is to obtain a clustering that reflects the structure of the data. A clustering that reflects well the properties of the data may require more work. Finally, various implementations of hierarchical clustering should not be judged simply by their speed. Many times, slower algorithms may simply be trying to do a better job of extracting the data features. Finally, and most importantly, hierarchical diagrams convey information only in their topology. The order of the genes within a given cluster and the order in which the clusters are plotted do not convey useful information and can be misleading.

11.3.3   Kohonen maps or self-organizing feature maps (SOFM)

The **Kohonen map**, also called **self-organizing feature map** (SOFM), was proposed by Teuvo Kohonen in the late 80s [187, 188]. The SOFM is a type of clustering. As any clustering algorithm, the SOFM will divide the input patterns into groups of similar patterns. In this respect, it is similar to k-means and hierarchical clustering. However, unlike the clustering produced by k-means and hierarchical clustering, the relationship between the clustered patterns actually conveys information about the relationships and reciprocal positions of the patterns in the original input space.

As we have seen, in k-means, the relative position of the patterns in the resulting clustering is not only uninformative but can also be misleading. This was illustrated in Fig. 11.9 in which two patterns very close in the input space were assigned to different clusters and appeared very far from each other in the resulting clustering. The hierarchical clustering is more informative than the k-means inasmuch large groups of patterns that are indistinguishable in k-means are further divided and organized into sub-trees which provide more information regarding the relative relationships between the respective patterns. However, even in a hierarchical clustering diagram, the elements of any particular subtree can still be plotted swapped around, bringing dissimilar patterns to be drawn nearby in the dendrogram (Fig. 11.21).

Unlike k-means and hierarchical, the SOFM clustering is designed to create a plot in which similar patterns are plotted next to each other. It is said that the SOFM maps the input space into a feature space in which the neighborhood relationship reflects the degree of similarity between patterns. Plotting the patterns in such a space creates a **feature map**. A feature map has the property that distances and relationships measured on the feature map are proportional to distances and relationships between patterns according to the similarity metric chosen.

The SOFM is actually a **neural network** technique [130, 138]. Neural networks are

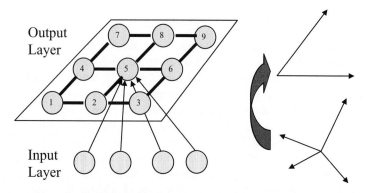

FIGURE 11.22: A two-dimensional self-organizing feature map (SOFM). Each unit in the SOFM is connected to all inputs. The map uses a distance measure and a neighborhood. The neighborhood of a unit is a set of nearby units. For instance the neighborhood of unit 5 can include units 2, 4, 6 and 8 (a 4-neighborhood) or units 1, 2, 3, 4, 6, 7, 8 and 9 (an 8-neighborhood). This SOFM implements a dimensionality reduction by projecting a space with 4 dimensions into a space with 2 dimensions.

a class of techniques inspired by the brain. The fundamental paradigm is to perform complex computations using networks of very simple elements. These simple elements are called units or neurons. In such networks, much like in the brain, the processing abilities come from the pattern and strength of the connections between units. A SOFM is usually a grid of such very simple elements. A SOFM can use a one-dimensional grid (like a string), a two-dimensional grid (an array) or a three-dimensional grid (a cube or parallelepiped). One-dimensional and two-dimensional SOFMs are usually the most widely used.

Fig. 11.22 illustrates a two-dimensional SOFM with 4 inputs. Each unit of the SOFM is connected to all inputs. Each such connection is characterized by a **weight** or **connection strength**. For any unit in the SOFM, all its weights form a vector. The size of the vector is equal to the number of dimensions of the input space since each unit has a link from each input. If the SOFM is used to analyze experiments, each experiment will be a pattern and the number of inputs (or the number of input features or the dimensionality of the input space) will be equal to the number of genes. If the SOFM is used to analyze genes, each gene will be a pattern and the dimensionality of the input space will be equal to the number of experiments. The values of the weight vector determine a point in the input space. It is said that each unit represents a **prototype**. The weights are initialized with random values which is equivalent to saying that the units of SOFM are randomly distributed in the input space.

The SOFM is constructed by **training**. The neural network training process is similar to the learning process that happens in the brain. The input patterns are presented repeatedly. Every such presentation of an input pattern modifies slightly the strength of some of the connections in the network. If the connections are modified randomly, nothing useful happens. However, if the connections are modified according to some **training rule**, the process leads to a gradual adaptation that is known as **learning**. The purpose of the learning in the training of the self-organizing feature map is to extract from the data the most important features and, based on them, to group the data into meaningful clusters that share such important features. In SOFM, this is achieved using two ideas: a **neighborhood** and a **winner-take-all approach**.

Let us assume that we are clustering 10,000 genes. Let us assume that each gene was measured in each of 4 experiments. The input space has 4 dimensions and there are 10,000 patterns, each pattern being described by a vector with 4 elements. The SOFM is trained as follows. Each gene is presented to the input of the network. Using the chosen distance, each unit in the SOFM calculates the distance between its weight vector (a vector of 4 numbers) and the current gene (another vector of 4 numbers). The unit which is found to be closest to the current gene is declared to be the winner. The weights of the winner are modified in such a way that the weight vector becomes more similar to the current gene. The exact amount of modification brought is determined by a small positive number called the **learning rate**. Essentially, the input winner is moving in the direction of the current input pattern with the magnitude of the move being determined by the learning rate. At the same time, all units in the neighborhood of the winner are also changed in the same way but to a lesser extent. The winning unit is practically pulling its neighbors closer to the current gene. This process can be visualized by imagining all SOFM units being connected to each other by rubber bands. When a particular unit is moved, its neighbors will also be moved a little bit. The process then continues for the next gene until all 10,000 genes have been processed. This represents one iteration of the algorithm. In order to ensure a convergence, the learning rate and sometimes the size of the neighborhood is gradually reduced over a number of iterations.

The result of this training algorithm is that different units in the SOFM become prototype genes representing profiles most often encountered in the genes analyzed. The SOFM provides three benefits. Firstly, each unit of the SOFM will contain a prototype. This prototype will represent the typical behavior of the genes triggering that particular unit. More generally, the prototype will represent the set of common features extracted from the input patterns by a given unit. Secondly, the SOFM yields a set of clusters. All input patterns activating a unit will be clustered together. Finally, the relationship between the units activated by specific genes will be closely related to the relationships between the genes. A gene will always be more similar to another gene in its immediate neighborhood than to a gene further apart. Also the placement of the clusters themselves will have the same feature. The landscape of the feature map constructed by the Kohonen network will represent faithfully the properties and features of the data. For this reason, it is said that the SOFM implements a dimensionality reduction: the n-dimensional input space is projected into a space with 1, 2 or 3 dimensions depending on the network used (see Fig. 11.22).

SOFMs can be used as clustering and/or visualization tools. One could simply perform the Kohonen training and then plot the clusters generated from the SOFM. An example of this usage is shown in Fig. 11.23.

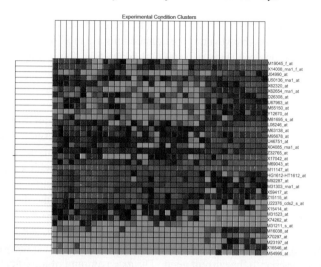

FIGURE 11.23:   The results of a 1D clustering in Euclidian distance on both genes and experiments for 35 selected genes from the ALL-AML data set [127]. Unlike the results of a k-means or hierarchical clustering, two genes that are plotted next to each other are necessarily similar according to the chosen distance.

Another way to use the SOFM is to plot the prototypes in the input space together with the topology of the network showing the links between the units. For example, Fig. 11.24 represents a 2D self-organizing feature map trained on 3-dimensional data. Each dot represents a gene. The 3 coordinates of each dot are the expression values measured in 3 different mRNA samples. The left panel represents an early stage during the training. The feature map has started to bend and stretch in order to capture the features of the data. The right panel represents the feature map at the end of the training. The feature map is bent and stretched such that it covers the entire data set. In this case, there were more units than genes so each unit models a gene and there are unused units. Note that similar genes are represented by neighboring nodes in the network. Since the plot is drawn in the input space, it is clear that this approach can be used directly only when the dimensionality of the input space is small (2-3 dimensions). However, this technique can be coupled nicely with PCA. The PCA can be used first to project the problem into a space with 2 or 3 dimensions and then SOFM can be used to create a feature map that would indicate similarities and relationships between the various data points.

Finally, SOFMs can be used for visualization. In this case, the plot will show the units of the trained Kohonen network as they are activated by various input patterns. This is truly a dimensionality reduction since any n-dimensional pattern can be seen through the activation determined by it on the 1, 2 or 3-dimensional feature map. Such an example is shown in Fig. 11.25. The patterns are 3D atomic structures of

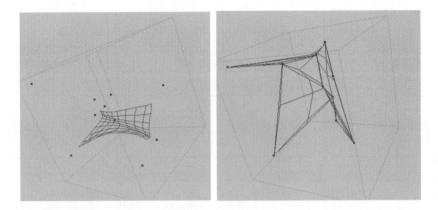

FIGURE 11.24: A 2D self-organizing feature map trained on 3-dimensional data. Each dot represents a gene. The 3 coordinates of each dot are the expression values measured in 3 different mRNA samples. The left panel represents an early stage during the training. The feature map has started to bend and stretch in order to capture the features of the data. The right panel represents the feature map at the end of the training. The feature map is bent and stretched such that is covers the entire data set. In this case, there were more units than genes so each unit models a gene and there are unused units. Note that similar genes are represented by neighboring nodes in the network.

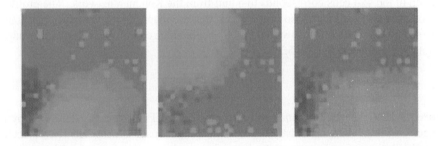

FIGURE 11.25: Examples of responses of a trained 20x20 Kohonen map when patterns from 3 different clusters were used as inputs. The patterns are structures of several HIV proteases resistant to Indinavir. Each HIV protease is described by a vector with 22 elements. The 22-dimensional input space is conveniently projected on the 2-dimensional space of the feature map while maintaining the relationships between patterns. Left to right: high resistance, medium resistance and low resistance patterns [94]. Images obtained with SNNS. From Drăghici, S., Predicting HIV drug resistance with neural networks, Bioinformatics, Vol. 19, No. 1, p. 102, 2003. With permission.

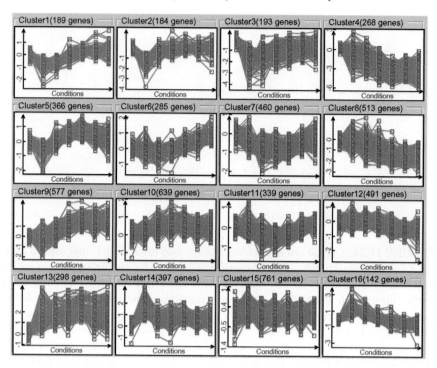

FIGURE 11.26: A 2D self-organizing feature map on a 4x4 network. Each profile represents the expression level of a yeast gene over the sporulation time series [66]. Each of the 4x4 cells corresponds to a unit from the Kohonen network. The profiles plotted in each cell are those of the genes that are mapped into that particular unit. Note that the profiles in any two neighboring cells share certain characteristics. Image obtained with GeneSight 3.5.2.

several HIV proteases resistant to Indinavir. Each HIV protease is described by a vector with 22 elements. The 22-dimensional input space is conveniently projected on the 2-dimensional space of the feature map while maintaining the relationships between patterns. From left to right, the figure shows the activation of the map when a high resistance, medium resistance and low resistance pattern is presented at the inputs [94].

Another type of visualization available with SOFM is to plot together all input patterns that activate a given unit from the Kohonen map. Fig. 11.26 shows a 2D self-organizing feature map constructed on a 4x4 network. Each profile represents the expression level of a yeast gene over the sporulation time series [66]. Each of the 4x4 cells corresponds to a unit from the Kohonen network. The profiles plotted in each cell are those of the genes that are mapped into that particular unit. Note that the profiles in any two neighboring cells share certain characteristics.

As for any other algorithm using a distance to assess similarity, the results obtained with the SOFM depend essentially on the choice of the distance. In order to illustrate this, we will consider again the data shown in Fig. 10.19. These data contain several genes having distinct profiles clearly visible in a 3D PCA plot such as that in Fig. 10.20. A self-organizing feature map of these data obtained using the Euclidean distance is shown in Fig. 11.27. The Euclidean distance groups together profiles with small differences in values regardless whether they occur at the same or different time points. The clusters group together profiles with expression values in the same range (e.g. row 1, column 1 and row 1, column 3) without distinguishing between the different shapes within the given range. Although informative, the clustering is not perfect: three of the resulting clusters are empty. A different initialization and a different range of parameters (neighborhood radius and learning rate) might produce a better clustering. Given the data, the results of the clustering are good but not exceptional. The SOFM was able to identify 3 of the profiles present in the data: row 1, column 2, row 2, column 2 and row 3, column 1. However, the "up-down-up" and "down-up-down" profiles were grouped together at low intensities (row 1, column 1). The same happened for the "up-down-up" and "down-up-down" profiles at high intensities (row 1, column 3). We would have probably preferred to see that the algorithm identifies each individual shape as well as their ranges.

Fig. 11.28 shows the same data clustered using the same 2D SOFM but with a correlation distance instead of the Euclidean distance. As expected for the correlation distance, the clusters group together the profiles with a similar shape over time regardless of their position on the vertical axis. For instance, the cell on row 2, column 1 contains all genes with an "up-down-up" behavior regardless of whether they vary around -2, 0 or 2. Like any microarray data set with many genes, this data set contains a lot of random noise. Many genes affected by random noise happen to have profiles similar to the meaningful genes and are picked up by various clusters. The cell on row 1, column 2 found the "low – steady increase – high" behavior but also captured some noisy genes.

Finally, Fig. 11.29 shows the self-organizing feature map obtained from the same data using the Chebychev distance. As discussed in Section 11.2, the Chebychev distance uses the maximum differences between any pair of coordinates. In this case, the Chebychev distance is most useful. The SOFM was able to extract in an unsupervised manner 8 out of the 9 different gene profiles present in the data. The network was able to distinguish between the similar "up-down-up" profiles in row 2, column 1 and row 3, column 3.

The self-organizing feature maps also have drawbacks. The random initialization makes the results non-deterministic: the same algorithm applied twice on the same data may produce different results. The initialization of the Kohonen network may be tricky. If some initial prototypes are initialized with values very unlike any of the data, they will never win the competition, will never be modified and will remain unused. In order to avoid this, one may always initialize the network with some random patterns from the data. This will ensure that all units will win at least once (for the pattern they have been initialized with), which means they will all be used. Another drawback of SOFM is related to the size (number of units) and layout

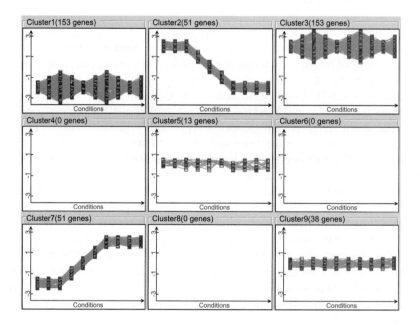

FIGURE 11.27:   A 2D SOFM clustering of the TIGR sample data shown in Fig. 10.19 using the Euclidean distance.   The Euclidean distance groups together profiles with small differences in values regardless whether they occur at the same or different time points.   The clusters group together profiles with expression values in the same range (e.g. row 1, column 1 and row 1, column 3) without distinguishing between the different shapes within the given range.   Image obtained with GeneSight 3.5.2.

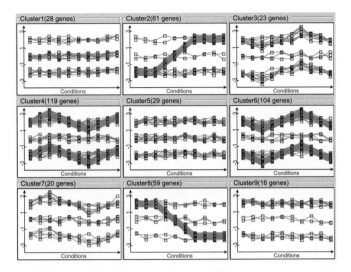

FIGURE 11.28:   A 2D SOFM clustering of the TIGR sample data shown in Fig. 10.19 using the correlation distance. The correlation distance groups together the profiles with the same shape disregarding the range. Image obtained with GeneSight 3.5.2.

(string, array or cube) of the network. Choosing the size and layout of the network has an heuristic component. In most cases, this is done by trial and error. Usually many very similar clusters may indicate a network that is too large for the given data and indistinct clusters lumping together many inhomogeneous patterns may indicate that the network is too small.

11.4   Summary

Two frequently posed problems related to microarrays are: i) finding groups of genes with similar expression profiles across a number of experiments and ii) finding groups of individuals with similar expression profiles within a population. This is the task of the cluster analysis or clustering.

A distance measure is necessary in order to assess the degree of similarity between patterns. This chapter discussed various distances, their definitions and their properties. The following distances have been discussed: Euclidean, standardized Euclidean, squared Euclidean, angle, correlation, jackknife, Manhattan (city block), Chebychev, Minkowski and Mahalanobis. Each distance has certain properties that can be used to emphasize certain characteristics of the data.

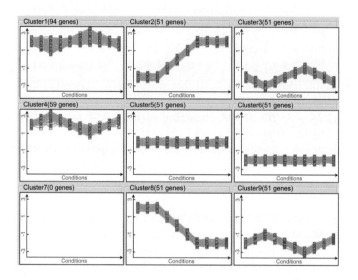

FIGURE 11.29: A 2D SOFM clustering of the TIGR sample data shown in Fig. 10.19 using the Chebychev distance. In this case, the Chebychev distance is most useful. The SOFM was able to extract in an unsupervised manner 8 out of the 9 different gene profiles present in the data. Note how the use of the Chebychev distance allowed the SOFM to separate between the similar up-down-up profiles in row 2, column 1 and row 3, column 3. Image obtained with GeneSight 3.5.2.

The distance is usually used to cluster the genes and/or experiments. The chapter presented the k-means clustering, hierarchical clustering and self-organizing feature maps (Kohonen maps). The k-means algorithm clusters patterns in a given number of groups. The relationship or ordering between different groups in a given k-means cluster diagram is meaningless. The relationship or ordering between the elements of the same group is also meaningless. The k-means clustering uses a random initialization and is therefore non-deterministic (different runs on the same data may produce different results).

The hierarchical clustering constructs a complete tree structure between patterns. It is more informative than k-means since each cluster is further divided into sub-trees showing the internal structure of the clusters. There are several approaches: top-down (by division), bottom-up (by agglomeration) and incremental. Each approach has different properties. The bottom-up may be slower but may provide a more informative tree structure. The information extracted by the hierarchical clustering algorithm is contained in the topology of the dendrogram (how the branches are connected). A particular dendrogram or tree diagram may not necessarily reflect the true relationships between patterns. For instance, patterns drawn next to each other may be more different than patterns drawn at large distances from each other.

The self-organizing feature map (SOFM) or Kohonen map is a regular grid of elements that store data prototypes. Usually, the Kohonen map is a one, two or three-dimensional array, i.e. a string, a rectangular grid or a parallelepiped. Each element in the grid is connected directly to its neighbors. The prototypes are a summation of certain common features shared by several pieces of data and are extracted from the data without user intervention. The process of extracting this information from the data is known as learning or training. Unlike the diagrams constructed by k-means or hierarchical clustering, the Kohonen feature map has the property that if the images of two data points are close on the feature map, those two data points are similar according to the distance used. Kohonen feature maps also have some drawbacks. They tend to be sensitive to the way they are initialized, are non-deterministic and the size and layout involve heuristic choices.

The existing literature is very rich in papers concerned with clustering methods and algorithms, as well as their applications [120, 132, 137, 180, 188, 212, 234, 262, 266, 289, 296, 295]. A particularly interesting approach was used by Cho et al. to discern whether hierarchical clusters are enriched in specific functional categories [65].

Chapter 12

Data pre-processing and normalization

If at first you don't succeed, transform your data set.

—Unknown [117]

12.1 Introduction

Pre-processing is a step that extracts or enhances meaningful data characteristics. Sometimes, pre-processing prepares the data for the application of certain data analysis methods. A typical example of pre-processing is taking the logarithm of the raw values. Normalization is a particular type of pre-processing done to account for systematic differences across data sets. A typical example of normalization is modifying the values in order to compensate for the different dye efficiency in the two channel microarray experiments using cy3 and cy5.

This chapter will discuss the main pre-processing and normalization procedures currently used, emphasizing their motivations, effects and limitations. Certain pre-processing steps such as the logarithmic transformation are equally applicable to cDNA and Affymetrix data while others such as background correction and probe level pre-processing are specific to a given technology.

12.2 General pre-processing techniques

12.2.1 The log transform

The logarithmic function has been used to pre-process microarray data from the very beginning [290, 292]. There are several reasons for this. Firstly, the logarithmic transformation provides values that are more easily interpretable and more meaningful from a biological point of view. Let us consider two genes that have background corrected intensity values of 1000 in the control sample. A subsequent measurement of the same two genes in a condition of interest registers background corrected intensity values of 100 and 10,000 respectively for the two genes (a 16 bit tiff file can

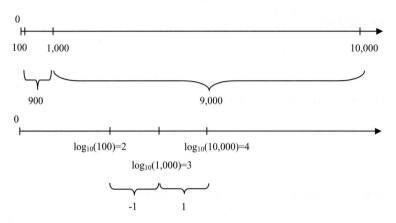

FIGURE 12.1: The effect of the logarithmic transform. The top panel shows the relative position of the values 100, 1000 and 10000. If one considers the absolute difference between these values, one is tempted to consider the difference to the right more important than the difference to the left since $10000 - 1000 = 9000 \gg 1000 - 100 = 900$. The bottom panel shows the relative position of the log transformed values 2, 3 and 4. The transformed numbers reflect the fact that both genes increased and decreased in a similar way (10 times).

contain values between 1 and 65,536). If one considers the absolute difference between the control values and the two experiment values, one would be tempted to consider that one gene is much more affected than the other since:

$$10000 - 1000 = 9000 \gg 1000 - 100 = 900$$

This effect is illustrated in Fig. 12.1. However, from a biological point of view the phenomenon is the same, namely both genes registered a 10-fold change. The only difference between the genes is that the 10-fold change was an increase for one gene and a decrease for the other one. It is very convenient to transform the numbers in order to eliminate the misleading disproportion between these two relative changes. The logarithmic transformation (henceforth log) accomplishes this goal. Using a log transform, in base 10 for instance, will transform the values into:

$$\log_{10}(100) = 2$$

$$\log_{10}(1000) = 3$$

and

$$\log_{10}(10000) = 4$$

reflecting the fact that the phenomena affecting the two genes are the same only that they happen in different directions. This time, the genes are shown to vary by:

$$2 - 3 = -1$$

for one gene and:

$$4 - 3 = 1$$

for the other gene. Note how the values now reflect the fact that the two genes change by the same magnitude in different directions. It is said that the log partially decouples the variance and the mean intensity, i.e. makes changes such as ± 10 fold in the example above more independent of where they happen. Thus, fold changes happening around small intensities values will be comparable to similar fold changes happening around large intensities values.

A second and very strong argument in favor of the log transformation is related to the shape of the distribution of the values. The log transformation makes the distribution symmetrical and almost normal [202, 253, 292]. This is illustrated in Fig. 12.2. In this figure, the top panel shows the histogram of the background corrected intensity values. Note that the intensity range spans a very large interval, from zero to tens of thousands. The distribution is very skewed having a very long tail towards high intensity values. The bottom panel in Fig. 12.2 shows the distribution of the same values after the log transformation.

Finally, a third argument in favor of using the log transformation is convenience. If the log is taken in base 2, the later analysis and data interpretation are greatly facilitated. For instance, selecting genes with a 4 fold variation can be done by cutting a ratio histogram at the value $\log_2(ratio) = 2$. Henceforth in this text, the base of the logarithm will be assumed to be equal to 2.

12.2.2 Combining replicates and eliminating outliers

As we have seen, due to the large amount of noise typically associated with microarray data, one must make repeated measurements. Chapter 7 showed how such repeated measurements can be used to estimate the amount of noise and compare the inter-experiment and within experiment variations. However, in certain situations it is convenient to combine the values of all replicates in order to obtain a unique value, representative for the given gene/condition combination. Such repeated measurements may in fact be, depending on the situation, different spots in cDNA arrays or different values measured on different arrays in both cDNA or Affymetrix arrays. Typically, such values are combined by calculating a measure of central tendency such as mean, median or mode. However, substituting a set of values by a unique value does imply a loss of information and must be done with care. Chapter 4 discussed several examples in which various measures of central tendency are misleading and represent poorly the set of values. Nevertheless, there are strong incentives to calculate a unique expression value for a given gene in a unique condition. Such incentives may include the ability to compare various genes across conditions or tissues, storage and retrieval in expression databases, etc.

Two approaches may be attempted in order to somehow alleviate the loss of information associated to the compression of many repeated measurements into condition representative values. The first approach is to store several parameters of the distribution of original values besides the measure of central tendency. Such values may

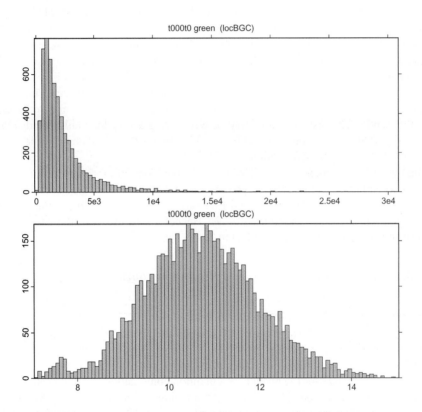

FIGURE 12.2:   The effect of the log transform upon the distribution of the intensity values. The top panel shows the histogram of the background corrected intensity values. The distribution is very skewed (elongated) towards high intensity values. The bottom panel shows the distribution of the same values after the log transformation.

include: the number of values, standard deviation as well as other parameters of the original distribution. Such additional parameters may be used to asses the confidence into a particular value such as a mean. For instance, a mean value obtained from 10 replicates with a low variance will be much more trustworthy than a mean value obtained from 3 replicates with a high variance. The second approach is to try to clean the data by eliminating the outliers. This can be done by calculating a mean and standard deviation σ from the original data and eliminating the data points situated outside some given interval (e.g. $\pm 3\sigma$). The remaining data are re-processed in the same way by calculating a new mean and standard deviation. The process is repeated until no more outliers are detected. Finally, the representative value may be taken as the mean of the remaining values.

12.2.3   Array normalization

The fundamental driving force behind the extensive use of microarrays is the hope that arbitrary comparisons between the gene expression levels in various conditions and/or tissues will be possible eventually. A crucial requirement before such arbitrary comparisons are possible is to normalize the data in such a way that the data are independent of the particular experiment and technology used [43, 44]. In an ideal world, the technology would allow the computation of the gene expression level in some universal reference system using some standard units such as number of mRNA copies per cell. At this time, there is no agreed upon way of normalizing microarray data in such a universal way. It is also still controversial whether data collected with different technologies such as oligonucleotide and cDNA arrays can be compared directly. For instance, Yuen et al. reported a fairly good correlation between expression data measured using the GeneChip with data measured using a cDNA chip [300] while Kuo et al. found no correlation [190]. The main difference between these two studies is that the latter compared data from two different labs (i.e., data based on the same 60 cell lines that were cultured in different labs), whereas the other group used data based on identical biological material. The most recent study by Li et al. showed a good correlation between the data obtained with the two technologies [197]. Interestingly, Yuen et al. found that the bias observed with the commercial oligonucleotide arrays was less predictable and calibration was unfeasible. The same study concluded that fold-change measurements generated by custom cDNA arrays were more accurate than those obtained by commercial oligonucleotide arrays after calibration. At the same time, Li et al. conclude that the data from oligonucleotide arrays is more reliable when compared to long cDNA array (Incyte Genomics) [197]. With such a lack of consensus in the literature, the next best thing to arbitrary cross-technology comparisons is to be able to compare data obtained with a given technology. Fortunately, this is a simpler problem and various approaches allow such comparisons.

The difficulty is related to the fact that various arrays may have various overall intensities. This can be due to many causes including different protocols, different amounts of mRNA, different settings of the scanner, differences between individual arrays and labelling kits, etc. For Affymetrix arrays, there is a difference between the

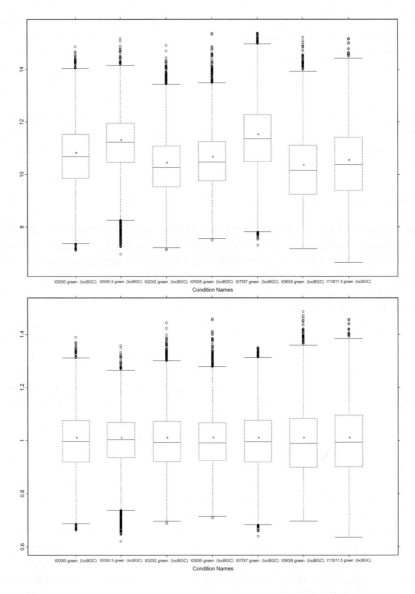

FIGURE 12.3:   The need for array normalization. The top panel shows a comparison of the green channels on the arrays in the yeast sporulation data [66]. The graph shows that there are important differences between the various arrays corresponding to different time points. A comparison of the uncorrected values would produce erroneous results. The bottom panel shows the same comparison after all values have been divided by the array mean. Figure obtained with GeneSight 3.5.2.

overall mean of each individual array. For cDNA arrays, there can also be a difference between each individual channel (dye) on the same array. Fig. 12.3 illustrates the need for an array normalization. In this figure, the top panel shows a comparison of the green channels on the arrays in the yeast sporulation data [66]. An examination of the box plots shows that there are important differences between the various arrays corresponding to different time points. A comparison of the uncorrected values would produce erroneous results. For instance, a gene expressed at an average level on array 5 (left to right) would appear as highly expressed when compared with values coming from array 6. If this gene is also expressed at an average level on array 6, the comparison of its un-normalized values between the two arrays might conclude that the gene is up-regulated in 5 with respect to 6. In fact, the difference in values may be due exclusively to the overall array intensity. The bottom panel shows the same comparison after all values have been divided by the mean of the array from which they were collected. Exactly the same phenomenon can be observed in the case of Affymetrix data.

For both oligonucleotide and cDNA arrays, the goal is to normalize the data in such a way that values corresponding to individual genes can be compared directly from one array to another. This can be achieved in various ways as follows.

12.2.3.1 Dividing by the array mean

The mean can be substituted with the median, mode or percentile if the distribution is noisy or skewed. One could also apply a Z transformation by subtracting the mean and divide by the standard deviation [241]. For cDNA arrays, this approach adjusts overall intensity problems but does not address dye non-linearity. This will be addressed with specific color normalization techniques, such as LOWESS, discussed below.

Variations of this approach eliminate the values in the upper and lower 10% of the distribution (thought to be regulated) and then divide by the mean of the remainder. The motivation behind this is that the number of genes regulated and the amount of regulation should not influence the normalization.

12.2.3.2 Subtracting the mean

Subtracting the mean is usually used in the context of the log transformed data. A well known property of the logarithmic function is:

$$\log_a \frac{x}{y} = \log_a x - \log_a y \qquad (12.1)$$

or log of ratio is the difference of the logs of the numerator and denominator. Therefore, dividing all values by the mean of the array and then taking the log of the normalized value should be analogous to applying the log of all values, then calculating the mean of the log values on the array and normalizing by subtracting this mean. This is indeed the case; the effect of the two processing sequences is similar. However, they are not quite the same. Let us consider an array A containing n genes. The values x_i are read from this array (e.g background corrected values for cDNA arrays or average differences for Affymetrix arrays). We can divide the values by the mean of the array to obtain:

$$x'_i = \frac{x_i}{\overline{x_i}} = \frac{x_i}{\frac{\sum_{i=1}^n x_i}{n}} \tag{12.2}$$

For reasons explained in Section 12.2.1, we can now take the log to obtain the normalized values:

$$x_{in} = \log(x'_i) = \log\left(\frac{x_i}{\overline{x_i}}\right) \tag{12.3}$$

where x_{in} are the normalized values of x_i according to the "divide by mean – log" sequence. Alternatively, we can choose to take the log of the raw values first and then to subtract the mean. Taking the log would provide the following values:

$$x'_i = \log(x_i) \tag{12.4}$$

We now subtract the mean of the log values on the whole array:

$$x_{in} = \log(x_i) - \overline{\log(x_i)} = \log(x_i) - \frac{\sum_{i=1}^n \log(x_i)}{n} \tag{12.5}$$

We can use the properties of the log function to re-write the last term as:

$$\frac{\sum_{i=1}^n \log(x_i)}{n} = \frac{1}{n}\log(x_i \cdot x_2 \cdots x_n) = \log \sqrt[n]{x_i \cdot x_2 \cdots x_n} \tag{12.6}$$

This last form emphasizes the fact that this is the geometric mean of the x_i values. Using the notation:

$$\overline{x_i} = \frac{\sum_{i=1}^n x_i}{n} \tag{12.7}$$

for the arithmetic mean of the values x_i and the notation:

$$\overline{x_{ig}} = \sqrt[n]{x_i \cdot x_2 \cdots x_n} \tag{12.8}$$

for the geometric mean of the values x_i, we see that the equations 12.3 and 12.5 are in fact exactly the same but for the type of the mean used. Thus, the sequence "divide by mean - log" uses the arithmetic mean:

$$x_{in} = \log\left(\frac{x_i}{\overline{x_i}}\right) \tag{12.9}$$

while the sequence "log - subtract mean" uses the geometric mean:

$$x_{in} = \log\left(\frac{x_i}{\overline{x_{ig}}}\right) \tag{12.10}$$

The geometric mean is the same (when all x_i are equal) or lower than the arithmetic mean. In general, the arithmetic mean is used when finding the average of numbers that are added to find the total, and the geometric mean is used when the items of interest are multiplied to obtain the total. For instance, if a company has a profit of 7, 9 and 11 million over 3 consecutive years, the average yearly profit would be calculated appropriately using the arithmetic mean. However, if a company's track record is described in terms of yearly growth such as 3%, 2.5% and 2%, the average growth would be calculated appropriately using the geometric mean. Since the intensities corresponding to various genes are additive, the arithmetic mean is probably more suitable than the geometric mean for array normalization purposes.

12.2.3.3   Using control spots/genes

This involves modifying the values of each channel/experiment such that certain control spots/genes in both experiments have the same or similar values. The control spots should span the whole intensity range. If the intensity of a control gene on array A is found to be c times higher than the intensity of the same control gene on array B, all genes with intensity values in the same range on array A should be divided by c in order to make them comparable with the values read from array B.

12.2.3.4   Iterative linear regression

This approach was proposed in the context of normalizing the two channels of a cDNA array [14]. However, the method can be used to align any two sets of values. The basic idea is that of an iterative linear regression. Performing a linear regression means fitting a straight line of the form $y = m \cdot x + n$ through the data in such a way that the errors (the differences between the y values predicted by using the straight line model above and the real values) are minimal in the least square sense. At each step, those genes that are not modelled well by the linear model are thrown out. Essentially, the approach assumes that there is a linear correspondence between the two sets of values and tries to fit the best slope and fit that would make the two sets match for those genes that are unchanged.

If we consider the sets of values x_{1i} and x_{2i} as coming from either two channels of a cDNA array or two different oligonucleotide arrays the processing can be described in the following steps.

1. Apply a log transform to obtain $\log(x_{1i})$ and $\log(x_{2i})$.

2. Perform a simple linear regression. This means fitting a straight line of the form: $\log(x_{1i}) = \log(x_{2i}) \cdot m + b$ through the data in such a way that the errors (residuals) are minimal.

3. Find the residuals $e = \log(x_{1i}) - \log(x_{2i})_c$ where $\log(x_{1i})_c$ is the value calculated as a function of $\log(x_{2i})$.

4. Remove all genes that have residuals greater than 2σ from 0.

5. Repeat the steps above until the changes between consecutive steps are lower than a given threshold.

6. Normalize $log(x_{2i})$ using the regression line found above: $log(x_{2i})_n = log(x_{2i}) \cdot m + b$

Note again that this normalization does not take into consideration, nor does it correct for any non-linear distortion introduced by the dyes, if such distortion exists. Normalization techniques able to correct for this are discussed in Section 12.3.3.

12.2.3.5 Other aspects of array normalization

An explicit array normalization is not always necessary. For instance, if an ANOVA approach is used (see Chapter 7) together with a suitable experiment design, the systematic difference between arrays will be extracted as the array bias term. Also, other techniques such as LOWESS achieve also the array normalization as part of their non-linear correction. Thus, if two channels of the same cDNA arrays are to be compared and LOWESS is used to compensate for the dye non-linearity, an explicit step of array normalization (e.g. divide by mean) may not be necessary.

12.3 Normalization issues specific to cDNA data

The cDNA experiment is discussed in detail in Chapter 2 but, for convenience, the process will be summarized here. In a typical multi-channel cDNA experiment, various samples are labelled with different dyes. The most usual experiment uses two dyes, or colors, such as cy3 and cy5. For instance, the control sample will be labelled with cy3 and the experiment sample will be labelled with cy5. Once labelled, the two samples will be mixed and hybridized on the array. For a gene expressed in both samples, the hybridization will involve a competitive aspect since cDNA from both samples will have to compete for the few complementary strands spotted on the array. We assume that each spot is represented by a value computed from the pixel intensities as it was discussed in Chapter 3. The process and phenomena are essentially the same even if more or fewer dyes are used.

12.3.1 Background correction

A first pre-processing step is the background correction. The idea behind the background correction is that the fluorescence of a spot is the effect of a summation between the fluorescence of the background and the fluorescence due to the labelled mRNA. Thus, the theory goes, in order to obtain the value proportional to the amount of mRNA, one needs to subtract the value corresponding to the background. This can be done in several ways as follows.

12.3.1.1 Local background correction

The intensity of the background is calculated in a local area around the spot. A measure of central tendency (e.g. mean, median or mode) is calculated and subtracted from the spot intensity. This method is preferred when the background intensity varies considerably from spot to spot. This method is to be avoided when the local neighborhood of the spots does not contain sufficiently many pixels. This may happen on very high density arrays when the spots may be separated by only a few pixels.

12.3.1.2 Sub-grid background correction

A measure of central tendency is calculated for all spots in a subgrid. This is an useful approach in high density arrays. A subgrid includes sufficiently many pixels to allow a more reliable estimate of a measure of central tendency while it is still smaller than the whole array and may be flexible enough to compensate for local variations in the background intensity. Furthermore, most current robots print a sub-grid using the same pin so a sub-grid should be homogeneous as far as the shape and size of the spots are concerned.

12.3.1.3 Group background correction

This is similar to the subgrid correction but uses a smaller number of spots. This method would use a neighborhood of the given spot (e.g. all spots situated in a circle of radius 3) and calculate a measure of local tendency using the background pixels in this neighborhood. This is more flexible than the subgrid, allowing a better adaptation to a non-uniform background while still estimating the background value using more than the few pixels around a single spot.

12.3.1.4 Background correction using blank spots

This method can be used when the design of the array included a few blank spots, i.e. spot locations where no DNA was deposited. Again, a measure of central tendency is calculated on a number of such blank spots.

12.3.1.5 Background correction using control spots

A particular criticism of the approach above is related to the assumption that the spot intensity is the result of a simple summation between the background intensity and the labelled DNA intensity. In fact, the background intensity depends on the properties of the interaction between the labelled target in the solution and the substrate. However, the spot intensity depends on the properties of the interaction between the labelled target and the DNA deposited in the spot. Some researchers have studied this and concluded that the labelled target may be more likely to stick to the substrate in the background of a spot than to hybridize non-specifically on a spot containing some DNA. If this is the case, subtracting any value characterizing the target-substrate interaction may be an over-correction. A possibility is to use some control spots using exogenous DNA and use the intensity of the non-specific hybridization on such spots as a better background correction. Such spots are called control spots.

12.3.2   Other spot level pre-processing

One may choose to discard those spots that have found to be unreliable in the image processing stage. This can be done by flagging those spots in the image processing stage and by discarding the values coming from those spots in the data pre-processing stage. If values are discarded, the missing values may or may not be substituted with some estimates. If the missing values are to be estimated, common estimates may be the mean, median or mode of other spots representing the same gene on the given arrays (if replicate spots are used, as they should) or the mean, median or mode of the same gene in other experiments. Care should be exercised if missing values are estimated because this affects the number of degrees of freedom of the sample (see Chapter 6).

12.3.3   Color normalization

One important problem is that various dyes have slightly different biochemical properties which may affect the data collected. The purpose of the color normalization is to eliminate the data artifacts introduced by the dyes. Fig. 12.4 shows an example of a scatter plot cy3 vs. cy5 of the raw data in a real cDNA experiment. As discussed in Section 10.4, in any given experiment, most of the genes of an organism are expected to be unchanged. In consequence, most of the points in a cy3-cy5 scatter plot are expected to appear along the diagonal cy3=cy5. A brief examination of the scatter plot in Fig. 12.4 shows that this is hardly the case. In this plot, most of the points are found below the diagonal suggesting that for most of the points, the values measured on the cy3 channel are higher than the values measured on the cy5 channel. In principle, this can have two possible causes: either the mRNA labelled with cy3 was more abundant for most of the genes or the cy3 dye is somehow more efficient and, for the same amount of mRNA, the average intensities read are higher.

The idea of a flip dye experiment was introduced in order to control such phenomena. In a flip dye experiment, the two samples of mRNA, A and B are labelled first with cy3 (A) and cy5 (B) and then with cy5 (A) and cy3 (B). Subsequent hybridization and image analysis will produce two sets of data which represent the same biological sample. A plot of the expression levels measured on the two channels for the same mRNA, for instance A labelled with cy3 vs. A labelled with cy5, should produce a straight line $cy3 = cy5$. One is prepared to accept small random variations from the reference but any general trend will indicate a non-negligible influence that should be corrected.

In the example above, the very same mRNA is labelled with two different colors so any differences will be exclusively due to the labelling process. Thus we expect the graph to be symmetric independently of the number of genes involved. Similar expectations hold even if each sample is labelled with a different dye *if the experiments*

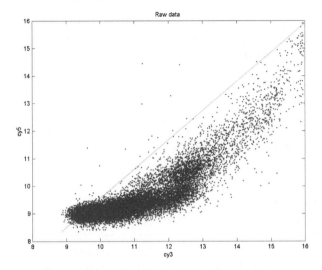

FIGURE 12.4: Typical graph of the cy3/cy5 scatter plot of the raw data obtained from a cDNA experiment. The straight line represents the first diagonal cy3=cy5.

involve all or most genes in a genome. In such situations, it is assumed that most genes will not change since across the board expression changes would probably kill the organism. However, such expectations are **not** warranted if a small subset of genes is considered, especially if the genes are known to be functionally related.

The dyes used may have different overall efficiencies due to many possible reasons. A non-linear dye effect with a stronger signal provided by one of the two dyes is certainly present if the commonly used cy3 and cy5 fluorescent dyes are used. However, the problem should be investigated and addressed for any dyes used, even if the claim of the manufacturer is that the dyes are equally efficient and perfectly linear. For convenience, we will assume that only two dyes are used and we will discuss the color normalization referring to these dyes as cy3 and cy5. However, any other dyes can be used in practice. Furthermore, the same methods can be used for any number of dyes (e.g. if performing a 4 color experiment).

The best way to control and compensate for the dye effect is perform a flip-dye experiment as described above. If the experiment design includes flipping the dyes (see Chapter 8), no extra hybridizations are needed. The essential aspect is to collect data from the same mRNA sample using both dyes. Once such data are available, one could draw a scatter plot of the same mRNA labelled with cy5 and cy3. Such a plot is shown in Fig. 12.4. Since the data were obtained from the same mRNA labelled differently, we know that any departure from the $x = y$ line is due either to the inherent random noise or to the dyes. The plot exhibits two major, non-random, features: i) the data are consistently off the diagonal and ii) the "cloud" of data points has a specific shape often described as "banana-like" or "comma-like." This particular shape is the effect of the non-linear distortion introduced by the dyes and is the main target of the color normalization. The color distortion can be corrected using one of the following approaches:

1. Curve fitting and correction

2. Lowess normalization

3. Piece-wise linear normalization

12.3.3.1   Curve fitting and correction

The color normalization may be achieved through a curve fitting followed by a corresponding data correction. It has been noted [150] that the color distortion introduced by the dyes has an exponential shape in the ratio-intensity scatter plot. Based on this observation, one can use the data to find the parameters (base and shift) of the exponential function that fits the data. This process is illustrated in Fig. 12.5. The data represent a single time point from the yeast cell cycle expression data [254]. In all graphs, the horizontal axis represents $\log(cy5)$ while the vertical axis represents the log of the ratio $\log(cy3/cy5)$. The data in the top left panel are normalized by subtracting the mean but not corrected for color distortion. The distribution exhibits a clear non-linear distortion. The data are divided into intensity intervals on the horizontal axis $\log(cy5)$. This is equivalent to dividing the graph into vertical slices corresponding to such intervals. A centroid of the data is calculated for each interval. An exponential curve of the form:

$$y = a + b \cdot e^{-cx} \tag{12.11}$$

is fitted through the centroids. The purpose of this fit is to find the best combination of values a, b and c that would make the exponential curve above to represent best the color distortion present in the data. Once this function is found, it is used to adjust the data: the vertical coordinate of each data point will be shifted by a value given by the fitted exponential curve at that location in such a way that the color distortion is compensated for. The bottom right panel shows the color corrected plot.

The example above used an exponential curve because this was suitable for the distortion introduced by the commonly used dyes cy3 and cy5. Different dyes may introduce distortions that are better compensated by using functions different from the exponential. However, the general idea behind this approach will remain the same: fit a suitable function through the data and thus obtain a model of the distortion; use this model to adjust the data and eliminate the dye effect. A related idea is used by the LOWESS transformation.

12.3.3.2   LOWESS/LOESS normalization

The LOWESS transformation, also known as LOESS, stands for LOcally WEighted polynomial regreSSion [69, 70]. In essence, this approach divides the data into a

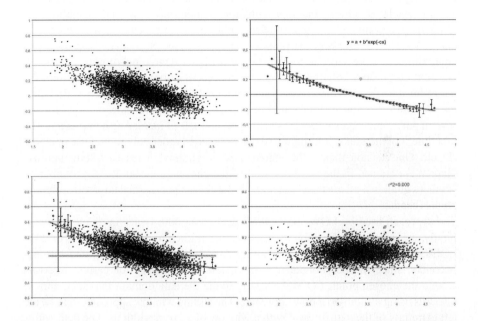

FIGURE 12.5: Exponential normalization. Top left: ratio-intensity plot of the raw data. Top right: the data are divided into groups based on the signal intensity $\log(cy5)$ and a centroid of the data is calculated for each such group. Bottom left: an exponential curve is fitted through the centroids. Bottom right: each data point is shifted with a value given by the fitted exponential. The normalized data is straight and centered around $\log \frac{cy3}{cy5} = 0$. The data are a subset (a single time point) of the yeast cell cycle [254]. In all graphs, the horizontal axis represents $\log(cy5)$ while the vertical axis represents the log of the ratio $\log(cy3/cy5)$.

number of overlapping intervals and fits a function in a way similar to the exponential normalization discussed above. However, the function fitted by LOWESS is a polynomial of the form:

$$y = a_0 + a_1 x + a_2 x^2 + a_3 x^3 + \cdots \qquad (12.12)$$

Polynomials are very nice mathematical objects in the sense that they can approximate a large category of functions.[1] However, the polynomial approximation has two general problems. Firstly, the approximation is good only in a small neighborhood of the chosen point and the quality of the approximation gets worse very quickly as one gets further away from the point of approximation. Secondly, the polynomial approximation is very prone to over-fitting if higher degree polynomials are used. Over-fitting produces highly non-linear functions that attempt to match closely the target function around the known data points but "wiggle" excessively away from them. Both problems are illustrated in Fig. 12.6. The figure shows a function and several polynomial approximations using polynomials of increasing complexity. The first order polynomial is a straight line (1), the second order polynomial is a parabola (2), etc. One can note that, as the degree of the polynomial increases, a better approximation is obtained in the neighborhood of the chosen point but the approximation degrades quickly far from the given point as the chosen polynomial has more inflexion points (bends).

The approach used by LOWESS/LOESS deals with both issues in an elegant way. Firstly, the degrees of the polynomials used are limited to 1 (in LOWESS) or 2 (in LOESS) in order to avoid the over-fitting and the excessive twisting and turning. Secondly, since the polynomial approximation is good only for narrow intervals around the chosen point, LO(W)ESS will divide the data domain into such narrow intervals using a sliding window approach. The sliding window approach starts at the left extremity of the data interval with a window of a given width w. The data points that fall into this intervals will be used to fit the first polynomial in a weighted manner. The points near the point of estimation will weigh more than the points further away. This is achieved by using a weight function such as:

$$w(x) = \begin{cases} \left(1 - |x|^3\right)^3 & , \ |x| < 1 \\ 0 & , \ |x| \geq 1 \end{cases} \qquad (12.13)$$

where x is the distance from the estimation point. Other weighting functions can also be used as long as they satisfy certain conditions [69].

The procedure continues by sliding the window to the right, discarding some data points from the left but capturing some new data points from the right. A new polynomial will be fitted with this local data set and the process will continue sliding the window until the entire data range has been processed. The result is a smooth curve

---

[1] For the mathematically minded reader, we can recall that any differentiable function can be approximated around any given point by a polynomial obtained by taking a finite number of terms from the Taylor series expansion of the given function.

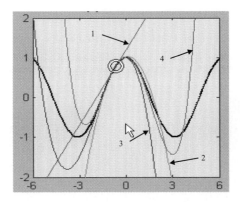

FIGURE 12.6: Polynomial approximation as used in LOWESS/LOESS. The sinusoid function is the target function to be approximated. Several approximations are shown: first order (1), second order (2), third order (3) and fourth order (4). As the degree of the polynomial increases, a better approximation is obtained in the neighborhood of the chosen point but the approximation degrades quickly far from the given point as the chosen polynomial "wiggles" more.

that provides a model for the data. The smoothness of the curve is directly proportional to the number of points considered for each local polynomial, i.e. proportional with the size of the sliding window. If there are n data points and a polynomial of degree d is used, one can define a smoothing parameter q as a user-chosen parameter between $\frac{d+1}{n}$ and 1. The LO(W)ESS will use $n \cdot q$ (rounded up to the nearest integer) points in each local fitting. Large values of q produce smooth curves that wiggle the least in response to variations in the data. Smaller values of q produce more responsive curves that follow the data more closely but are less smooth. Typically, useful values of q for microarray data range between 0.05 and 0.5 for most data sets. These values tend to be lower than the values of q used in other applications.

The effects of the LOWESS normalization are illustrated in Fig. 12.7 and Fig. 12.8. Fig. 12.7 shows the ratio-intensity plot before (left panel) and after (right panel) the LOWESS correction. In this plot, the horizontal axis represents the sum of the log intensities $\log_2(cy3 \cdot cy5) = \log_2(cy3) + \log_2(cy5)$ which is a quantity directly proportional to the overall intensity of a given spot. The vertical axis represents $\log_2(cy3/cy5) = \log_2(cy3) - \log_2(cy5)$ which is the usual log-ratio of the two samples. Similar plots are obtained if the horizontal axis is chosen to represent $1/2(\log_2(cy3) + \log_2(cy5))$ or $\log_2(cy3)$. Note the strong non-linear dye distortion in the left panel and how this is corrected by LOWESS in the right panel. The left panel also shows the non-linear regression curve calculated by LOWESS. Fig. 12.8 shows the same data as a scatter plot cy5 vs. cy3. The horizontal axis represents $\log_2(cy3)$ and the vertical axis represents $\log_2(cy5)$. In this case, the overall mean of the data was normalized before applying LOWESS.

The biggest advantage of LO(W)ESS is that there is no need to specify a particular

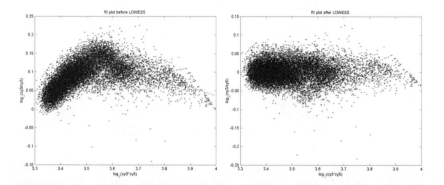

FIGURE 12.7: LOWESS normalization. A ratio-intensity plot before (left panel) and after (right panel) the LOWESS correction. The horizontal axis represents $\log_2(cy3) + \log_2(cy5)$ which is a quantity proportional to the overall signal intensity. The vertical axis represents the log-ratio of the two samples $\log_2(cy3/cy5)$. The left panel also shows the non-linear regression curve.

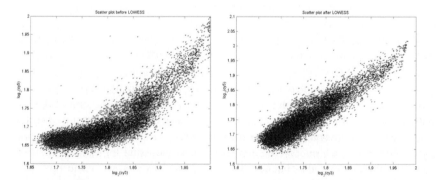

FIGURE 12.8: LOWESS normalization. A scatter plot before (left panel) and after (right panel) the LOWESS correction. The horizontal axis represents $\log_2(cy3)$ and the vertical axis represents $\log_2(cy5)$.

type of function to be used as a model (e.g exponential function in the exponential normalization). The only parameters that need to be specified by the user are the degree of the polynomials d and the smoothing factor q. Furthermore, LO(W)ESS methods use least squares regression which is very well studied. Existing methods for prediction, calibration and validation of least squares models can also be applied to LO(W)ESS.

Disadvantages of LO(W)ESS include the fact that it does not produce a regression function, or model, that is easily representable as a mathematical formula. In particular, the color distortion model found on a particular data set cannot be transferred directly to another data set or group of researchers. LO(W)ESS needs to be applied every time, on every data set and will produce a slightly different model in each case. Another disadvantage is related to the fact that the procedure is very computationally intensive. As an example, a set of 10,000 genes processed with $q = 0.1$ means that each individual least square curve fitting will be done on a subset of 1,000 data points. Furthermore, there will be approximately 10,000 such computations, as the sliding window scrolls through the entire data range. However, this is a relatively minor problem in the context of all other challenges related to microarray data analysis and taking into consideration the availability of powerful and cost effective computer hardware. A more important disadvantage is related to LO(W)ESS' susceptibility to noise and outliers. More robust versions of LO(W)ESS have been proposed [69] but even the robust version may be disrupted by extreme outliers. This is one of the reasons for which outliers (e.g. flagged spots) should be removed from the data before attempting a LO(W)ESS normalization.

12.3.3.3 Piece-wise normalization

The piece-wise normalization is closely related to LO(W)ESS attempting to preserve its advantages while improving on the computational aspects [87]. The idea is that LO(W)ESS performs a lot of computations which, for microarray data, are often redundant. Let us consider again the example of a set of 10,000 genes on which the size of the sliding window is chosen to be $nq = 1,000$. The classical LO(W)ESS requires approximatively 10,000 curve fittings. In turn, each such curve fitting is an iterative process requiring of the order of $1,000$ computations. In practice, for microarray data, the difference between two adjacent models each calculated on 1,000 points of which 999 are common is very likely to be minimal. The piece-wise normalization substitutes the sliding window approach with a fixed set of overlapping windows. In each such interval, the data are approximated by a linear function. A quadratic function could also be used as in LOESS. The user controls the smoothness of the resulting curve by choosing the number of such intervals and the degree of overlap.

The results of the piece-wise normalization are very comparable to the results of the LO(W)ESS normalization. Fig. 12.9 shows a comparison of the two methods on the same data set. The top-left panel in this figure shows the scatter plot of the uncorrected average log values cy5 vs. cy3. The top-right panel shows the same plot after piece-wise linear normalization. The bottom-left panel shows the same data

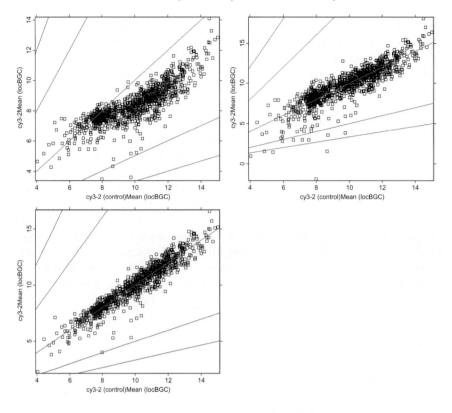

FIGURE 12.9: A comparison between piece-wise linear and LOWESS normalization. Top-left: scatter plot of the uncorrected average log values cy5 vs. cy3. Top-right: same plot after piece-wise linear normalization. Bottom-left: same data corrected with LOWESS. Note that both the piece-wise linear and the LOWESS use linear functions for the local regression.

corrected with LOWESS. Note that both the piece-wise linear and the LOWESS use linear functions for the local regression.

Advantages of the piece-wise normalization include a much better speed.[2] Perhaps more importantly, the piece-wise normalization can produce a compact mathematical description of the non-linearity that can be stored and used on different data sets. This description will be a piece-wise linear or quadratic function with a number of pieces equal to the number of intervals specified by the user. In principle, this could be done with LOWESS as well, only that the number of individual functions to be stored would be the same order of magnitude as the size of the data set which is unfeasible.

---

[2]For the computer scientist reader, the algorithm complexity is reduced from $O(n^2)$ to $O(n)$.

As LOWESS and perhaps a bit more so, the piece-wise normalization is susceptible to the effect of the outliers. A more robust version uses adaptive bins that take into consideration the change in parameters between adjacent intervals [87].

12.3.3.4 Other approaches to cDNA data normalization

It is important to note that the measured variance is dependent on the mean intensity having high variance at low intensity levels and low variance at high intensities. This can be corrected by using an iterative algorithm that adjusts gradually the parameters of a probabilistic model [64]. This approach can be further refined using a Gamma-Gamma-Bernoulli model [215]. Various other approaches to the normalization of cDNA data are discussed in the literature [36, 98, 109, 291, 299, 136, 177, 276, 290, 292].

12.4 Normalization issues specific to Affymetrix data

The Affymetrix technology as well as the data processing techniques implemented in the Data Mining Tool software package and other related software packages provided by the company are proprietary. Because of this, there is relatively little variation among the way Affymetrix data are processed in different laboratories. Also, there is relatively little research on the pre-processing and normalization of such data [194, 195, 235, 280]

The normalization of the oligonucleotide arrays designed for gene expression is slightly different from the normalization of cDNA data. Here, a gene is represented by a number of probe pairs (short oligonucleotide sequences) with each pair containing a perfect match and a mismatch (the same sequence with a different nucleotide in the middle). The terminology is illustrated in Fig. 12.10. A brief description of the major pre-processing steps and a discussion of the most important issues will be included here for reference. However, as with any proprietary technology, the particulars are subject to changes. For most up to date algorithms and pre-processing details we refer the reader to the Affymetrix technical documentation [4, 5, 6, 164].

12.4.1 Background correction

The image of the chip is captured in .CEL files. Also the .CEL file contains the raw intensities for all probe sets. Unlike cDNA, the Affymetrix array does not have a background as such. Therefore, the background correction of the probe sets intensities is performed using the neighboring probe sets. For this purpose, the array is split into a number of K rectangular zones (at the moment $K = 16$). Control cells and masked cells are ignored in this computation. The cells in each zone are ranked and the lowest 2% are chosen to represent the background value in the given zone. The background value is computed as a weighted sum of the background values of the neighboring zones with the weight being inversely proportional to the square of the distance to a given zone (see Fig. 12.10). Specifically, the weighting of zone k for a cell situated at the chip coordinates (x, y) is:

$$w_k(x, y) = \frac{1}{d_k^2(x, y) + c} \qquad (12.14)$$

where c is a smoothing constant that ensures the denominator never gets too close to zero. For every cell, the weighted background of the cell at coordinates (x, y) will be:

$$b(x, y) = \frac{1}{\sum_{k=1}^{K} w_k(x, y)} \sum_{k=1}^{K} w_k(x, y) b_{Z_k} \qquad (12.15)$$

where b_{Z_k} is the background of zone Z_k

As for cDNA arrays, one would like to correct the values by subtracting the background intensity value from the cell intensity values. However, it is very possible for a particular cell to have an intensity lower than the background value calculated according to Eq. 12.15. This would produce a negative background corrected value which in turn would produce problem in the subsequent processing (e.g. the log function is not defined for negative values). In order to address this problem, one can use the same approach to calculate a local noise value:

$$n(x, y) = \frac{1}{\sum_{k=1}^{K} w_k(x, y)} \sum_{k=1}^{K} w_k(x, y) n_{Z_k} \qquad (12.16)$$

where n_{Z_k} is the local noise value in zone Z_k calculated as the standard deviation of the lowest 2% of the background in that zone.

The individual cell intensities can be adjusted using a threshold and a floor:

$$I'(x, y) = \max(I(x, y), 0.5) \qquad (12.17)$$

where $I(x, y)$ is the raw intensity at location (x, y). Finally, the background corrected intensity values can be calculated as follows:

$$I_c(x, y) = \max\left(I'(x, y) - b(x, y), NF \cdot n(x, y)\right) \qquad (12.18)$$

where NF is a selected fraction of the global background variation (the default is currently 0.5).

12.4.2   Signal calculation

As for cDNA arrays, the goal of the pre-processing is to modify the gene specific intensity values in order to obtain a value that reflects as accurately as possible the amount of transcript in the solution used. As noted before, Affymetrix arrays use several probes with both perfect matches (PM) and mismatched (MM) cells. The challenge is to combine these values in a unique, meaningful value, proportional to the transcript. The amount of hybridization on the mismatched sequence is thought to

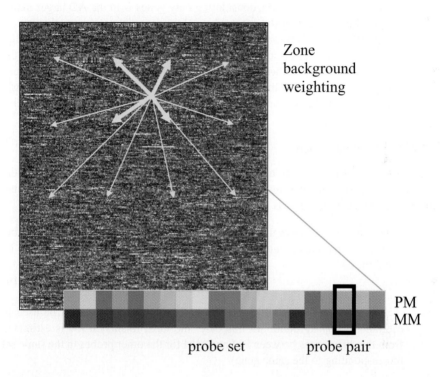

FIGURE 12.10: Affymetrix terminology and background zone weighting computation. A gene is represented by a number of perfect match (PM) and mismatch (MM)
probe pairs. Each pair is composed of two cells. The background value is computed
as a weighted sum of the nearby zone background values. The weight of a zone is
inversely proportional to the distance to it. The background value for a given zone is
calculated from the cells with the lowest 2% intensities in that zone.

be representative for non-specific hybridization and its intensity is usually subtracted from the intensity of the perfect match. Subsequently, such differences are averaged for all probes corresponding to a gene and an average difference (AD) is calculated. The higher this AD, the higher the expression level of the gene. The software provided by Affymetrix also calculates a *call*, i.e. a ternary decision about the gene: absent (A), marginally present (M) and present (P). A conundrum can occur if genes with an A call have AD higher than P genes. Options include ignoring the calls and using only the AD, considering only P genes, considering only genes with the AD larger than a threshold [127] or calculating the expression values according to some other model (e.g. dChip [194]). Furthermore, a typical problem often encountered with earlier versions of the pre-processing software is related to obtaining negative values for the average difference. This is rather difficult to interpret since usually the PM sequence should hybridize stronger than the MM sequence.[3] Recent versions of the Affymetrix software use a modified algorithm that eliminates the possibility of obtaining negative signal values.

12.4.2.1   Ideal mismatch

The very reason for including a MM probe is to provide a value that estimates the background non-specific hybridization as well as any other stray signals affecting the PM probe. There are 3 cases as follows:

1. If the MM value is lower than the PM value, the MM is considered a reasonable estimate of the background and the ideal mismatch (IM) will be taken to be equal to MM:

$$IM_{i,j} = MM_{i,j} \qquad (12.19)$$

2. If the MM is larger than PM for a given probe, clearly it cannot be used to estimate the background. In this case, the ideal mismatch IM is estimated from the differences between PM and MM for the other probes in the same set (corresponding to the same gene):

$$SM_i = T_{bi}(\log_2(PM_{i,j}) - \log_2(MM_{i,j}) , j = 1, \ldots, n \qquad (12.20)$$

where T_{bi} is a one-step Tukey's biweight estimate similar to a weighted mean. Full details about the computation of this estimate can be found in [7]. If the SB_I value thus calculated is greater than a threshold τ_c called *contrast tau*, the IM is calculated as:

$$IM_{i,j} = \frac{PM_{i,j}}{2^{SB_i}} \qquad (12.21)$$

3. Finally, if the SB_i value calculated above is smaller than the contrast, one cannot consider it an accurate estimate of MM. In these situations, the IM will be estimated by a value slightly lower than the PM:

---

[3]The sequences have been chosen to be specific to the target gene.

$$IM_{i,j} = \frac{PM_{i,j}}{2^{1+\frac{\tau_c}{\tau_c - SB_i}{\tau_s}}} \tag{12.22}$$

where τ_s is another threshold called *scale tau*. The current values for these thresholds are $\tau_c = 0.03$ and $\tau_s = 10$ [7].

12.4.2.2 Probe values

Once the ideal match value IM has been calculated for every probe, the probe value is calculated as the difference between the PM and the IM:

$$V_{i,j} = PM_{i,j} - IM_{i.j} \tag{12.23}$$

This is the classical idea of calculating the difference between the PM and MM values for every probe pair in the probe set. The log function is again useful and probe values *PV* can be calculated as:

$$PV_{i,j} = \log(V_{i,j}), \, j = 1, \ldots, n \tag{12.24}$$

Finally, the value proportional to the amount of transcript in the solution is the **signal log value** which is calculated again using the one-step Tukey's biweight estimate T_{bi}:

$$SLV = T_{bi}(\log V_{i.j}, \ldots, \log V_{i,n_i}) \tag{12.25}$$

Once again, the Tukey's biweight estimate is just a weighted mean of the values $\log V_{i.j}, \ldots, \log V_{i,n_i}$. Here, we considered a probe set with n_i probes.

It is possible that the values $PM_{i,j}$ and $IM_{i.j}$ are really close for a given probe pair. This will make their difference $V_{i,j} = PM_{i,j} - IM_{i.j}$ very close to zero and the log of the difference will be a very large negative value that may disrupt the weighted mean T_{bi}. In order to avoid this, the software uses another threshold $\delta = 2^{-20}$ for the $V_{i.j}$ values. Any $\log V_{i.j}$ value lower than δ will be set equal to it. This will set a floor of -20 for all values in the weighted mean.

12.4.2.3 Scaled probe values

A trimmed mean is calculated in order to improved the reliability of the values obtained. A trimmed mean is obtained by first eliminating the lowest and highest 2% of the values and then calculating the mean of the remaining values. However, since the signal values obtained above were logs, this computation needs to be performed on the anti-log (or exponential) values: 2^{SLV}. We will use TM to denote the trimmed mean:

$$TM(2^{SLV}, 0.02, 0.98) \tag{12.26}$$

where the parameters 0.02 and 0.98 specify where the tails of the distribution have been cut (lowest and highest 2%).

Finally, the values RV_i reported by the software are calculated as:

$$RV_i = nf \cdot sf \cdot 2^{SLV} \tag{12.27}$$

where nf is a normalization factor and sf is a scaling factor.

The scaling factor sf is calculated such as a target signal TS value is obtained for the trimmed mean:

$$sf = \frac{TS}{TM(2^{SLV}, 0.02, 0.98)} \qquad (12.28)$$

The normalization factor nf is calculated in order to allow the direct comparison of two arrays, usually called baseline (or reference) and experiment. The normalization factor is computed as a simple ratio of the two trimmed means corresponding to the two arrays:

$$nf = \frac{TM(2^{SLV_{baseline}}, 0.02, 0.98)}{TM(2^{SLV_{experiment}}, 0.02, 0.98)} \qquad (12.29)$$

Since a comparison analysis is done at the probe pair level, the individual probe pair values are also modified by the scaling and normalization factors. The scaled probe values SPV are calculated as:

$$SPV_{i,j} = PV_{i,j} + \log_2(nf \cdot sf) \qquad (12.30)$$

This rather complex sequences of steps has been shown to provide good results. The main ideas of this pre-processing sequence can be summarized as follows:

1. The cell intensities are corrected for background using some weighted average of the backgrounds in the neighboring zones.

2. An ideal mismatch value IM is calculated and subtracted from the PM intensity. If the mismatch MM is lower than the PM, the IM will be taken to be the difference $PM - MM$.

3. The adjusted PM values are log-transformed.

4. A robust mean of these log transformed values is calculated. The signal value is calculated as the exponential of this robust mean.

5. The signal value obtained is scaled using a trimmed mean.

12.4.3   Detection calls

The idea of the detection call is to characterize the gene as either present (P), which means that the expression level is well above the minimum detectable level, absent (A), which means the expression level is below this minimum, or marginal (M), which means that the expression level of the gene is somewhere near the minimum detectable level. The most recent version of the Affymetrix software uses a classical hypothesis testing approach using a Wilcoxon rank test which is a non-parametric test based on the rank order of the values. The p-value reported by the software is the probability of the null hypothesis being true (see Chap. 5).

In order to calculate the calls, the software first removes the saturated probe pairs. If all probe pairs corresponding to a gene are saturated, the gene is reported as present and the p-value is set to zero. The probe pairs for which the PM and MM values are very close (within a limit τ) are also discarded. A discrimination score R is calculated for the remaining probe pairs:

$$R_i = \frac{PM_i - MM_i}{PM_i + MM_i} \tag{12.31}$$

The hypotheses are:

$$H_0 : median(R_i - \tau) = 0$$
$$H_a : median(R_i - \tau) > 0$$

The default value of τ is 0.015. Clearly, increasing τ will reduce the number of false positives (false P calls) but will also reduce the number of true detected calls. This is a clear situation in which reducing the probability of a Type I error also reduces the power of the test. The calls are decided based on the p-value using two thresholds α_1 and α_2:

$$\begin{array}{ccc} \text{Present} & \text{Marginal} & \text{Absent} \\ p < \alpha_1 & \alpha_1 \leq p < \alpha_2 & p \geq \alpha_2 \end{array}$$

It is important to note that in this case, the true probability of a Type I error does not correspond directly to the chosen alpha level. This is due to two different reasons. Firstly, the hypotheses formulation involves a constant τ. For a given data set, changing the value of τ can change the calls made and thus affect the false-positive rate. Secondly, the results will also depend on the number of probe-pairs used. This number will be constant for a given array type but may change from one array type to another.

12.4.4 Relative expression values

When two conditions are to be compared directly, it is best to make the comparison at the individual probe level. If there are differences between the hybridization efficiencies of various probes selected to represent the same gene, these differences will be automatically cancelled out in a probe level comparison. In order to do this, the software first calculates a probe log ratio PLR for each probe j in the probe set i on both baseline and experiment arrays:

$$PLR_{i,j} = SPV_{i,j}^{experiment} - SPV_{i,j}^{baseline} \tag{12.32}$$

In this equation, the SPV values are calculated as in Eq. 12.30. Once these probe log ratios are available, they can be combined using the same one-step Tukey's biweight estimate in a unique, gene specific, signal log ratio:

$$SLR_i = T_{bi}(PLR_{i,1}, \ldots, PLR_{i,n_i}) \tag{12.33}$$

The fold change can now be calculated as the exponential of the log ratio:

$$FC = \begin{cases} 2^{SLR_i}, & SLR_i \geq 0 \\ -2^{-SLR_i}, & SLR_i < 0 \end{cases} \tag{12.34}$$

12.5 Other approaches to the normalization of Affymetrix data

Another approach for the pre-processing and normalization of high-density oligonucleotide DNA arrays relies on the idea that the probe-specific biases are significant
but they are also reproducible and predictable. Thus, one could develop a statistical
model of the phenomena happening at the probe level and use such model to calculate estimates of the gene expression. Furthermore, the same model can be used to
detect cross-hybridizations, contaminated regions and defective regions of the array
[194, 195].

The model proposed by Wong et al. assumes that the intensity value of a probe j
increases linearly with the expression of a gene in the i-th sample, θ_i. Furthermore,
it is assumed that this happens for both PM and MM, only that the PM intensity will
increase at a higher rate than the MM intensity. These ideas can be formalized in the
following equations:

$$MM_{ij} = \nu_j + \theta_i \alpha_j + \epsilon \tag{12.35}$$

and

$$PM_{ij} = \nu_j + \theta_i \alpha_j + \theta_i \phi_j + \epsilon \tag{12.36}$$

where PM_{ij} and MM_{ij} denote the PM and MM intensity values for the i-th array
and the j-th probe pair for the given gene, ν_i is the baseline response of the j-th probe
pair due to non-specific hybridization, α_j is the rate of increase of the MM response
of the j-th probe pair, ϕ_j is the additional rate of increase in the corresponding PM
response and ϵ is the random error. It has been shown that this model fits the data
well yielding a residual sum of squares (see Chapter 7) of only 1%, much better than
a simple additive model. From Eq. 12.35 and 12.36 one can extract an even simpler
model for the $PM - MM$ differences:

$$y_{ij} = PM_{ij} - MM_{ij} = \theta_i \phi_j + e_{ij} \tag{12.37}$$

This model can be extended to all genes in the sample and used to detect and handle
cross-hybridizations, image contaminations and outliers from other causes [194, 195].
This technique has been implemented in the package dChip which is available free
of charge for academic use[4].

12.6 Useful pre-processing and normalization sequences

It is useful to try to discuss the pre-processing as a whole as opposed to focusing on
the details of every single pre-processing step. In order to do this, we may imagine

---

[4]The dChip software can be downloaded from http://www.dchip.org at the time of going to press.

the pre-processing as a pipeline, with raw data entering at one end and the normalized data coming out at the other end. As discussed, the specific pre-processing and normalization steps may be combined in a large number of ways to produce useful sequences. For instance, the normalization for the overall array level can be achieved by performing a division of the values by the mean of each array before taking the logs or by subtracting the mean after the log is taken. Also, the values of the replicate spots may be combined at various stages during the pre-processing.

A pre-processing and normalization sequence that might be useful for cDNA data would include the following steps:

1. Background correction. This can be done using:

   (a) Local background correction

   (b) Subgrid background correction

   (c) Local group background correction

   (d) Background correction using blank spots

2. Eliminate the spots flagged in the image processing stage.

3. Substitute missing values (optional).

4. Normalize overall array intensity. This may be optional if the normalization for color distortion equalizes the array levels, as well. The alternatives include:

   (a) Divide by mean

   (b) Iterative regression

   The normalization of overall array intensity may be performed after eliminating the outliers and/or differentially regulated genes.

5. Normalize for color distortion. This can be done using:

   (a) Curve fitting and correction

   (b) LO(W)ESS

   (c) Piece-wise normalization

6. Calculate the ratio of the two channels cy3/cy5 or cy5/cy3.

7. Apply a logarithmic transformation (base 2 recommended).

8. Combine the replicate values to calculate a mean/median/mode value and the standard deviation.

For Affymetrix data, we recommend the use of the Affymetrix software for the pre-processing and normalization. A viable alternative for the pre-processing is the dChip package implementing the model proposed by Wong et al. (Eq. 12.37). Both packages do background correction, probe level normalization and array intensity normalization.

If the data are available as average differences (as from older versions of the Affymetrix DMT software), the following sequence may be useful:

| Detection calls | | Magnitude of exp/base ratio | Biological meaning | Interesting gene |
exp	base			
M/A	M/A	meaningless; can be anything	the gene is probably not expressed in either condition	no
M/A	P	meaningless; close to zero	gene expressed on the baseline array, not expressed in the experiment	yes
P	M/A	meaningless; very large	gene expressed in the experiment, not expressed in the baseline array	yes
P	P	meaningful; can be anything	gene expressed on both channels, useful ratio	depends on the ratio

TABLE 12.1: The interplay between the A/M/P calls and the expression values when an experiment array is compared to a baseline array. The only genes that can be safely ignored are those that are absent (A) or only marginally detected (M) in both arrays. All other combinations are or may be interesting from a biological point of view.

1. Apply a floor function to bring all negative values to a small positive value.

2. Apply a logarithmic transformation

3. Normalize overall array intensity.

In many cases, it is useful to use the detection calls provided by the Affymetrix package. However, it is not a good idea to eliminate completely the genes that are absent (A) or marginally detected (M). Table 12.1 shows that the only genes that can probably be safely ignored are those that are absent (A) or only marginally detected (M) in both arrays. All other combinations are or may be interesting from a biological point of view. For instance a gene that is absent in the baseline and present in the experiment is probably a gene whose activity may be up-regulated in the condition studied. Even though it may not be possible to report a meaningful fold change for this gene, its further study may be well warranted.

12.7 Summary

The aim of the normalization is to account for systematic differences across different data sets (e.g. quantity of mRNA) and eliminate artifacts (e.g. non-linear dye effects). The normalization is crucial if results of different experimental techniques are to be combined. This chapter discussed several issues related to the pre-processing and normalization of microarray data. A few general pre-processing techniques useful for both cDNA and Affymetrix data include the logarithmic transformation, combining replicates and eliminating outliers and array normalization. The log transform partially decouples the variance and the mean intensity and makes the distribution almost normal. Combining the replicates should be done with care since in many cases this involves losing some information. Outliers can be eliminated in an iterative process involving calculating the mean and standard deviation σ and eliminating values outside $\pm 3\sigma$. Array normalization can be done by dividing by the mean before the log or by subtracting the mean after the log. The former is equivalent to a correction using the arithmetical mean while the latter is equivalent to a correction using the geometrical mean.

Many other pre-processing techniques are specific to a given technology and they were discussed separately. The cDNA data is usually pre-processed by correcting for the background, correcting for color non-linearities or using more sophisticated statistical models. The background correction can be done locally (only the background local to the spot is considered), using a group of spots such as a sub-grid (suitable for high density arrays where there are not enough pixels around a single spot to produce a reliable value) or blank spots. The color normalization can be performed using a model based approach (e.g. the exponential normalization), a LO(W)ESS approach or a piece-wise linear normalization. The exponential normalization has been shown to work well for the usual cy3/cy5 dyes. The LO(W)ESS approach should work well for any dyes but is computationally intensive and does not provide a normalization model that can be stored for reference or used for other data sets. The piece-wise linear normalization is similar to LO(W)ESS but is faster and does provide a normalization model.

The Affymetrix data is pre-processed as follows. The cell intensities are corrected for background using some weighted average of the backgrounds in the neighboring zones. An ideal mismatch value IM is calculated and subtracted from the PM intensity. If the mismatch MM is lower than the PM, the IM will be taken to be the difference $PM - MM$. The adjusted PM values are log-transformed and a robust mean of these log transformed values is calculated. The signal value is calculated as the exponential of this robust mean and scaled using a trimmed mean. Alternatively, model-based normalizations approaches also exist (e.g. dChip) and have been shown to perform very well.

12.8   Appendix

12.8.1   A short primer on logarithms

The log is the inverse of the exponential function. Thus, $y = \log_a x$ is defined as the power to which one needs to raise a in order to obtain x: $a^y = x$. For instance, $\log_2 8 = 3$ since $2^3 = 8$ and $\log_{10} 100 = 2$ since $10^2 = 100$. The quantity a is

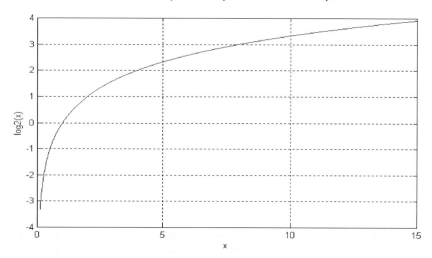

FIGURE 12.11: The logarithmic function: $y = \log_2 x$.

called the base of the logarithm. Since the exponential and the logarithm are each other's inverse, the following identities hold true:

$$a^{\log_a x} = \log_a a^x = x \tag{12.38}$$

Fig. 12.11 shows the graphic of the logarithmic function $y = \log_2 x$.
The main properties of the logarithms are as follows:

1. $\log_a a^x = x$ which is to say that the power to which one needs to raise a in order to obtain a^x (a to the power x) is the very same x.

2. $a^{\log_a x} = x$ which is to say that a raised to the power to which one needs to raise a in order to obtain x is indeed x.

3. $\log_a (x \cdot y) = \log_a x + \log_a y$ or log of a product is the sum of the logs of the factors.

4. $\log_a \frac{x}{y} = \log_a x - \log_a y$ or log of ratio is the difference of the logs of the numerator and denominator.

5. $\log_a x^y = y \cdot \log_a x$. This can be proven easily by applying repeatedly the product rule above.

Chapter 13

Methods for selecting differentially regulated genes

Everything should be made as simple as possible, but not simpler.

—Albert Einstein

13.1 Introduction

In many cases, the purpose of the microarray experiment is to compare the gene expression levels in two different specimens.[1] In most cases, one sample is considered the reference or control and the other one is considered the experiment. Obvious examples include comparing healthy vs. disease or treated vs. untreated tissues. Sample comparison may be done using different arrays (e.g. oligonucleotide arrays) or multiple channels on the same array (e.g. cDNA arrays). In all such comparative studies, a very important problem is to determine those genes that are differentially expressed in the two samples compared.

Although simple in principle, this problem becomes more complex in reality because the measured intensity values are affected by numerous sources of fluctuation and noise [92, 241, 285]. For spotted cDNA arrays, there is a non-negligible probability (about 5%) that the hybridization of any single spot containing complementary DNA will not reflect the presence of the mRNA. Furthermore, the probability that a single spot will provide a signal even if the mRNA is not present is even greater (about 10%) [193]. Most cDNA arrays address this problem by having several replicate spots for each gene.

The Affymetrix technology tries to respond to the challenge of a poor reliability for single hybridizations by representing a gene through a set of probes. The probes correspond to short oligonucleotide sequences thought to be representative for the given gene. Each oligonucleotide sequence is represented by two probes: one with the exact sequence of the chosen fragment of the gene (perfect match or PM) and one with a mismatch nucleotide in the middle of the fragment (mismatch or MM).

---

[1]Some of the material in this chapter is reprinted from S. Drăghici, "Statistical intelligence: effective analysis of high-density microarray data" published in Drug Discovery Today, Vol. 7, No. 11, p. S55-S63, Copyright (2002), with permission from Elsevier.

For each gene, the value that is often taken as representative for the expression level of the gene is the average difference between PM and MM (see Chapters 2, 3 and 12). In principle, this value is expected to be positive because the hybridization of the PM is expected to be stronger than the hybridization of the MM. However, many factors including non-specific hybridizations and a less than optimal choice of the oligonucleotide sequences representative for the gene may determine a MM hybridization stronger than the PM hybridization for some probes. In such cases, older versions of the software reported negative average differences. The latest versions have corrected this by using a more sophisticated pre-processing method (see Chap. 12). However, the numerical manipulations performed in order to avoid such negative values still introduce non-linearities in the expression values and make the gene selection task difficult even for Affymetrix data.

In this context, distinguishing between genes that are truly differentially regulated and genes that are simply affected by noise becomes a real challenge. All methods discussed here are completely independent of the technology used to obtain the data (e.g. cDNA or Affymetrix). The main difference between the different types of data is the pre-processing as discussed in Chap. 12. The Affymetrix data will be pre-processed by combining the fluorescence levels of individual probes between match and mismatch to yield average differences (or expression indexes) and detection calls (present, absent, marginal, etc.). The cDNA data are minimally processed by subtracting the background from the fluorescence values of the spots and correcting for color non-linearity. Furthermore, when data from different arrays are compared, such comparisons must be first made meaningful by bringing the arrays at comparable levels of intensity. This is usually done by some global normalization such as dividing the values on each array by their mean over the whole array. Finally, in most cases, one would like to apply a log transform in order to improve the characteristics of the distribution of expression values.

In the following, we will exemplify several methods using the very simple example of a comparison between two conditions: experiment and control.

13.2   Criteria

The performance of a gene selection method can be calculated in terms of positive predicted value (PPV), negative predicted value (NPV), specificity and sensitivity. In general, for any diagnosis or classification method, one could compare the truth with the results reported by the method. In a binary decision situation such as changed/unchanged, the results can always be divided into 4 categories: truly changed that are reported as changed (**true positives**), unchanged that are reported as changed (**false positives**), truly changed that are reported as unchanged (**false negatives**) and truly unchanged that are reported as such (**true negatives**). Based on these, one can define the 4 quantitative criteria: **positive predicted value** (PPV),

negative predicted value (NPV), **specificity** and **sensitivity**. These are defined as follows:

$$PPV = \frac{TP}{TP + FP} \tag{13.1}$$

$$NPV = \frac{TN}{TN + FN} \tag{13.2}$$

$$Specificity = \frac{TN}{TN + FP} \tag{13.3}$$

$$Sensitivity = \frac{TP}{TP + FN} \tag{13.4}$$

$$Accuracy = \frac{TP + TN}{N} \tag{13.5}$$

where TP is (the number of) true positives, TN is true negatives, FP is false positives, FN is false negatives and N is the total number of instances $N = TP + FN + TN + FN$. These quantities range from 0 to 1. Sometimes, they are expressed as percentages with 1 being equal to 100%. A perfect method would yield no false positives and no false negatives. In this case, the accuracy, specificity, sensitivity, PPV and NPV would all be equal to 1 or 100%. These measures also depend on the proportion of truly changed $(TP + FN)$ with respect to the total number of instance N. The proportion $\frac{TP + FN}{N}$ is called **prevalence**. The prevalence does influence the other measures as well as the usefulness of any classification or diagnosis method. Example 4.6 in Chap. 4 showed that a diagnosis method with a specificity of 90% and a PPV of 99.5% may still be insufficient for practical purposes for a disease with a prevalence of only 1 in 5,000 individuals (0.02%).

If the numbers are reported for up-regulated genes, TP would be the number of genes truly up-regulated and TN would be the number of genes not up-regulated. It has to be noted, that in this case, the set of genes that are not up-regulated includes the unchanged genes, as well as the down-regulated genes. In consequence, FP would be the number of genes reported as up-regulated where in fact they are either not regulated or down-regulated and FN would be the number of genes that are truly up-regulated and are not reported as such.

13.3   Fold change

13.3.1   Description

The simplest and most intuitive approach to finding the genes that are differentially regulated is to consider their fold change between control and experiment. Typically, an arbitrary threshold such as 2 or 3 fold is chosen and the difference is considered as

Reported	True			
	changed	unchanged		
changed	TP	FP	Positive predicted value	$\frac{TP}{TP+FP}$
unchanged	FN	TN	Negative predicted value	$\frac{TN}{TN+FN}$
	Sensitivity $\frac{TP}{TP+FN}$	Specificity $\frac{TN}{TN+FP}$		

TABLE 13.1: The definitions of positive predicted value (PPV), negative predicted value (NPV), specificity and sensitivity. A perfect method will produce PPV=NPV=specificity= sensitivity=1.

significant if it is larger than the threshold [166, 81, 80, 265, 260, 281, 284]. Sometimes, this selection method is used in parallel on expression estimates provided by several techniques such as radioactive and fluorescent labelling [226]. A convenient way to select by fold change is to calculate the ratio between the two expression levels for each gene. This is more traditional for the cDNA data but can be easily done for Affymetrix data, as well. Such ratios can be plotted as a histogram (Fig. 13.1).

As discussed previously, in a screening experiment involving many genes, most genes will not change. Thus, the experiment/control ratio of most genes will be grouped around 1 which means that their logs will be grouped around 0. The horizontal axis of such a plot represents the log ratio values. In consequence, selecting differentially regulated genes can be simply done by setting thresholds on this axis and selecting the genes outside such thresholds. For instance, in order to select genes that have a fold change of 4 and assuming that the log has been taken in base 2, one would set the thresholds at $+/-2$. Note the strong resemblance between this figure and Fig. 5.5 in Chap. 5. This indicates already that the process used by the fold selection method is in principle the same as that used in a classical hypothesis testing situation. The difference is that in a hypothesis testing situation, the thresholds are chosen very precisely in order to control the probability of the Type I error (calling a gene differentially regulated by mistake) while in the fold change method, the thresholds are chosen arbitrarily.

If the log expression levels in the experiment are plotted against the log expression levels in control in a scatter plot (left panel in Fig. 13.2), the genes selected will be at a distance of at least 2 from the diagonal that corresponds to the expression being the same in control and experiment. The fold change method reduces to drawing lines parallel to the diagonal at a distance corresponding to the chosen threshold and selecting the genes outside the central area defined by these thresholds. In a ratio-intensity plot the reference line corresponding to unchanged genes will be the horizontal line $y = 0$ (right panel in Fig 13.2). Similarly, the method works by establishing selection boundaries parallel to the reference line and selecting the genes outside the boundaries, away from the reference.

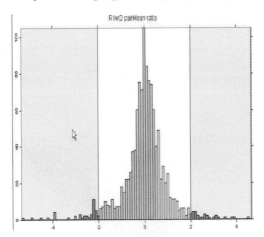

FIGURE 13.1: Fold-change on a histogram. Experiment-control ratios can be plotted as a histogram showing the number of genes (vertical axis) for every ratio value (horizontal axis). The horizontal axis is graded in fold change units. Selecting differentially regulated genes based on fold change corresponds to selecting the genes in the tails of the histogram (blue areas) by setting thresholds at the desired minimum fold change. Note the resemblance of this figure with Fig. 5.5 in Chap. 5.

13.3.2 Characteristics

The fold change method is often used because it is simple and intuitive. However, the method has important disadvantages. The most important drawback is that the fold threshold is chosen arbitrarily and may often be inappropriate. For instance, if one is selecting genes with at least 2 fold change and the condition under study does not affect any genes to the point of inducing a 2 fold change, no genes will be selected resulting in zero sensitivity. Reciprocally, if the condition is such that many genes change dramatically (or if a lower threshold is chosen), the method will select too many genes and will have a low specificity. In this respect, the fold change method is nothing but a blind guess.

Another important disadvantage is related to the fact that the microarray technology tends to have a bad signal/noise ratio for genes with low expression levels. On a scatter plot and ratio-intensity plot this is illustrated by a funnel shape of the distribution. This is caused by a larger variance of the values measured at the low end of the scale (to the left) and a low variance of values measured at the high end of the scale (to the right) and by certain numerical phenomena discussed in Chap. 10 (see Fig. 10.6). A gene that is closer to the diagonal at a high expression level might be more reliable than a gene that is a bit further from the diagonal at a low level. Since the fold change uses a constant threshold for all genes, it will introduce false positives at the low end, thus reducing the specificity, while missing true positives at the high end and thus reducing the sensitivity.

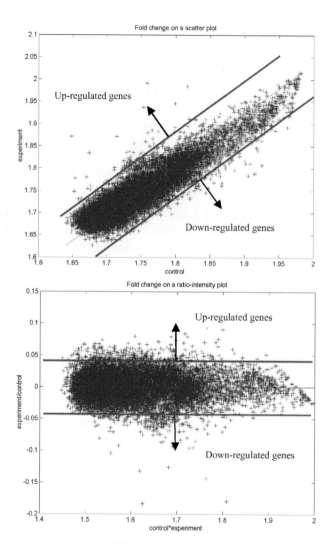

FIGURE 13.2: Fold-change on a scatter plot of log values (left) and ratio-intensity plot (right). The fold method reduces to drawing two linear selection boundaries at a constant distance from the $exp = control$ line and selecting the genes as shown. The line $exp = control$ is the diagonal $y = x$ in a scatter plot and the line $y = 0$ in a ratio-intensity plot.

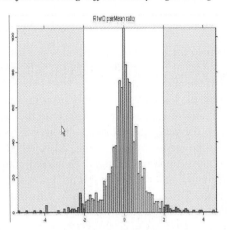

FIGURE 13.3: Selecting by the unusual ratio criterion. The frequencies (vertical axis) of the z-scores of the genes (horizontal axis) are plotted in a histogram. The horizontal axis is now graded in standard deviations. Setting thresholds at the +/-2 marks corresponds to selecting genes with a ratio more than 2 standard deviations away from the mean log ratio.

13.4   Unusual ratio

13.4.1   Description

The second widely used selection method involves selecting the genes for which the ratio of the experiment and control values is a certain distance from the mean experiment/control ratio [263, 238, 239]. Typically, this distance is taken to be $\pm 2\sigma$ where σ is the standard deviation of the ratio distribution. In other words, the genes selected as being differentially regulated will be those genes having an experiment/control ratio at least 2σ away from the mean experiment/control ratio. In practice, this can be achieved very simply by applying a z-transform to the log ratio values. The z-transform essentially subtracts the mean and divides by the standard deviation. In consequence, a histogram of the transformed values will still be centered around 0 (most genes will have a ratio close to the mean ratio) but the units on the horizontal axis will represent standard deviation (see Fig. 13.3). Thus, setting thresholds at +/-2 will correspond to selecting those genes which have an unusual ratio, situated at least 2 standard deviations away from the mean ratio.

13.4.2   Characteristics

This method is superior to the fold change method while still simple and intuitive. The advantage of the unusual ratio method is that it will automatically adjust the cut-off threshold even if the number of genes regulated and the amount of regulation vary considerably. Thus, the unusual ratio method uses thresholds on how different the experiment/control ratio of a gene is with respect to the mean of all such ratios instead of thresholds on the values of the ratios themselves. No matter how many genes are regulated and no matter by how much, this method will always pick the genes that are affected most. In particular, as we have seen, if the ratio distribution is close to a normal distribution and the thresholds are set at +/-2 standard deviations, this method will select the 5% most regulated genes. This is because the probability of having a Z lower than -2 is $P(Z < -2) = 0.0228$ whereas the probability of having a Z value higher than 2 is $P(Z > 2) = 1 - P(Z < 2) = 1 - 0.9772 = 0.0228$. Cumulative, the probability of having a Z value more extreme than ±2 is $0.0228 + 0.0228 = 0.0456$.

However, the unusual ratio method still has important intrinsic drawbacks. Thus, the method will report 5% of the genes as differentially regulated *even if there are no differentially regulated genes*. This happens because in all microarray experiments there is a certain amount of variability due to noise. Thus, if the same experiment is performed twice, the expression values measured for any particular gene will likely not be exactly the same. If the method is applied to study differential regulated genes in two control experiments, the unusual ratio method will still select about 5% of the genes and report them as "differentially" regulated. This is because different measurements for the same gene will still vary a little bit due to the noise. The method will dutifully calculate the mean and standard deviation of this distribution and will select those genes situated +/-2 standard deviations away from the mean. In this case, the null hypothesis *is true* and the method will still reject (incorrectly) the null hypothesis 5% of the time, i.e. will report 5% of the genes as being differentially regulated.

Furthermore, the method will still select 5% of the genes *even if much more genes are in fact regulated*. Thus, while the fold method uses an arbitrary threshold and can provide too many or too few genes, the unusual ratio method uses a fixed proportion threshold that will always report the same proportion of the genes as being differentially regulated. On a scatter plot (such as the one in Fig. 13.2), the ratio method continues to use cut-off boundaries parallel to the diagonal which will continue to overestimate the regulation at low intensity and underestimate it at high intensity.

A variation of the unusual ratio method selects those genes for which the absolute difference in the average expression intensities is much larger than the estimated standard error ($\hat{\sigma}$) computed for each gene using array replicates. For duplicate experiments the absolute difference has to be larger than $4.3\hat{\sigma}$ and $22.3\hat{\sigma}$ for the 5% and 1% significance levels, respectively [68]. For triplicate experiments the requirements can be relaxed to $2.8\hat{\sigma}$ and $5.2\hat{\sigma}$ for the 5% and 1% significance levels, respectively.

A number of other *ad hoc* thresholding and selection procedures have also been used that are equivalent to a selection based on the unusual ratio method. For instance,

[238, 239] only considered genes for which the difference between the duplicate measurements did not exceed half their average. Furthermore, the genes considered as differentially regulated were those genes that exhibited at least a 2-fold change in expression. Although this criterion seems to use the fold method, it can be shown [68] that the combination of the duplicate consistency condition and the differentially regulated condition can be expressed in terms of mean and standard deviations and therefore it falls under the scope of the unusual ratio method.

13.5   Hypothesis testing, corrections for multiple comparisons and resampling

13.5.1   Description

Another possible approach to gene selection is to use univariate statistical tests (e.g. t-test) to select differentially expressed genes [19, 68, 98]. This approach essentially uses the classical hypothesis testing approach discussed in Chap. 5 in conjunction with some correction for multiple comparisons discussed in Chap. 9. This approach will be briefly reviewed here.

Let us consider that the log ratios follow a distribution like the one illustrated in Fig. 13.4. For a given threshold and a given distribution the confidence level or p-value is the probability of the measured value being in the shaded area by chance. The thinking is that a gene whose log ratio falls in the shaded area is far from the mean log ratio and will be called differentially regulated (upregulated in this case). However, the measured log ratio may be there just due to random factors such as noise. The probability of the measurement being there just by chance is the p-value. In this case, calling the gene differentially regulated will be a mistake (Type I error) and the p-value is the probability of making this mistake.

Regardless of the particular test used (e.g. t-test if normality is assumed), one needs to consider the fact that when many genes are analyzed at one time, some genes will appear as being significantly different just by chance [149, 146, 244, 245, 283]. In the context of the selection of differentially regulated genes, a Type I error (or rejecting a true null hypothesis) is manifested as reporting an un-regulated gene as differentially regulated (false positive). In an individual test situation (the analysis of a single gene), the probability of a Type I error is controlled by the significance level chosen by the user. On a high density array containing many genes, the significance level at gene level does not control the overall probability of making a Type I error anymore. The overall probability of making at least a mistake, or family-wise significance level, can be calculated from the gene level significance level using several approaches discussed in Chap. 9. Bonferroni [41, 42] and Šidák corrections [272] are quite simple but very conservative. For arrays involving thousands or tens of thousands of genes, these methods are not practical. The Holm step-down group of methods [149, 244, 148, 146], false discovery rate (FDR) [30, 31], permutations

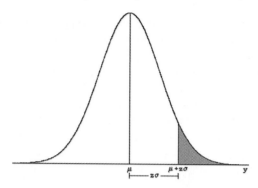

FIGURE 13.4: For a given threshold and a given distribution the p-value is the probability of the measured value being in the shaded area by chance. If the method is applied to the distribution of the log ratios, the p-value is the probability of making a mistake when calling a gene differentially regulated if its measured log ratio is in the shaded area.

[283] and significance analysis of microarray (SAM) [268] are all suitable methods for multiple comparison corrections in the context of microarray data. SAM in particular is more than a method for multiple comparison correction and can probably be considered a gene selection method on its own.

A univariate testing procedure (e.g. t-test or Wilcoxon) followed by a Westfall and Young adjustment for multiple testing [283] has been used by Dudoit et al. [98].

13.5.2   Characteristics

A drawback of the methods based on hypothesis testing is that they tend to be a bit conservative. As discussed in Chap. 5, not being able to reject a null hypothesis and call a gene differentially regulated does not necessarily mean that the gene is not so. In many cases, it is just that insufficient data do not provide sufficient statistical proof to reject the null hypothesis. However, those genes that are found to be differentially regulated using such methods will most likely be so.

The classical hypothesis testing approach also has the disadvantage of assuming that the genes are independent which is clearly untrue in the analysis of any real data set. Fortunately, combining a classical hypothesis testing approach with a re-sampling or bootstrapping approach (e.g. in step-down or SAM) tends to lose the conservative tendencies and also takes into consideration dependencies among genes. If the experiment design and the amount of data available allows it, the use of these methods is recommended.

13.6 ANOVA

13.6.1 Description

A particularly interesting approach to microarray data analysis and selecting differentially regulated genes is the ANalysis Of VAriance (ANOVA) [8, 51, 143]. This was discussed in detail in Chap. 7 but will be reviewed briefly here. The idea behind ANOVA is to build an explicit model about the sources of variance that affect the measurements and use the data to estimate the variance of each individual variable in the model.

For instance, Kerr and Churchill [183, 182, 180] proposed the following model to account for the multiple sources of variation in a microarray experiment:

$$\log\left(y_{ijkg}\right) = \mu + A_i + D_j + G_g + (AD)_{ij} + (AG)_{ig} + (VG)_{kg} + (DG)_{jg} + \epsilon_{ijkg}$$
$$(13.6)$$

In this model, μ is the overall mean signal of the array, A_i is the effect of the ith array, D_j represents the effect of the jth dye, G_g is the variation of the gth gene, $(AD)_{ij}$ is the effect of the array-dye interaction, $(AG)_{ig}$ is the effect of a particular spot on a given array (array-gene interaction), $(VG)_{kg}$ represents the interaction between the kth variety and the gth gene, $(DG)_{jg}$ is the effect of the dye-gene interaction and ϵ_{ijkg} represents the error term for array i, dye j, variety k and gene g. In this context a variety is a condition such as healthy or disease. The error is assumed to be independent and of zero mean. Finally, $\log\left(y_{ijkg}\right)$ is the measured log-ratio for gene g of variety j measured on array i using dye j.

Sums of squares are calculated for each of the factors above. Then mean squares will be obtained by dividing each sum of squares by its degrees of freedom. This will produce a variance-like quantity for which an expected value can be calculated if the null hypothesis is true. These mean squares can be compared the same way variances can: using their ratio and an F distribution. Each ratio will have the MS_E as the denominator. Essentially, each individual test asks the question whether a certain component MS_x has the variance significantly different from the variance of the noise MS_E. The differentially regulated genes will be the genes for which the $(VG)_{kg}$ factor representing the interaction between the variety and gene is significant.

13.6.2 Characteristics

The advantage of ANOVA is that each source of variance is accounted for. Because of this, it is easy to distinguish between interesting variations such as gene regulation and side effects such as differences due to different dyes or arrays. The caveat is that ANOVA requires a very careful experiment design [182, 180] that must ensure a sufficient number of degrees of freedom. Thus, ANOVA cannot be used if the experiments have not been designed and executed in a manner consistent with the ANOVA model used.

13.7 Noise sampling

13.7.1 Description

A full blown ANOVA requires a design that blocks all variables under control and randomizes the others. In most cases, this requires repeating several microarrays with various mRNA samples and swapping dyes if a multi-channel technology is used. A particular variation on the ANOVA idea can be used to identify differentially regulated genes using spot replicates on single chips to estimate the noise and calculate confidence levels for gene regulation. The noise sampling method [88, 92, 93] modifies the Kerr-Churchill model as follows:

$$\log R(gs) = \mu + G(g) + \epsilon(g, s) \tag{13.7}$$

where $\log R(gs)$ is the measured log ratio for gene g and spot s, μ is the average log ratio over the whole array, $G(g)$ is a term for the differential regulation of gene g and $\epsilon(g, s)$ is a zero-mean noise term.

In the model above, one can calculate an estimate $\hat{\mu}$ of the average log ratio μ:

$$\hat{\mu} = \frac{1}{n \cdot m} \sum_{g,s} \log\left(R\left(g, s\right)\right) \tag{13.8}$$

which is the sum of the log ratios for all genes and all spots divided by the total number of spots (m replicates and n genes). An estimate $\widehat{G(g)}$ of the effect of gene g can also be calculated as:

$$\widehat{G(g)} = \frac{1}{m} \sum_{g} \log\left(R\left(g, s\right)\right) - \hat{\mu} \tag{13.9}$$

where the first term is the average log ratio over the spots corresponding to the given gene. Using the estimates above, one can now calculate an estimate of the noise as follows:

$$\widehat{\epsilon(g, s)} = \log\left(R\left(g, s\right)\right) - \hat{\mu} - \widehat{G(g)} \tag{13.10}$$

For each spot (g, s), this equation calculates the noise as the difference between the actual measured value $\log\left(R\left(g, s\right)\right)$ and the estimated array effect $\hat{\mu}$ plus the gene effect $\widehat{G(g)}$. This will provide a noise sample for each spot. The samples collected from all spots yield an empirical noise distribution.[2] A given confidence level can

---

[2]Note that no particular shape (such as Gaussian) is assumed for the noise distribution or for the distribution of the gene expression values, which makes this approach very general.

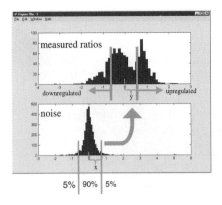

FIGURE 13.5: Using the empirical noise distribution to find differentially expressed genes. The confidence intervals found on the noise distribution (lower panel) can be mapped onto confidence intervals on the distribution of the expression values (upper panel). The confidence interval can be different between the up and down-regulated parts of the distribution if the noise is skewed. Reprinted from Drăghici, S. et al., Statistical intelligence: effective analysis of high-density microarray data, Drug Discovery Today, Vol. 7, pp. S55–S63, 2002. With permission from Elsevier.

be associated with a deviation from the mean of this distribution. To avoid using any particular model, the distance from the mean can be calculated by numerically integrating the area under the distribution. This distance on the noise distribution can be put into correspondence to a distance y on the measured distribution (Fig. 13.5) by bootstrapping [93, 283]. Furthermore, the dependency between intensity and variance can be taken into account by constructing several such models covering the entire intensity range and constructing non-linear confidence boundaries similar to those in Fig. 13.6.

13.7.2 Characteristics

The noise sampling method has the important advantage that its non-linear selection boundaries adapt automatically both to various amounts of regulation and different amounts on noise for a given confidence level chosen by the user. Fig. 13.6 illustrates the difference between the selection performed by the fold-change method (1) and the selection performed by the noise sampling method (2). Using the non-linear boundaries (2) takes into consideration the fact that the variance varies with the intensity (lower variance at higher intensities and higher variance at low intensities). A full blown ANOVA requires a special experimental design but provide error estimates for all variables considered in the model. The noise sampling method does not require such a special experiment design but only provides estimates for the log ratios of the genes (see Section 13.12.1 in Appendix). It has been shown that the noise sampling method provides a better sensitivity than the unusual ratio method and a much better specificity than the fold change method [93]. The method can be

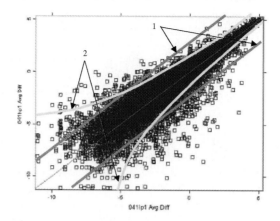

FIGURE 13.6:   A scatter plot representing the experiment values plotted against the control values. Unchanged genes will appear on the diagonal as the two values are similar. Selecting genes with a minimum fold change is equivalent to setting linear boundaries parallel to the y=x diagonal at a distance from it equal to the minimum fold change requested (1). The noise sampling method performs a better selection by using non-linear boundaries (2) that adapt to the increased noise variance at low intensities.

used equally well for Affymetrix data as well as for experiments involving several arrays.

13.8   Model based maximum likelihood estimation methods

In order to explain this approach, we will consider a simpler problem. Instead of considering two conditions and try to identify genes that are up-regulated, down-regulated and unchanged in the experiment vs. the control, we will consider a single array and will use this approach in order to distinguish the genes that are expressed from the genes that are not expressed [193]. The description of the method will assume the use of cDNA arrays using j replicate spots for each gene. However, this approach is by no means limited to this type of arrays or this specific problem [64, 121, 233, 193].

13.8.1   Description

We will consider two events. $\mathcal{E}_g$ will represent the event that mRNA for gene g in the array is contained in the target sample tissue, i.e. the gene is expressed. $\overline{\mathcal{E}_g}$ will represent the complement of $\mathcal{E}_g$, i.e. the event in which the gene in unexpressed. Let us denote by p the *a priori* probability of a gene being expressed. This is in fact equal to the fraction of expressed genes present in the mRNA sample. Since there are only two possibilities, either the gene is expressed or not, the probability of the gene being unexpressed is $1 - p$.

The method considers the probability density function (pdf) f_{E_j} associated with the genes that are expressed and the pdf f_{U_j} associated with the genes that are unexpressed. The probability density functions will describe the probability of measuring a given intensity value[3] y for a gene. If the gene is expressed, the intensity y will follow f_{E_j} whereas if the gene is not expressed, the intensity y will follow f_{U_j}. This is illustrated in Fig. 13.7. The horizontal axis represents intensity values; the vertical axis represents the probability. The graph to the left represents f_{U_j}. The graph to the right represents f_{E_j}. The placement of the two graphs corresponds to our expectations: the maximum probability (the peak of the pdf) for unexpressed genes is lower than the maximum probability for the expressed genes. The intersection of the two graphs corresponds to those occasional spots that belong to expressed genes but have intensities lower than those of some unexpressed genes and, conversely, spots that belong to unexpressed genes that happen to have higher intensities than some of those expressed genes (area A in Fig. 13.7). The two distributions can be combined in the following statistical model:

$$f_j(y) = p \cdot f_{E_j}(y) + (1 - p) \cdot f_{U_j}(y) \qquad (13.11)$$

where $f_j(y)$ is the pdf of observed intensity value y. This is **a mixture model** for the distribution of observed log ratios. This is called a mixture model because the observed intensity values are assumed to be distributed as a mixture of two distributions.

Equation 13.11 is straightforward to interpret. The first term in Eq. 13.11 corresponds to the case in which the observed intensity comes from an expressed gene. This will happen with probability p and in this case, the observed intensity will follow the pdf of an expressed gene which is $f_{E_j}(y)$. The second term in Eq. 13.11 corresponds to the case in which the observed intensity comes from an unexpressed gene. This will happen with probability $1 - p$ and in this case, the observed intensity will follow the pdf of an expressed gene which is $f_{U_j}(y)$. A gene is either expressed or not expressed so the probabilities are added together.

The goal of the approach is to decide whether a given spot j corresponds to an expressed or unexpressed gene. In other words, given that the measured intensity Y_{gj} of the gene g on spot j has the value y, we are to calculate the probability of gene g being expressed at spot j.[4] This is exactly the definition of the conditional probability (see Chap. 4):

---

[3]More precisely, the value is the log ratio of the background corrected mean of the pixels in a given spot. See Chap. 12 for a more complete discussion of the pre-processing and normalization.

[4]It may seem intuitive that if a gene g is expressed, this will happen for all its spots. However, as we will see, there is a non-zero probability that an expressed gene will not appear as such when examining a single spot.

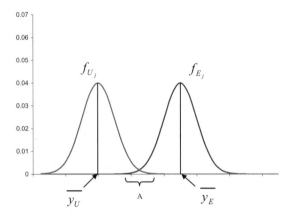

FIGURE 13.7:   A mixture model is the result of superposing two distributions. The horizontal axis represents the values of the observed intensity y. The vertical axis represents the probability of observing a given y value. In this case, the distribution to the left corresponds to the unexpressed genes and the distribution to the right corresponds to the expressed genes. Note that most unexpressed genes will have an observed intensity around $\bar{y}_U$ while most expressed genes will have an observed intensity around $\bar{y}_E$. In region A, the two distribution overlap.

$$Pr\left\{\mathcal{E}_g|Y_{gj} = y\right\} \tag{13.12}$$

From Bayes theorem (see Chap. 4), this can be expressed as:

$$Pr\left\{\mathcal{E}_g|Y_{gj} = y\right\} = \frac{p \cdot f_{E_j}(y)}{f_j(y)} \tag{13.13}$$

Eq. 13.13 states that the probability of the gene being expressed given that the observed intensity is y is the probability of a gene being expressed p times the probability that an expressed gene has the intensity y and divided by the probability of observing the intensity y.

If we knew the exact expressions of p, $f_{E_j}(y)$ and $f_{U_j}(y)$ we could use Eq. 13.11 to calculate $f_j(y)$, and then Eq. 13.13 to calculate the probability of a gene being expressed from its observed intensity values.

In order to obtain $f_{E_j}(y)$ and $f_{U_j}(y)$ we can just assume that they are normal distributions. In doing so, the model becomes **a mixed normal probability density function**. In this case, each distribution is completely described by only two values: a mean μ and a variance σ^2 (see Chap. 4 and in particular Eq. 4.61 for the analytical expression of the normal distribution). The values p, μ_{E_j}, $\sigma^2_{E_j}$, μ_{U_j} and $\sigma^2_{U_j}$ can be estimated numerically, using **a maximum likelihood** approach. The maximum likelihood method searches various combinations of the parameters such as the obtained equation fits the data as best as possible. Once obtained, the maximum likelihood estimates are used in conjunction with Eq. 13.11 and Eq. 13.13 to calculate the probability of a gene being expressed from its observed intensity values.

This approach has been used in conjunction with a special set of 288 genes, of which 32 genes were expected to appear highly expressed, in order to study the importance of replication in cDNA microarray gene expression studies [193]. Notably, this work proved several important facts:

1. Any individual spot can provide erroneous results. However, the probability of 3 or more spots being consistently wrong is negligible.

2. The intensities do seem to be normally distributed. This indicates that a mixed normal probability model may be useful in the analysis of other experiments.

3. The probability of a false negative (an expressed gene that either fails to be represented as probe or, if it is represented as probe, fails to be hybridized to the cDNA that are deposited on the slide) is as high as 5% for any single replicate.

4. The probability of a false positive (an unexpressed gene for which the fluorescence intensity suggests that the gene is expressed) is as high as 10% for any single replicate.

13.8.2   Characteristics

This approach is very general and very powerful. Advantages include the fact that the maximum likelihood estimators (MLE) become unbiased minimum variance estimators as the sample size increases and that the likelihood functions can be used to test hypotheses about models and parameters.

There are also several disadvantages. The first one is that they require solving complex non-linear equations. This is very computationally intensive and requires specialized software. Furthermore, a more realistic situation involving two or more experiments in which the genes have to be divided into up-regulated, unchanged and down-regulated requires a more complex mixture model:

$$f_j(y) = p_1 \cdot f_{UP_j}(y) + p_2 \cdot f_{DOWN_j}(y) + p_3 \cdot f_{UNCHANGED_j}(y) \qquad (13.14)$$

involving 3 a priori probabilities p_1, p_2 and p_3 corresponding to the up-regulated genes, down-regulated genes and unchanged genes, respectively. Also, the computation of the MLE becomes more complex due to the increased number of parameters. Finally, another disadvantage of the maximum likelihood estimate approach is that the results become quickly unreliable as the sample size decreases. Indeed, for small samples, the estimates may be very much different from the real underlying distribution. Furthermore, MLE estimates can become unreliable when data deviates considerably from normality.

13.9 Affymetrix comparison calls

The comparison calls calculated by the Affymetrix software are designed to answer the question whether a particular transcript on one array is changed significantly with respect to another array. Usually, the two arrays are called the experiment array and the baseline array. The calls are categorical classifications into one of the following classes: "Decreased," "Marginally Decreased," "No Change," "Marginally Increased" and "Increased." As in hypothesis testing, "No Change" does not actually mean that the gene has not changed but rather that the amount of change in the context of the measured variance does not allow a rejection of the null hypothesis that the gene is unchanged. In other words, the data are not sufficient to prove that the gene has changed, which may not necessarily mean that the gene has not changed.

If all probe pairs of a probe set are saturated (PM or MM $\geq$ 46000), the software will report that no comparative calls can be made. If any one of the 4 cells, PM and MM in baseline and PM and MM in experiment, are saturated the corresponding probe pair is not used in further computations. The remaining probe pairs are used to form two vectors. The first vector will contain the differences between PM and MM:

$$q = (PM_1 - MM_1, PM_2 - MM_2, \cdots, PM_n - MM_n) \qquad (13.15)$$

The second vector will contain the differences between the PM and the background level computed as described in Sec. 12.4.1:

$$z = (PM_1 - b_1, PM_2 - b_2, \cdots, PM_n - b_n) \qquad (13.16)$$

However, the distribution of q and z values over all probe pairs on a given array are different from each other. Two balancing factors are used to match these two distributions [7]. Finally, a Wilcoxon signed rank test is used to test a two-tail set of hypotheses on the median of the adjusted differences between the vectors q and z in the baseline and experiment, respectively:

$$f_1 \cdot q_E[i] - q_B[i] \qquad (13.17)$$

and

$$C \cdot (f_2 z_E[i] - z_B[i]) \qquad (13.18)$$

where f_1 and f_2 are empirically determined balancing factors and C is a scaling constant ($C = 0.2$ by default).

The hypotheses tested are:

$H_0:$ $median(v) = 0$
$H_a:$ $median(v) > 0$

The cut-off values γ_1 and γ_2 for the p-values are adjusted for multiple comparisons using a Bonferroni-like dividing factor only that the exact value of the factor is determined empirically. It is important to note that the p-values produced by this approach are actually overestimates of the probability of a Type I error because the two values used $PM - MM$ and $PM_i - b_i$ are not independent. However, this is acceptable because we only seek an ordering of the values and the thresholds between increase and marginally increase and decrease and marginally decrease, respectively, are more or less arbitrary.

13.10   Other methods

Another maximum likelihood estimation approach for two color arrays is described in [64]. This approach is based on the hypothesis that the level of a transcript depends on the concentration of the factors driving its selection and that the variation for any particular transcript is normally distributed and in a constant proportion relative to most other transcripts. This hypothesis is then exploited by considering a constant coefficient of variation c for the entire gene set and constructing a 3rd degree polynomial approximation of the confidence interval as a function of the coefficient of variation c. This approach is also interesting because it provides the means to deal with signals uncalibrated between the two colors through an iterative algorithm that compensates for the color difference.

Sapir et al. [233] present a robust algorithm for estimating the posterior probability of differential expression based on an orthogonal linear regression of the signals obtained from the two channels. The residuals from the regression are modelled as a mixture of a common component and component due to differential expression. An expectation maximization algorithm is used to deconvolve the mixture and provide estimates of the probability that each gene is differentially regulated as well as estimates of the error variance and proportion of differentially expressed genes.

Two hierarchical models (Gamma-Gamma and Gamma-Gamma-Bernoulli) for the two channel (color) intensities are proposed in [215]. One advantage of such an approach is that the models constructed take into consideration the variation of the posterior probability of change on the absolute intensity level at which the gene is expressed. This particular dependency is also considered in [229] where the values measured on the two channels are assumed to be normally distributed with a variance depending on the mean. Such intensity dependency reduces to defining some curves in the green-red plane corresponding to the two channels and selecting as differentially regulated the genes that fall outside the equiconfidence curves (see Fig. 13.6).

Another multiple experiment approach is to identify the differentially expressed genes by comparing their behavior in a series of experiments with an expected expression profile [127, 118]. The genes can be ranked according to their degree of similarity to a given expression profile. The number of false positives can be controlled through random permutations that allow the computation of suitable cut-off thresholds for the degree of similarity. Clearly, these approaches can only be used in the context of a large data set including several microarrays for each condition considered.

Other methods used for the selection of differentially regulated genes include gene shaving [132], assigning gene confidence [207] or significance [268], bootstrap [108, 181] and Bayesian approaches [21, 202, 282].

Finally, more elaborate methods for the data analysis of gene expression data exist. Such methods include singular value decomposition [11], independent component analysis [198] and many others. The goals of such methods go well beyond the selection of differentially regulated genes and, as such, they are outside the scope of this chapter.

13.11 Summary

A plethora of refined methods is available for the selection of differentially regulated genes. Although still in widespread use, the early methods of selection by fold change and unusual ratio are clearly inadequate. Using a fold change without a clear biological justification is just a blind guess. The unusual ratio method will always report some genes as regulated even if two identical tissues are studied (false positives). These two methods suffer from severe drawbacks and their use should be discontinued as methods for selecting differentially regulated genes. However, studying the fold change of genes of known function is and will continue to remain important. In order words, computing statistics as required by biological reasons is fully justified (e.g. how do apoptosis related genes change in immortalization?). However, drawing biological conclusions based on an arbitrary choice of fold change is not (for instance concluding that gene X is relevant to immortalization because it has a fold change of 2).

Selecting the differentially regulated genes can be done using a classical hypothesis testing approach. When using univariate statistical tests for hundreds or thousands of genes (e.g. with data coming from most commercial chips) the correction for multiple comparisons is absolutely crucial. A number of methods use this approach.

ANOVA is a general data analysis approach that can also provide information able to discern the differentially regulated genes. Kerr and Churchill have proposed several ANOVA models and experiment designs (see also Chapters 7 and 8).

The noise sampling method yields only $2n$ residuals (as opposed to $4n$ residuals when ANOVA is applied on individual channels) and no estimates for the other factors (e.g. array effect, gene effect, etc.). However, the approach does provide the differential expression terms for all genes, as well as noise estimates for calculating confidence intervals. Furthermore, the noise sampling method is suitable for use on individual arrays which has two important consequences. Firstly, the application of this technique does not require repeating experiments[5] and thus is advantageous

---

[5]However, repeating experiments is very strongly recommended.

when either the number of arrays and/or the amount of mRNA are limited. Secondly, because the noise estimates come from a single array, they can be used to construct an overall noise distribution which will characterize the experimental quality of the array and thus provide the experimental biologist with a very useful feedback regarding the quality of the laboratory process.

Current statistical methods offer a great deal of control and the possibility of selecting genes within a given confidence interval. However, all such methods rely essentially on a careful experiment design and the presence of replicate measurements. A good way to obtain reliable results is, arguably, some version of the ANOVA method. However, in most cases, this will probably mean involving a statistician from the very beginning and designing the experiment in such a way that enough degrees of freedom are available in order to answer the relevant biological questions.

Model based maximum likelihood methods use a statistical model involving one or several known distributions. A maximum likelihood approach is used to estimate the parameters of the model in order to fit it best to the given data. Advantages include flexibility and large applicability while disadvantages include the amount of computation and bad results for small sample sizes.

The chapter also included a short description of the approach used to calculate the fold changes for Affymetrix arrays. The methods used include a number of empirical estimates and adjustments that have been shown to work well with real data. The decisions are ultimately based on a non-parametric Wilcoxon signed-rank test for paired data where the pairing is done at probe level between the baseline and experiment arrays.

13.12 Appendix

13.12.1 A comparison of the noise sampling method with the full blown ANOVA approach

The noise sampling method presented here is a particular application of the analysis of variance (ANOVA) method. Another ANOVA variation was proposed by Kerr-Churchill [182, 183]. It is informative to compare the two approaches. Let us consider the following model proposed by Kerr and Churchill:

$$\log(y_{ijkg}) = \mu + A_i + D_j + V_k + G_g + (AG)_{ig} + (VG)_{kg} + \epsilon_{ijkg} \quad (13.19)$$

In this equation, i indexes the array, j indexes the dye, k indexes the mRNA and g indexes the genes. Let us assume that there are two arrays in a flip dye experiment. A flip dye experiment involves two arrays. In the first array, the experimental mRNA is labeled with one dye, typically cy3, and the control mRNA is labeled with a second dye, typically cy5. In the second array, the colors are reversed: cy5 for experiment and cy3 for control. In this case, we have $(i = 1, 2, j = 1, 2, k = 1, 2, g = 1..n)$.

The flip dye experiment will provide the following $4n$ values: y_{111g}, y_{122g}, y_{212g} and y_{221g}. According to the Kerr-Churchill approach, the values for μ, A_i, D_j, V_k, G_g, VG_{kg} and AG_{ig} are obtained using a least-square fit. In consequence, there will be $4n$ residuals ϵ_{ijkg}. Subsequently, the mean differential expression can be calculated as:

$$\overline{dE_g} = \frac{(y_{111g} - y_{122g}) + (y_{221g} - y_{212g})}{2} - bias \qquad (13.20)$$

and the $4n$ residuals can be used to add error bars. In contrast, the approach proposed here goes straight to ratios which correspond to difference of log values, as follows:

$$y_{111g} = \mu + A_1 + D_1 + V_1 + G_g + VG_{1g} + AG_{1g} + \epsilon_{111g} \qquad (13.21)$$

$$y_{122g} = \mu + A_1 + D_2 + V_2 + G_g + VG_{2g} + AG_{1g} + \epsilon_{122g} \qquad (13.22)$$

which can be subtracted to yield:

$$y_{111g} - y_{122g} = (D_1 - D_2) + (V_1 - V_2) + (VG_{1g} - VG_{2g}) + (\epsilon_{122g} - \epsilon_{122g}) \quad (13.23)$$

Taking $(D_1 - D_2) + (V_1 - V_2) = \mu$, $(VG_{1g} - VG_{2g}) = G(g)$ and $(\epsilon_{122g} - \epsilon_{122g}) = \epsilon(g, s)$ we obtain Eq. 13.7. In conclusion, the approach presented here yields only $2n$ residuals (as opposed to $4n$ residuals when ANOVA is applied on individual channels) and no estimates for μ, A_i, G_g and AG_{ig}. However, the approach does provide the differential expression terms for all genes $(VG_{1g} - VG_{2g})$ as well as noise estimates for calculating confidence intervals.

Chapter 14

Functional analysis and biological interpretation of microarray data

There are no facts, only interpretations.

—Friedrich Wilhelm Nietzsche

The whole is simpler than the sum of its parts.

—Willard Gibbs

14.1 Introduction

This text has focused so far on the numerical analysis of the microarray data and the nitty-gritty of the number crunching. However, the ultimate purpose of the gene expression experiments is to produce biological knowledge, not numbers.[1] Independently of the methods used, the result of a microarray experiment is, in most cases, a set of genes found to be differentially expressed between two or more conditions under study. The challenge faced by the researcher is to translate this list of differentially regulated genes into a better understanding of the biological phenomena that generated such changes. Although techniques aimed at this goal have started to appear (e.g. inferring gene networks [80, 82, 229, 264], function prediction [112, 119, 189, 288], etc.) this approach is substantially more difficult. A good first step in this direction can be the translation of the list of differentially expressed genes into a functional profile able to offer insight into the cellular mechanisms acting in the given condition. Even if our information about the genes were complete and accurate, the mapping of lists of tens or hundreds of differentially regulated genes to biological functions, molecular functions and cellular components is not a trivial matter.

In this chapter, we will shift the focus from the exclusive numerical analysis performed so far to a more integrated bioinformatics approach that will try to accomplish the goal stated above. While the means will continue to consist of statistical and data analysis tools similar to the ones discussed so far, the emphasis will shift

---

[1] Some of the material in this chapter is reprinted from Genomics, Vol. 81, No. 2, Drăghici et al., "Global functional profiling of gene expression", p. 98–104, Copyright (2003), with permission from Elsevier.

towards refining the data analysis results into biological knowledge. We will discuss a number of problems related to this higher level analysis as well as some tools designed to address such problems.

14.2 The Gene Ontology

14.2.1 The need for an ontology

Many biologists now believe that probably there is a limited set of genes and proteins, many of which are conserved in most or all living cells. Recognizing this unification of biology and the information about genes and proteins shared by different organisms led to the development of a comparative genomics approach. Knowledge of a biological role of such a shared gene or protein in one organism may be used to infer its role in other organisms. The ability to infer biological roles provides an opportunity that annotation from one organism can be transferred to another organism in an automated fashion. However, the problem with any automated processing is that the existing databases use different formats and, most importantly, different vocabularies in order to describe functional annotation [16].

One way of meeting these challenges is to develop an ontology for gene annotations. The immediate advantage of such an ontology is the ability to explore functional annotations of genomes of different organisms in an automatic way. An ontology is also helpful in describing attributes of genes. For example, genes may have more than one product. Gene products possess one or more biochemical, physiological or structural functions. Often gene products are located in specific cellular compartments. The structured, controlled vocabulary of an ontology provides a picture that is bigger than just sequences of nucleotides or amino acids [204]. Such an ontology has been developed over the past few years by the Gene Ontology (GO) Consortium [16, 17, 126].

14.2.2 What is the Gene Ontology (GO)?

The stated goal was to produce a dynamic, structured, precisely defined, common, controlled vocabulary useful to describe the roles of genes and gene products in any organism even as knowledge of gene and protein roles in cells is accumulating and changing [16, 17, 126]. The main features of the GO are as follows:

1. It has a controlled vocabulary and therefore it is machine readable.

2. It is multidimensional, with more than one axis of classification.

3. It can be used across species.

4. It allows multiple classes of relationships, e.g. "is a" and "part of" relationships.

5. It allows expression of regulatory and reaction relationships.

6. It allows representation of incomplete knowledge.

Our knowledge of what genes and their proteins do is very incomplete and is changing rapidly everyday. To keep up with the new information being available everyday, the GO consortium provides links for each node in the GO ontologies, to related databases, that includes GenBank, EMBL, DDBJ, PIR, MIPS, YPD, WormPD, SWISS-PROT, ENZYME and other databases [156]. However, providing links to the external databases is not sufficient. The ontologies themselves must be updated continuously as more information becomes available.

14.2.3 What does GO contain?

GO includes three independent ontologies for **biological process**, **molecular function** and **cellular component**. They are all attributes of a gene, a gene product or a gene product group. A gene product can have one or more molecular functions, be used in one or more biological processes and may be associated with one or more cellular components. The biological process is defined as a biological objective to which the gene or gene product contributes. A process is the result of one or more ordered assemblies of molecular functions. The molecular function is defined as the biochemical activity of a gene product. The molecular function only describes what is done without specifying where or when the actual activity takes place. The cellular component refers to the place in the cell where the gene product is active. It must be noted that not all terms are applicable to all organisms [16].

GO is a database independent of any other. GO itself is not populated with gene products of any organism, but rather GO terms are used as attributes of genes and gene products by related databases. These databases use GO terms to annotate objects such as genes or gene products stored in their repositories, provide references and the kind of evidence that is available to support the annotations. They also provide tables of cross-links between GO terms and their database objects to GO, which are then made available publicly [159]. The list of all crossed-referenced databases along with their abbreviated names in GO is available at [161].

14.2.3.1 GO data representation

Each term in GO is a node of a Directed Acyclic Graph (DAG). A DAG is very similar to a tree with the only difference that in a DAG it is possible for a node to have more than one parent. In GO, the relationship between a parent and a child can be either "instance of" or "part of". If a child has more than one parent, it may have different relationships with its different parents. GO uses "%" to denote the "instance of" relationship and "<" to denote the "part of" relationship. Each GO term is given a unique identifier of the form "*GO:nnnnnnn*," where *nnnnnnn* is a zero-padded integer of seven digits. The unique identifier is used for cross-referencing in the databases using GO.

Gene Ontology is dynamic and changes as more information is made available. If a term is retired, it is removed from the DAG and made a child of the meta term "obsolete." When two terms, for example term A and term B are merged, for example as term A, the GO ID of term B is made a secondary GO ID of term A. Similarly, if a term is split each new term is assigned a new GO ID and the original GO ID is made a secondary GO ID of the new terms.

Each GO term is also cross-referenced with a number of databases. The links to such external databases are provided as "database-abbreviation:id," where id is an identifier in the given database [161].

One of the goals of GO is to provide an ontology that is applicable across species. However, some of the functions, processes and components are specific to species. In those cases when a text string has a different meaning in different species, GO differentiates them lexically [159]. Definitions are provided in a single text file [158]. External databases that choose to cross-reference GO by providing links to the genes and gene products in their repository must also provide information about the type of evidence supporting the functional assignments used. The GO Consortium divides the types of evidence in the following categories [157, 160]:

1. IMP inferred from mutant phenotype

2. IGI inferred from genetic interaction

3. IPI inferred from physical interaction

4. ISS inferred from sequence similarity

5. IDA inferred from direct assay

6. IEP inferred from expression pattern

7. IEA inferred from electronic annotation

8. TAS traceable author statement

9. NAS non-traceable author statement

10. ND no biological data available

11. IC inferred by curator

Among these evidence codes, IEA (Inferred from Electronic Annotation) is of the lowest quality.

14.2.4   Access to GO

GO Consortium provides the AmiGO browser that allows the user to browse the GO database in tree or directed acyclic graph view. AmiGO allows searches using a GO term or a gene product. The results of the query performed with a GO term will contain all gene products annotated with the term used for the query. If a query is made for a gene product, the results will provide the gene and all its associations [162]. Some of the related databases also provide browsers to search for GO terms in their own database. Mouse Genome Informatics provides a GO browser [163] that allows search on a GO term and displays all mouse genes annotated to the term. European BioInformatics Institute (EBI) provides the QuickGO browser [154] integrated into InterPro [155] and the EP:GO browser [153] integrated into the Expression Profiler [152].

14.3   Other related resources

A tremendous amount of genetic data is available on-line from several public databases (DBs). NCBI provides sequence, protein, structure and genome DBs, as well as a taxonomy and a literature DB. Of particular interest are UniGene (non-redundant set of gene-oriented clusters) and LocusLink (genetic loci). SWISS-PROT is a curated protein sequence DB that provides high level annotation and a minimal level of redundancy [20, 205]. Kyoto Encyclopedia of Genes and Genomes (KEGG) contains a gene catalogue (annotated sequences), a pathway DB containing a graphical representation of cellular processes and a LIGAND DB [170, 171, 217]. GenMAPP is an application that allows the user to create and store pathways in a graphic format, includes a multiple species gene database and allows a mapping of a user's expression data on existing pathways [73]. Other related databases and on-line tools include: PathDB (metabolic networks [277]), GeneX (NCGR) (source independent microarray data DB [208]), Arrayexpress [101], SAGEmap [191], μArray [101], ArrayDB [216], ExpressDB [2] and Stanford Microarray Database [247, 255]. Two meta-sites containing information about various genomic and microarray on-line DBs are [248] and [71].

Data format standardization is necessary in order to automate data processing [47]. The Microarray Gene Expression Data Group (MGED) is working to standardize the Minimum Information About a Microarray Experiment (MIAME), the format (MAGE) and ontologies and normalization procedures related to microarray data [48, 102].

14.4   Translating lists of differentially regulated genes into biological knowledge

The availability of GO and many cross-referencing databases seems to solve the problem of interpreting the results of a microarray experiment from a biological point of view. Most databases provide efficient search mechanisms that return quickly all annotation information associated to any specific gene or gene product of interest. However, the problem is that most relevant databases are oriented towards a manual, gene by gene querying. If the processing of the list of differentially regulated genes were to be done manually, one would take each accession number corresponding to a regulated gene, search various public databases and compile a list with, for instance, the biological processes that the gene is involved in. The same type of analysis could be carried out for other functional categories such as biochemical function, cellular role, etc. This task can be performed repeatedly, for each gene, in order to construct a master list of all biological processes in which at least one gene was involved. Further processing of this list can provide a list of those biological processes that are common between several of the regulated genes. It is intuitive to expect that those biological processes that occur more frequently in this list would be more relevant to the condition studied. For instance, if 200 genes have been found to be differentially regulated and 160 of them are known to be involved in, let us say, mitosis, it is intuitive to conclude that mitosis is a biological process important in the given condition. As we shall see in the following example, this intuitive reasoning is incorrect and a more careful analysis must be done in order to identify the truly relevant biological processes.

Example 14.1

Let us consider that we are using an array containing 2000 genes to investigate the effect of ingesting a certain substance X. Using some of the classical statistical and data analysis methods discussed in Chap. 13, we conclude that 200 of these genes are differentially regulated by substance X. For each of these 200 genes, one can query the available public databases containing information about the biochemical function, biological process, cellular role, cellular component, molecular function and chromosome location. Let us focus on the biological process for instance, and assume that the results for the 200 differentially regulated genes are as follows: 160 of the 200 genes are involved in mitosis, 80 in oncogenesis, 60 in the positive control of cell proliferation and 40 in glucose transport.

If we now look at the functional profile described above, we might conclude that substance X may be related to cancer since mitosis, oncogenesis and cell proliferation would all make sense in that context. However, a reasonable question is: what would happen if all the genes on the array used were part of the mitotic pathway? Would mitosis continue to be significant? Clearly, the answer is no. Therefore, in order to draw correct conclusions, it is necessary to compare the actual number of occurrences with the expected number of occurrences for each individual category.

This comparison is shown in Table 14.1 for the example considered. Now, the functional profile appears to be completely different. There are indeed 160 mitotic genes but, in spite of this being the largest number, we actually expected to observe 160 such genes so this is not better than chance alone. The same is true for oncogenesis. The positive control of cell proliferation starts to be interesting because we expected 20 and observed 60. This is 3 times more than expected. However, the most interesting is the glucose transport. We expected to observe only 10 such genes and we observed

biological process	genes found	genes expected	
mitosis	160	160	not better than chance
oncogenesis	80	80	not better than chance
positive control of cell proliferation	60	20	better than chance
glucose transport	40	10	much better than chance

TABLE 14.1:   The statistical significance of the data mining results. The number of genes that are involved in a given biological process can be misleading. Mitosis may appear to be the most important process affected since 160 of the 200 differentially regulated genes are involved in mitosis. In fact, this is no better than chance alone. In comparison, there are only 40 genes involved in glucose transport but this is 4 times more than expected by chance alone.

40, which is 4 times more than expected. Taking into consideration the expected numbers of genes radically changed the interpretation of the data. In light of these data, we may want to consider the correlation of X with diabetes instead of cancer. □

This example illustrates that the simple frequency of occurrence of a particular functional category among the genes found to be regulated can be misleading. In order to draw correct conclusions, one must analyze the observed frequencies in the context of the expected frequencies. The problem is that an event such as observing 40 genes when we expect 10 can still occur just by chance. This is unlikely, but it can happen. The next section explains how the significance of these categories can be calculated based on their frequency of occurrence in the initial set of genes M, the total number of genes N, the frequency of occurrence in the list of differentially regulated genes x and the number of such differentially regulated genes K. The statistical confidence thus calculated will allow us to distinguish between significant events and possibly random events.

14.4.1   Statistical approaches

Several different statistical approaches can be used to calculate a p-value for each functional category F. Let us consider there are N genes on the array used. Any given gene is either in category F or not. In other words, the N genes are of two categories: F and non-F (NF). This is similar to having an urn filled with N balls of two colors such as red (F) and green (not in F). M of these balls are red and N-M are green. The researcher uses their choice of data analysis methods to select which genes are regulated in their experiments. Let us assume that they picked a subset of K genes. We find that x of these K genes are red and we want to determine the probability of this happening by chance. So, our problem is: given N balls (genes) of which M are red and N-M are green, we pick randomly K balls and we ask what is the probability of having picked exactly x red balls. This is sampling without replacement because once we pick a gene from the array, we cannot pick it again.

The probability that a category occurs exactly x times just by chance in the list of differentially regulated genes is appropriately modelled by a hypergeometric distribution with parameters (N, M, K) [59]:

$$P(X = x \mid N, M, K) = \frac{\binom{M}{x}\binom{N-M}{K-x}}{\binom{N}{K}} \tag{14.1}$$

Based on this, the probability of having x *or fewer* genes in F can be calculated by summing the probabilities of picking 1 or 2 or ... or $x - 1$ or x genes of category F [264]:

$$p_u = P(X = 1) + P(X = 2) + \cdots + P(X = x) = \sum_{i=0}^{x} \frac{\binom{M}{i}\binom{N-M}{K-i}}{\binom{N}{K}} \tag{14.2}$$

This corresponds to a one-sided test in which small p-values correspond to underrepresented categories. The p-value for over-represented categories can be calculated as $p_o = 1 - p_u$ when $p_u > 0.5$:

$$p_o = 1 - \sum_{i=0}^{x} \frac{\binom{M}{i}\binom{N-M}{K-i}}{\binom{N}{K}} \tag{14.3}$$

The hypergeometric distribution is difficult to calculate when the number of genes is large (e.g., arrays such as Affymetrix HGU133A contain 22,283 genes). However, when N is large, the hypergeometric distribution tends to the binomial distribution [59]. If a binomial distribution is used, the probability of having x genes in F in a set of K randomly picked genes is given by the classical formula of the binomial probability in which the probability of extracting a gene from F is estimated by the ratio of genes in F present on the array M/N and the corresponding p-value can be respectively calculated as:

$$P(X = x \mid K, M/N) = \binom{K}{x}\left(\frac{M}{N}\right)^{x}\left(1 - \frac{M}{N}\right)^{K-x} \tag{14.4}$$

and

$$p = \sum_{i=0}^{x} \binom{K}{i}\left(\frac{M}{N}\right)^{i}\left(1 - \frac{M}{N}\right)^{K-i} \tag{14.5}$$

Alternative approaches include a χ^2 (chi-square) test for equality of proportions [110] and Fisher's exact test [206]. For the purpose of applying these tests, the data can be organized as shown in Fig. 14.1. The dot notation for an index is used to represent the summation on that index (see Sec. 7.1.2). In this notation, the number of genes on the microarray is $N = N_{.1}$, the number of genes in functional category F is $M = n_{11}$, the number of genes selected as differentially regulated is $K = N_{.2}$ and the number of differentially regulated genes in F is $x = n_{12}$. Using this notation, the chi-square test involves calculating the value of the χ^2 statistic as follows:

$$\chi^2 = \frac{N_{..}\left(|n_{11}n_{22} - n_{12}n_{21}| - \frac{N_{..}}{2}\right)^2}{N_{1.}N_{2.}N_{.1}N_{.2}} \tag{14.6}$$

	Genes on array	Diff. regulated genes	
having function F	n_{11}	n_{12}	$N_{1.} = \sum_{j=1}^{2} n_{1j}$
not having F	n_{21}	n_{22}	$N_{2.} = \sum_{j=1}^{2} n_{2j}$
	$N_{.1} = \sum_{i=1}^{2} n_{i1}$	$N_{.2} = \sum_{i=1}^{2} n_{i2}$	$N_{..} = \sum_{i,j} n_{ij}$

FIGURE 14.1: The significance of a particular functional category F can be calculated using a 2x2 contingency table and a chi-square or Fisher's exact test for equality of proportions. The N genes on an array can be divided into genes that are involved in the functional category of interest F ($n_{11} = M$) and genes that are not involved in F (n_{21}). The K genes found to be differentially regulated can also be divided into genes involved ($n_{21} = x$) and not involved (n_{22}) in F.

where $\frac{N_{..}}{2}$ in the numerator is a continuity correction term that can be omitted for large samples [125]. The value thus calculated can be compared with critical values obtained from a χ^2 distribution with $df = (2-1) \cdot (2-1) = 1$ degree of freedom. However, the χ^2 test for equality of proportion can not be used for small samples. The rule of thumb is that all expected frequencies: $E_{ij} = \frac{N_{i.} \cdot N_{.j}}{N_{..}}$ should be greater than or equal to 5 for the test to provide valid conclusions. If this is not the case, Fisher's exact test can be used instead [110, 176, 258]. Fisher's exact test considers the row and column totals $N_{1.}$, $N_{2.}$, $N_{.1}$, $N_{.2}$ fixed and uses the hypergeometric distribution to calculate the probability of observing each individual table combination as follows:

$$P = \frac{N_{1.}! \cdot N_{2.}! \cdot N_{.1}! \cdot N_{.2}!}{N_{..}! \cdot n_{11}! \cdot n_{12}! \cdot n_{21}! \cdot n_{22}!} \tag{14.7}$$

Using this formula, one can calculate a table containing all the possible combinations of $n_{11}n_{12}n_{21}n_{22}$. The p-value corresponding to a particular occurrence is calculated as the sum of all probabilities in this table lower than the observed probability corresponding to the observed combination [206].

Finally, Audic and Claverie have used a Poisson distribution and a Bayesian approach [18] to calculate the probability of observing a given number of tags in SAGE data. As noted by Man et al. [206], this approach can be used directly to calculate the probability of observing n_{12} genes of a certain functional category F in the selected subset given that there are n_{11} such genes on the microarray:

$$P(n_{12}|n_{11}) = \left(\frac{N_{.2}}{N_{.1}}\right)^{n_{12}} \cdot \frac{(n_{11}+n_{12})!}{n_{11}! \cdot n_{12}! \cdot \left(1 + \frac{N_{.2}}{N_{.1}}\right)^{n_{11}+n_{12}+1}} \tag{14.8}$$

The p-values are calculated as a cumulative probability distribution function (cdf) as follows [18, 206]:

$$p = \min \left\{ \sum_{k=0}^{k \leq n_{12}} P(k|n_{11}), \sum_{k=n_{12}}^{\infty} P(k|n_{11}) \right\} \tag{14.9}$$

Extensive simulations performed by Man et al. compared the chi-square test for equality of proportions with Fisher's exact test and Audic and Claverie's test and showed that the chi-square test has the best power and robustness [206].

14.5 Onto-Express

Onto-Express (OE) is a database and a collection of tools designed to facilitate the analysis described above [90, 185]. This is accomplished by mining known data and compiling a functional profile of the experiment under study. OE constructs a functional profile for each of the Gene Ontology (GO) categories [16]: cellular component, biological process and molecular function as well as biochemical function and cellular role, as defined by Proteome [222]. The precise definitions for these categories and the other terms used in OE's output have been discussed in Section 14.2. More details can be found in GO [16]. As biological processes can be regulated within a local chromosomal region (e.g. imprinting), an additional profile is constructed for the chromosome location. OE uses a database with a proprietary schema implemented and maintained in our laboratory [89]. Onto-Express uses data from GenBank, UniGene, LocusLink, PubMed, and Proteome.

14.5.1 Implementation

Onto-Express provides implementations of the χ^2 test, Fisher's exact test as well as the binomial test discussed in Section 14.4.1. Fisher's exact test is required when the sample size is small and the chi-square test cannot be used. For a typical microarray experiment with $N \simeq 10,000$ genes on the array and $K \simeq 100 = 1\%N$ selected genes, the binomial approximates very well the hypergeometric and is used instead. For small, custom microarrays (fewer than 200 genes), the χ^2 is used. The program calculates automatically the expected values and uses Fisher's exact test when χ^2 becomes unreliable (expected values less than 5). Thus, the choice between the three different models is automatic, requiring no statistical knowledge from the end-user. We did not implement Audic and Claverie's test because: i) it has been shown that χ^2 is at least as good [206] and ii) while very appropriate for the original problem involving ESTs, the use of a Poisson distribution may be questionable for our problem.

The current version of Onto-Express is implemented as a typical 3-tier architecture. The back-end is a relational DB implemented in Oracle 9i and running on a SunFire V880 with 4CPUs, 8 GB RAM memory, 200 GB of internal disk and accessing a 500 GB RAID array and tape jukebox backup.[2] The application performing the data

---

[2]A Sun Microsystems Equipment Grant that made possible the acquisition of this equipment is gratefully acknowledged here.

mining and statistical analysis is written in Java and runs on a separate Sun Enterprise 3500 server with 4CPUs, 4 GB RAM memory and 200 GB of internal disk storage. The front end is a Java applet served by a Tomcat/Apache web server running on a Sun Fire V100 web server appliance.

14.5.2 Graphical input interface description

The input of Onto-Express is a list of genes specified by either accession number, Affymetrix probe IDs or UniGene cluster IDs. In fact, the utility of this approach goes well beyond DNA microaarrays since such a list can be also constructed using any alternative technology such as protein arrays, SAGE, Westerns blots (e.g. high throughput PowerBlots [37]), Northerns blots, etc. At present[3], our database includes the human (*Homo sapiens*), fruit fly (*Drosophila melanogaster*), rat (*Rattus norvegicus*) and mouse (*Mus musculus*) genomes. Clearly, the utility of this analysis depends considerably on the amount of annotation data available for a given organism. The more data are available, the more useful this analysis will be. The content of the Onto-Express database is updated bi-weekly or whenever one of the primary databases (GenBank, UniGene, GO, etc.) are updated. Thus, the results of the analysis will constantly improve, as more annotation data becomes available.

The input interface is shown in Fig. 14.2. The analysis requires the following pieces of information:

1. The list of differentially regulated genes identified in the given condition.

2. The organism studied.

3. The list of genes on the array used. The database contains information about most mainstream commercial arrays. If such an array was used, the user can simply pick the appropriate array from a drop-down menu.

4. The method to be used for the correction for multiple comparisons (Bonferroni, Šidák, bootstrapping).

5. The type of input (accession, Affymetrix probe IDs, UniGene clusters).

6. A specification of the output required by the user. Currently, the user can select any (or all) of: cellular component, biological process, molecular function, biochemical function, cellular role and chromosome location.

The results are provided in graphical form and e-mailed to the user on request. By default, the functional categories are sorted in decreasing order of number of genes as shown in Fig. 14.3. The functional categories can also be sorted by confidence (see details about the computation of the p-values below) with the exception of the results for chromosomes, where the chromosomes are always displayed in their natural order.

---

[3]This book went to press in January 2003.

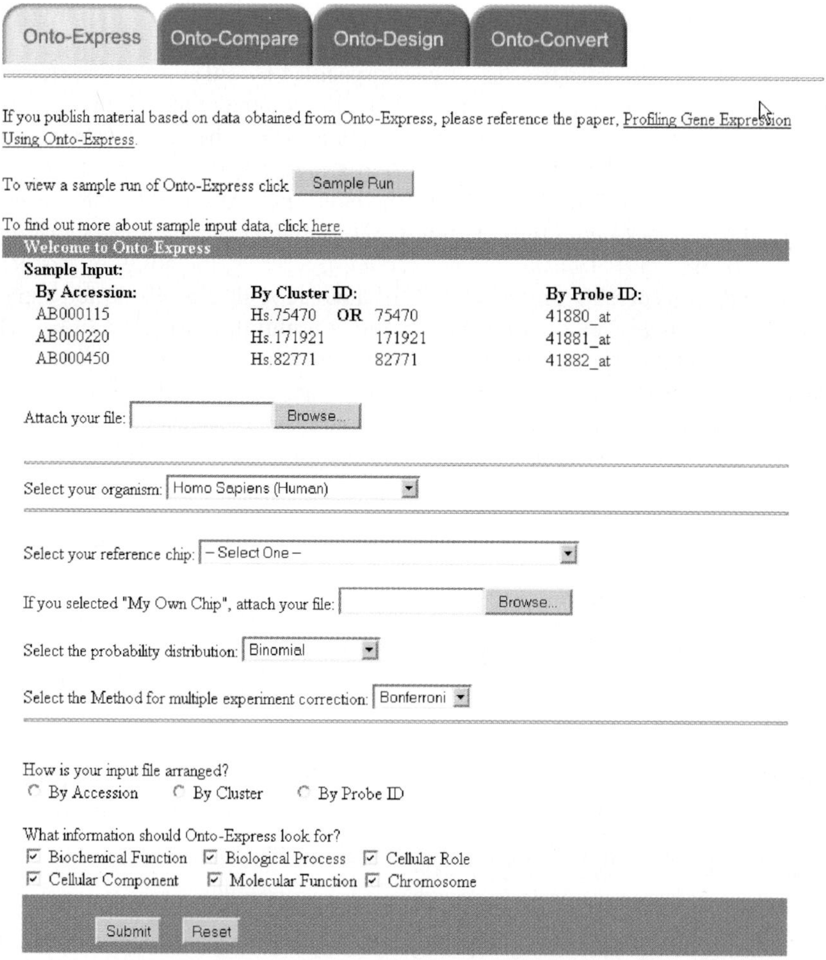

FIGURE 14.2: The input necessary for the Onto-Express analysis. Reprinted from Drăghici, S., Global functional profiling of gene expression, Genomics, Vol. 81, No. 2, pp. 98–104, 2003. With permission from Elsevier.

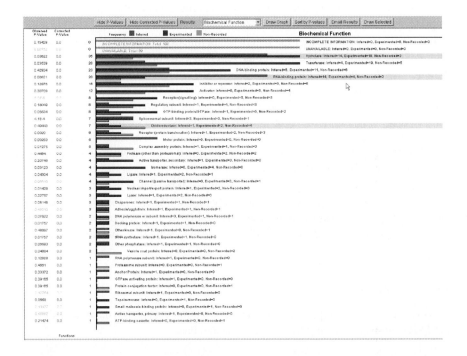

FIGURE 14.3: The main features of the Onto-Express output. The functional categories can be sorted by number of genes (shown) or by p-values (with the exception of the results for chromosomes where the chromosome are displayed in their order). Requesting a specific graph is done by choosing the desired category from the pull-down menu and subsequently clicking on "Draw graph." Right-clicking allows the user to select specific items from each functional category. These items that can be drawn in a composite graph by clicking on "Draw selected." Reprinted from Drăghici, S., Global functional profiling of gene expression, Genomics, Vol. 81, No. 2, pp. 98–104, 2003. With permission from Elsevier.

A particular functional category can be assigned to a gene based on specific experimental evidence or by theoretical inference (e.g. similarity with a protein having a known function). For human data, Onto-Express explicitly shows how many genes in a category are supported by experimental evidence (labelled with "experimented") and how many are predicted ("predicted"). Those genes for which it is not known whether they were assigned to the given functional category based on a prediction or experimental evidence are reported as "non-recorded." For other organisms, Onto-Express uses the GO evidence codes described in Section 14.2.3.1.

There is one graph for each of the biochemical function, biological process, cellular role, cellular component and molecular function categories. A specific graph can be requested by choosing the desired category from the pull-down menu and subsequently clicking the "Draw graph" button. Specific items within one functional category can be selected by right-clicking. A composite graph including items from various functional categories (e.g. biological process and cellular component and chromosome location) can be obtained by clicking the button "Draw selected". This is useful when generating publication quality graphs that have to include only a smaller number of highly relevant items.

Clicking on a category displays a hyper-linked list of the genes in that category. The list contains the UniGene cluster IDs uniquely identifying the genes, their Locus ID, the gene name, the accession number and the source of information (e.g. a reference to a publication). Clicking on a specific gene provides more information about that gene.

14.5.3   Some real data analyses

Onto-Express has been applied to a number of publicly available data sets [90, 185, 218]. Two examples from [90] will be presented here.

Example 14.2

A microarray strategy was recently used to identify 231 genes (from an initial set of 25,000) that can be used as a predictor of clinical outcome for breast cancer [270]. Using a classical approach based on putative gene functions and known pathways, van't Veer et al. identified several key mechanisms such as cell cycle, cell invasion, metastasis, angiogenesis and signal transduction as being implicated in cases of breast cancer with poor prognosis. The 231 genes found to be good predictors of poor prognosis were submitted to OE using the initial pool of 24,481 genes as the reference set. We concentrated on those functional categories significant at 5% ($p < 0.05$) and represented by two or more genes, These functional categories are presented in Fig. 14.4. It is interesting to note that Onto-Express' results included most of the biological processes postulated to be associated with cancer including the positive control of cell proliferation and anti-apoptosis. The spectacular aspect is related to the fact that these results have been obtained in a matter of minutes while the original analysis required many months.

Interestingly, oncogenesis, cell cycle control and cell growth and maintenance are not

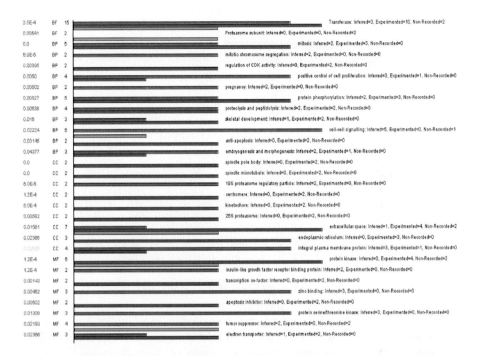

FIGURE 14.4: Significant correlations were observed between the expression level and poor breast cancer outcome for 231 genes [270]. This subset of genes was processed by Onto-Express to categorize the genes into functional groups as follows: BF= biochemical function, BP – biological process, CC – cellular component, MF – molecular function. The figure shows the 30 different functional groups associated with poor disease outcome in a significant way ($p < 0.05$ in left column). Red bar graphs represent genes for which the function was inferred, blue graphs represent genes for which the function was proved experimentally and green graphs represent genes for which this type of information was not recorded in the source database. Reprinted from Drăghici, S., Global functional profiling of gene expression, Genomics, Vol. 81, No. 2, pp. 98–104, 2003. With permission from Elsevier.

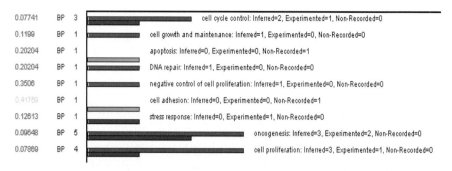

FIGURE 14.5: Some interesting biological processes that are not significant at the 5% significance level. Note that processes commonly associated with cancer such as: cell proliferation, cell cycle control and oncogenesis *are* significant at 10% significance level. Furthermore, the statistical analysis for apoptosis, cell growth and maintenance, etc. should be interpreted cautiously since they are represented by a single gene. Red bar graphs represent genes for which the function was inferred, blue graphs represent genes for which the function was proved experimentally and green graphs represent genes for which this type of information was not recorded in the source database. Reprinted from Drăghici, S., Global functional profiling of gene expression, Genomics, Vol. 81, No. 2, pp. 98–104, 2003. With permission from Elsevier.

significant at 5% but do become significant if the significance threshold is lowered to 10%. Fig. 14.5 shows these and a few other interesting functional categories. Note that the apoptosis, cell growth and maintenance, etc. do not appear to be significant although they are expected to be related to the condition studies. Two observations are required here. Firstly, the results of these categories should be interpreted cautiously because they are represented by a single gene. Secondly, even if they were represented by several genes the lack of significance cannot be interpreted as a proof of insignificance (see Sec. 5.4).

Onto-Express also identified a host of novel mechanisms. Protein phosphorylation was one of these additional categories significantly correlated with poor prognostic outcome. Apart from its involvement in a number of mitogenic response pathways, protein phosphorylation is a common regulatory tactic employed in cell cycle progression. PCTK1 [211] and STK6 [186] are among the cell cycle regulatory kinases identified as corollaries to prognostic outcome. Similarly, anti-apoptotic factors, surivin [124, 196] and BNIP3 [45], were identified. Both mechanisms are believed to be intimately linked and active in regulating cell homeostasis and cell cycle progression.   □

Example 14.3

The second data set used to validate our methods was focused around the link between BRCA1 mutations and tumor suppression in breast cancer. The expression of 373 genes was found to be significantly and consistently altered by BRCA1 induction [279].

FIGURE 14.6: Functional categories significantly ($p < 0.05$) stimulated by BRCA1 overexpression in breast cancer [279]: BF – biochemical function, BP – biological process, CC – cellular component, CR – cellular role, MF – molecular function. Red bar graphs represent genes for which the function was inferred, blue graphs represent genes for which the function was proved experimentally and green graphs represent genes for which this type of information was not recorded in the source database. Reprinted from Drăghici, S., Global functional profiling of gene expression, Genomics, Vol. 81, No. 2, pp. 98–104, 2003. With permission from Elsevier.

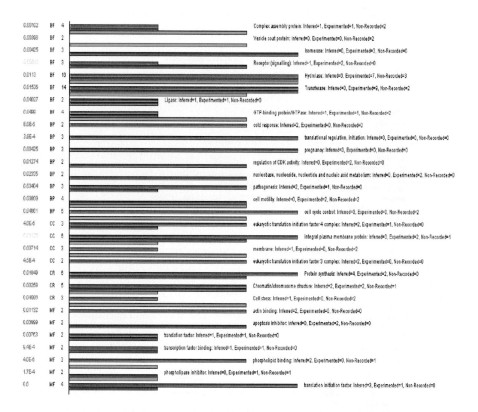

FIGURE 14.7: Functional categories significantly ($p < 0.05$) inhibited by BRCA1 overexpression in breast cancer [279]: BF – biochemical function, BP – biological process, CC – cellular component, CR – cellular role, MF – molecular function. Red bar graphs represent genes for which the function was inferred, blue graphs represent genes for which the function was proved experimentally and green graphs represent genes for which this type of information was not recorded in the source database. Reprinted from Drăghici, S., Global functional profiling of gene expression, Genomics, Vol. 81, No. 2, pp. 98–104, 2003. With permission from Elsevier.

We submitted this set to OE using the genes represented on the HuGeneFL microarray (aka HU6800; Affymetrix, Santa Clara, CA) as the reference set. This array contains approximately 6800 human ESTs. We divided the genes into up-regulated and down-regulated. The functional categories significantly represented in the set of up-regulated genes are stimulated by BRCA1 over-expression (Fig. 14.6). Functional categories significantly represented in the set of down-regulated genes are inhibited by BRCA1 over-expression (Fig. 14.7). Once again, our approach was validated by the fact that the biological processes found to be significantly affected included several processes known to be associated with cancer: mitosis, cell cycle control, and the control of the apoptotic programme.

This analysis also showed that BRCA1 had somewhat of a homeostatic effect on the cells, promoting many cell survival and maintenance pathways (e.g. mRNA processing, splicing, protein modification and folding). BRCA1 is known to be involved in the cell cycle checkpoint control (i.e. acting as a tumor suppressor [271]) and significantly downregulates several genes that normally promote transition through the cell cycle, including CDC2, CDC25B and the c-Ha-ras1 proto-oncogene.

□

14.5.4   Interpretation of the functional analysis results

Several other factors need to be considered when interpreting the result of this type of functional analysis. Firstly, the data analysis methods used to obtain the list of differentially regulated genes have different error rates. Thus, Onto-Express' *input* can contain false positives (genes reported as being differentially regulated when they are not). Since the presence and number of such false positives can influence the results, it is important to take this into consideration when interpreting the results. If the selection of differentially regulated genes has been done with methods that offer an explicit control of the false positive rates (confidence levels or probability of Type I error), one could repeat the analysis selecting genes at various confidence levels and compare the results.

Secondly, if a custom array is purposefully enriched with a certain type of genes, the significance of those specific genes will be artificially lowered. This biological bias has to be also taken into consideration when interpreting OE's results. For instance, an apoptosis array will contain many genes related to apoptosis. Because of this, any such genes found to be regulated may have large p-values and may not appear as significant. In interpreting this, it is useful to remember that a high p-value means that *the data are such that the null hypothesis cannot be rejected* which is different from saying that the null hypothesis is true. We refer the reader to the extensive discussion on this topic in Section 5.4. In other words, having a low p-value is a sufficient, but not necessary, condition for a gene to be interesting. It is very possible for a biological meaningful gene to have a large p-value in certain circumstances (e.g. when using focused arrays).

The exact biological meaning of the calculated p-values depends on the list of genes submitted as input. For example, if the list contains genes that are up-regulated and mitosis appears more often than expected, the conclusion may be that the condition under study stimulates mitosis (or more generally, cell proliferation) in a statistically significant way. If the list contains genes that are down-regulated and mitosis appears more often than expected (exactly as before), then the conclusion may be that the condition significantly inhibits mitosis.

Finally, microarray data are typically obtained from several repeated experiments. If a certain biological process is found to be affected in repeated, independent experiments, it is likely that the process is indeed so, independently of the number of genes representing that process on the array.

14.6   Summary

The ultimate goal of gene expression experiments is to gain biological knowledge. Many databases exist that contain a wealth of information about various genes, proteins and their interactions. It is useful to exploit the availability of such data to interpret the results of the microarray data analysis. In contrast to the approach of looking for key genes of known specific pathways or mechanisms, global functional profiling is a high-throughput approach that can reveal the biological mechanisms involved in a given condition. Onto-Express is a tool that translates *gene expression profiles* showing how various genes are changed in specific conditions into *functional profiles* showing how various functional categories (e.g., cellular functions) are changed in the given conditions. Such profiles are constructed based on public data and Gene Ontology categories and terms. Furthermore, Onto-Express provides information about the statistical significance of each of the pathways and categories used in the profiles allowing the user to distinguish between cellular mechanisms significantly affected and those that could be involved by chance alone.

Chapter 15

Focused microarrays – comparison and selection

If the only tool you have is a hammer, everything starts to look like a nail...

—Unknown

"The secret is to know your customer. Segment your target as tightly as possible. Determine exactly who your customers are, both demographically and psychographically. Match your customer with your medium. Choose only those media that reach your potential customers, and no others. Reaching anyone else is waste."

—Robert Grede, Naked Marketing, the Bare Essentials

15.1   Introduction

Microarrays have been introduced as powerful tools able to screen a large number of genes in an efficient manner.[1] The typical result of a microarray experiment is a number of gene expression profiles, which in turn are used to generate hypotheses, and locate effects on many, perhaps apparently unrelated pathways. This is a typical hypothesis generating experiment. For this purpose, it is best to use comprehensive microarrays that represent as many genes of an organism as possible. Currently, such arrays include tens of thousands of genes. For example, the HGU133 (A+B) set from Affymetrix Inc. contains 44,928 probes that represent 42,676 unique sequences from GenBank database, corresponding to 30,264 UniGene clusters.

Typically, after conducting a microarray experiment, one would select a small number of genes (e.g. 10-50) that are found to be differentially expressed. These genes are analyzed from a functional point of view, either going through online databases manually or by using an automated data mining tool such as Onto-Express [90, 185].

---

[1]Some of the material in this chapter is reprinted with permission from Eaton Publishing from the article: "Assessing the Functional Bias of Commercial Microarrays Using the Onto-Compare Database" by Drăghici et al. published in the BioTechniques supplement "Microarrays and Cancer: Research and Applications", March 2003.

This step identifies the biological processes, molecular functions, biochemical function and gene regulatory pathways impacted in the condition under study and generates specific hypotheses involving them. In many cases, only a small number of such pathways are identified. The next logical step is to focus on such specific pathways.

In many cases, it is desirable to construct a molecular classifier able to diagnose or classify samples into different categories based on their gene expression profiles [15, 35, 127, 249, 224, 256, 262]. This involves a training process that suffers from a "curse of dimensionality" [27]. In short, the curse of dimensionality refers to the fact that the difficulty of building such a classifier increases exponentially with the dimensionality of the problem, i.e. the number of genes involved. Furthermore, constructing a classifier requires many more training examples (i.e. samples or patients) than variables (genes). Both issues strongly suggest that the number of genes used to build the classifier has to be reduced to a minimum. In other words, it is best if the set of genes is restricted to strictly relevant genes.

Therefore, focusing on a smaller number of genes is both: i) the logical step that follows the initial screening experiment that generated the hypotheses as well as ii) a step required if molecular classifiers are to be constructed. Unlike the first step of exploratory search in which hypotheses are generated, the second steps should be a "hypothesis driven experiment" in which directed experiments are performed in order to test a small number of very specific hypotheses. However, specific hypotheses and a small number of pathways may still involve hundreds of genes. This is still too many for RT-PCRs, Western blotting and other gene specific techniques and therefore the microarray technology is still the preferred approach.

Many commercial microarray manufacturers have realized the need for such **focused arrays** and have started to offer many such arrays. For instance, ClonTech currently sells focused human microarrays for the investigation of the cardiovascular system, cell cycle, cell interaction, cytokines/receptors, hematology, neurobiology, oncogenes, stress, toxicology, tumors, etc. Many other companies have picked the same trend and offer focused arrays: Perkin-Elmer, Takara Bio, SuperArray Inc., Sigma Genosys, etc. Literally tens of focused arrays are available on the market with several companies offering customized arrays for the same pathways. Typically, a focused array includes a few hundreds of genes covering the biological mechanism(s) of choice. However, two microarrays produced by different companies are extremely unlikely to use the same set of genes. In consequence, various pathways will be represented to various degrees on different arrays even if the arrays are all designed to investigate the same biological mechanisms. This is an unavoidable **functional bias**. Such a bias will be associated with each and all arrays including less than the full genome of a given organism.

15.2 Criteria for array selection

The general criteria used to select an array usually include several categories of reasons. One such category includes reasons related to the **availability** of a particular array. For instance, one laboratory or core facility may have certain arrays readily available because they have been purchased in a larger lot, or because they remained from previous experiments, etc. Another large category of reasons is related to the **cost**. Even if the cost of the array itself is a relatively small component of the overall cost of the experiment, smaller arrays do tend to cost less and spending more money for a larger array is usually expected to be associated with a benefit. Another set of factors influencing the choice of the array is related to **technological preferences**. For instance, a researcher might prefer cDNA vs. oligonucleotide arrays, filters vs. glass, etc. Finally, the choice might be influenced by **data analysis issues**. For instance, certain normalization techniques, such as dividing by the mean of all genes on a given array, assume most genes do not change. This is probably true for an array containing thousands of genes but may not be true for smaller arrays. Furthermore, data analysis may be easier and more reliable if fewer genes are present on the array. Considerably fewer genes means fewer difficulties related to correction for multiple experiments in statistical hypothesis testing, much less computation in model fitting (e.g. expectation maximization approaches), less of a curse of dimensionality in the construction of a classifier for diagnosis purposes and in general a better ratio between number of dimensions (number of genes) and number of data points (mRNA samples).

The interplay of these factors eventually decides the choice of the array when, in fact, the choice of the array should be made primarily based on the scientific question at hand. If array A contains 10,000 genes but only 80 are related to a given pathway and array B contains only 400 genes but 200 of them are related to the pathway of interest, the experiment may provide more information if performed with array B instead of A. Furthermore, using a smaller array can also translate into significant cost savings. Since an array can contain thousands of probes and a typical user would compare several arrays at once, annotating the probes and comparing the arrays becomes anal task. Onto-Compare is a to

Onto-Compare is a tool that allows a researcher to perform this task in a fast and convenient way using terms from the Gene Ontology (GO) [16, 17].

15.3 Onto-Compare

In order to assess the biological bias of various arrays, we designed and implemented a custom database. This Onto-Compare (OC) database was populated with data collected from several online databases, as well as lists of genes (GenBank accession numbers) for each microarray as provided by their manufacturers. From the list of accession numbers, a list of unique UniGene cluster identifiers is prepared for each microarray and then a list of LocusLink identifiers is created for each microarray in our database. UniGene is a system for automatically partitioning GenBank mRNA sequences and ESTs into a non-redundant set of gene-oriented clusters with identical $3'$ untranslated regions ($3'$ UTRs)[243]. LocusLink[223] is a database of official gene names and other gene identifiers. Each locus in the LocusLink database is annotated using ontologies from the Gene Ontology Consortium[156] and ontologies from other researchers and companies. Gene Ontology Consortium provides ontologies for biological process, molecular function and cellular component. The data from these databases and gene lists are parsed and entered into our Onto-Compare relational database. The OC database is implemented in Oracle using a schema designed to allow for efficient querying. A group of Java programs and Perl routines are used to download, parse and enter the data into the OC database as well as to update the database on a regular basis[2] with minimal human intervention.

After creating a list of locus identifiers for each array, the list is used to generate the following profiles: biochemical functions, biological process, cellular role, cellular component and molecular function. The profile of each microarray is stored in the database. The list of genes deposited on a microarray is static, as long as the manufacturer maintains the array in production, but the annotations for those genes keep changing and are updated automatically. A Java program is used to facilitate the update of the database, which recreates these profiles as more information becomes available.

Since an array can contain thousands of probes and a typical user would compare several arrays at once, annotating the probes becomes a daunting computational task. For this reason, we precalculate the functional annotations for each array and store the results in the database. Thus, this very time consuming computation is only done after database updates. During user interaction, the data are merely queried from the database. However, a researcher might choose to use a set of arrays (e.g. the Affymetrix HGU133 is actually a set of two arrays). In order to accommodate for this, we allow the user to merge arrays and calculate functional profiles of the sets of arrays. Since a user can merge an arbitrary number of arrays, we cannot realistically precalculate the functional annotation for every possible union of arrays. Thus, this computation is done every time a user merges arrays.

Onto-Compare runs as a Java applet in a web browser on a user computer. The input screen is shown in Fig. 15.1. The results are presented as a table, in which the first column displays an ontology term and the rest of the column corresponds to one of the arrays from the set of selected arrays (see Fig. 15.2). The first row of the table represents the total number of unique GenBank sequences represented on each of the selected arrays. For each ontology term, Onto-Compare displays the total number of unique GenBank sequences found on each of the selected arrays. Onto-Compare also displays the total number of unique UniGene clusters found on each of the selected

---

[2]Currently this is done every two weeks.

FIGURE 15.1: The input screen of Onto-Compare. The arrays are organized by manufacturer and organism. The user can select any number of arrays to be compared. Reprinted from Drăghici, S. et al., Microarrays and cancer: research and applications, Biotechniques, p. 55–61, 2003. With permission from Eaton Publishing.

arrays for each ontology term in square brackets. Clicking an array name sorts the entire table by the total number of sequences for that array. Clicking on a value for total number of sequences displays accession numbers of the sequences, their corresponding cluster id and locus id along with the cluster's official gene symbol. Selecting check boxes and then clicking "Show selected functions" only displays the selected terms. The user can merge two or more arrays, by selecting the check boxes next to the array name and clicking the "Merge selected arrays" button.

15.4   Some comparisons

In order to illustrate the utility of Onto-Compare, we can consider the example of an anti-cancer drug candidate that inhibits *bcl2* which is an anti-apoptotic factor. The drug candidate was obtained from a large scale screening so the exact mechanism through which this drug promotes apoptosis is not yet known. We would like to use a microarray approach to study the interaction between the various genes and their respective proteins on the apoptotic pathway as this pathway is affected by the drug. This is a typical example of a hypothesis driven research. We have strong reasons to believe that a certain pathway or cellular mechanism is involved and we would like to focus our experiments on it.

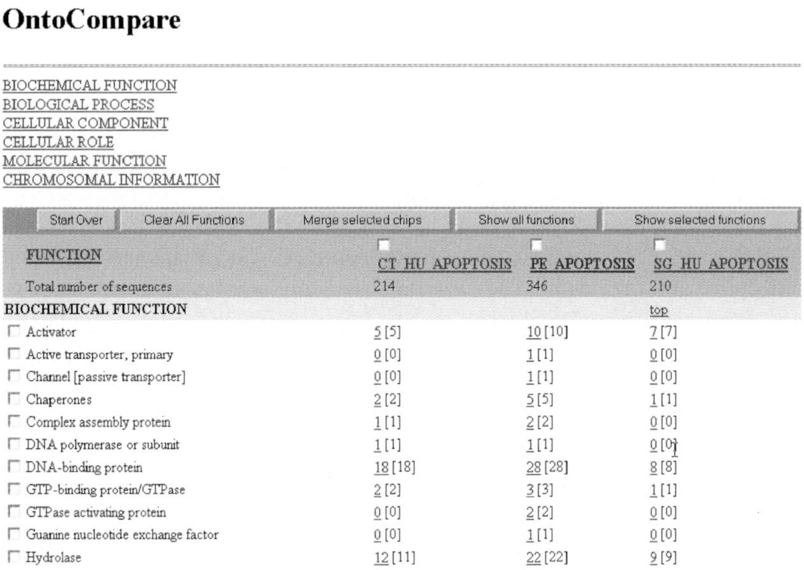

OntoCompare

BIOCHEMICAL FUNCTION
BIOLOGICAL PROCESS
CELLULAR COMPONENT
CELLULAR ROLE
MOLECULAR FUNCTION
CHROMOSOMAL INFORMATION

Start Over	Clear All Functions	Merge selected chips	Show all functions	Show selected functions

FUNCTION	CT HU APOPTOSIS	PE APOPTOSIS	SG HU APOPTOSIS
Total number of sequences	214	346	210
BIOCHEMICAL FUNCTION			top
☐ Activator	5 [5]	10 [10]	7 [7]
☐ Active transporter, primary	0 [0]	1 [1]	0 [0]
☐ Channel [passive transporter]	0 [0]	1 [1]	0 [0]
☐ Chaperones	2 [2]	5 [5]	1 [1]
☐ Complex assembly protein	1 [1]	2 [2]	0 [0]
☐ DNA polymerase or subunit	1 [1]	1 [1]	0 [0]
☐ DNA-binding protein	18 [18]	28 [28]	8 [8]
☐ GTP-binding protein/GTPase	2 [2]	3 [3]	1 [1]
☐ GTPase activating protein	0 [0]	2 [2]	0 [0]
☐ Guanine nucleotide exchange factor	0 [0]	1 [1]	0 [0]
☐ Hydrolase	12 [11]	22 [22]	9 [9]

FIGURE 15.2: A sample output screen of Onto-Compare. The functional categories analyzed are: biochemical function, biological process, cellular component, cellular role, molecular function and chromosomal location. The user can further merge arrays (e.g. for HG133A and B, not shown here), or focus the analysis on a subset of functional categories. Reprinted from Drăghici, S. et al., Microarrays and cancer: research and applications, Biotechniques, p. 55–61, 2003. With permission from Eaton Publishing.

There are several companies, such as ClonTech, Sigma-Genosys and Perkin-Elmer, that provide arrays designed for research involving apoptosis. The ClonTech human apoptosis array contains probes for 206 UniGene clusters; the Perkin-Elmer apoptosis microarray contains probes for 322 UniGene clusters and the Sigma-Genosys human apoptosis microarray contains probes for 200 UniGene clusters. There are 74 clusters common between all arrays. There are 167 clusters common between ClonTech and Perkin-Elmer array, 92 clusters common between Perkin-Elmer and Sigma-Genosys array and 82 clusters common between Clontech and Sigma-Genosys array. Comparison of these three apoptosis specific arrays using Onto-Compare is shown in Table 15.1. A biological process such as induction of apoptosis and a molecular function such as caspases are clearly relevant to the apoptosis [140, 298, 62]. Indeed, the table shows that all three arrays have similar numbers of genes representing induction of apoptosis, caspases and tumor necrosis factor receptors. Various interleukins are also reported to be mechanistically associated with apoptosis at both protein and gene levels [13, 147, 242]. However, neither Clontech nor PerkinElmer microarray contains any interleukin related gene. On the other hand, the Sigma-Genosys microarray contains 14 genes related to various interleukins. Clearly, among the 3 arrays considered, the Sigma-Genosys would be a better choice for testing any hypothesis involving the role of interleukins in apoptosis. Other processes such as immune response, cell-cell signalling, cell surface receptor linked signal transduction are also better represented on the Sigma-Genosys array. However, it is important to emphasize that the Sigma-Genosys array is not necessarily better than the other two arrays. In fact, since the Sigma-Genosys and the ClonTech arrays have almost the same number of genes, there must exist some functional categories that are represented better on the ClonTech array. Examples include processes such as cell cycle control, oncogenesis and negative control of cell proliferation.

As another example, we compared commercially available oncogene arrays. Such arrays are available from ClonTech (Atlas Select Human Oncogene 7831-1 and Human Oncogene/Tumor Suppressor 7745-1), Perkin-Elmer and SuperArray. The numbers of distinct sequences available on these arrays are 514, 199, 335, 26, respectively. A comparison between these arrays is shown in Table 15.2. It is interesting to note that the PE array contains 73 sequences (71 UniGene clusters) related to oncogenesis. At the same time, the ClonTech Oncogene array contains only 22 sequences (21 UniGene clusters) representing the same biological process. The results are remarkable in light of the fact that that PE array contains only 335 genes compared to the ClonTech Oncogene array which contains 514 genes. The same holds true for protein phosphorylation represented by 19 genes on the PE array and 6 genes on the ClonTech Oncogene array and negative control of cell proliferation represented by 15 genes on the PE array and only 6 genes on the ClonTech Oncogene array.

Finally, we performed another analysis in which we contemplated the use of both ClonTech Oncogene and ClonTech Tumor suppressor as a two-array solution and compared this ClonTech selection of genes with the genes used on the unique PE Oncogene/Tumor suppressor array. Note that when sets of arrays are used, many genes may be present on more than one array. Thus, the coverage of a given pathway or biological process cannot be inferred by simply summing the number of genes

Ontology Term	ClonTech	PerkinElmer	Sigma-Genosys
Total genes on array	214	346	210
induction of apoptosis	17 [17]	28 [27]	24 [24]
anti-apoptosis	15 [15]	20[20]	23[23]
immune response	0 [0]	1 [1]	19 [19]
cell-cell signaling	9 [9]	9 [9]	18 [18]
cell surface receptor linked signal transduction	5 [5]	9 [9]	17 [17]
oncogenesis	23 [23]	29 [29]	16 [16]
cell cycle control	30 [30]	31 [31]	12 [12]
positive control of cell proliferation	5 [5]	5 [5]	12 [12]
negative control of cell proliferation	16 [16]	20 [20]	10 [10]
induction of apoptosis by DNA damage	2 [2]	3 [3]	3 [3]
induction of apoptosis by extracellular signals	8 [7]	12 [12]	7 [7]
induction of apoptosis by hormones	1 [1]	1 [1]	1 [1]
induction of apoptosis by intracellular signals	2 [2]	2 [2]	2 [2]
induction of apoptosis by oxidative stress	0 [0]	0 [0]	1 [1]
induction of apoptosis via death domain receptors	4 [4]	5 [5]	7 [7]
caspase	1 [1]	3 [3]	2 [2]
caspase activator	1 [1]	3 [3]	2 [2]
caspase-1	1 [1]	1 [1]	1 [1]
caspase-10	1 [1]	2 [2]	1 [1]
caspase-2	1 [1]	1 [1]	1 [1]
caspase-3	1 [1]	1 [1]	1 [1]
caspase-4	1 [1]	1 [1]	1 [1]
caspase-8	2 [2]	1 [1]	1 [1]
caspase-9	1 [1]	1 [1]	1 [1]
caspase-activated deoxyribonuclease	1 [1]	2 [2]	2 [2]
tumor necrosis factor receptor	2 [2]	2 [2]	2 [2]
tumor necrosis factor receptor ligand	1 [1]	1 [1]	1 [1]
tumor necrosis factor receptor, type I	1 [1]	1 [1]	1 [1]
interleukin receptor	0 [0]	0 [0]	2 [2]
interleukin-1 receptor	0 [0]	0 [0]	2 [2]
interleukin-1, Type I, activating receptor	0 [0]	0 [0]	1 [1]
interleukin-10 receptor	0 [0]	0 [0]	2 [2]
interleukin-12 receptor ligand	0 [0]	0 [0]	2 [2]
interleukin-2 receptor	0 [0]	0 [0]	3 [3]
interleukin-2 receptor ligand	0 [0]	0 [0]	1 [1]
interleukin-4 receptor	0 [0]	0 [0]	2 [2]
interleukin-4 receptor ligand	0 [0]	0 [0]	1 [1]
interleukin-7 receptor	0 [0]	0 [0]	1 [1]
Total distinct genes in categories above	99 [98]	133 [132]	129 [129]

TABLE 15.1:   A comparison of three apoptosis specific microarrays: ClonTech human apoptosis, PerkinElmer apoptosis and Sigma-Genosys human apoptosis. Biological processes such as induction of apoptosis and molecular functions such as caspases and tumor necrosis are almost equally represented on each of the chips, but none of the interleukins are represented on ClonTech as well as PerkinElmer microarray. Processes such as immune response, cell-cell signalling, cell surface receptor linked signal transduction are better represented on the Sigma-Genosys array. Processes such as cell cycle control, oncogenesis and negative control of cell proliferation are better represented on the ClonTech array. The numbers represent sequences present on the arrays; the numbers in brackets represent distinct UniGene clusters. Reprinted from Drăghici, S. et al., Microarrays and cancer: research and applications, Biotechniques, p. 55–61, 2003. With permission from Eaton Publishing.

Biological Process	ClonTech Oncogene	ClonTech Tumor Suppressor	PE Onco-gene Tumor Suppressor	SuperArray Oncogene
Total genes on array	514	199	335	26
oncogenesis	22 [21]	40 [40]	73 [71]	8 [8]
signal transduction	46 [46]	39 [39]	54 [54]	6 [6]
cell proliferation	23 [23]	19 [19]	24 [24]	8 [8]
protein phosphorylation	6 [6]	17 [17]	19 [19]	5 [5]
cell cycle control	20 [20]	31 [31]	19 [19]	1 [1]
negative control of cell proliferation	6 [6]	14 [14]	16 [15]	0 [0]
cell-cell signalling	10 [9]	7 [7]	13 [13]	0 [0]
Total genes in categories above	91 [90]	109 [109]	144 [142]	15 [15]

TABLE 15.2: A comparison of the ClonTech (Atlas Select Human Oncogene 7831-1 and Human Oncogene/Tumor Suppressor 7745-1), Perkin-Elmer and Super-Array (Human Cancer/Oncogene) arrays. PE is a better choice for conditions potentially involving oncogenesis, protein phosphorylation and negative control of cell proliferation. The numbers represent sequences present on the arrays; the numbers in brackets represent distinct UniGene clusters. Reprinted from Drăghici, S. et al., Microarrays and cancer: research and applications, Biotechniques, p. 55–61, 2003. With permission from Eaton Publishing.

covering the given pathway on each array. Table 15.3 shows the comparison between the set of two ClonTech arrays and the PE array. When the two ClonTech arrays are used together, there is a good representation of general signal transduction, protein phosphorylation and negative control of cell proliferation. Remarkably, oncogenesis is still better represented on the PE array in spite of the fact that the set of two ClonTech arrays deploy 676 genes while the unique PE array uses only 335 genes.

The examples above show that each array or set of arrays has a certain biological bias. These examples should not be interpreted as proving that one particular array is better or worse than other similar arrays. However, these examples do show that in those situations in which a certain hypothesis exist, the choice of the array must be made based on a comprehensive functional analysis of the biological processes, biochemical functions, cellular components and chromosomal locations of the genes represented on each of the arrays considered.

15.5 Summary

The first step of an experiment involving microarrays will probably be of an exploratory nature aimed at generating hypotheses. Comprehensive arrays including as many genes as possible are useful at this stage. In most cases, the hypothesis generating phase will be followed by successive steps of focused research. Such hypothesis

Biological Process	ClonTech Oncogene and Tumor Suppressor	PE Oncogene Tumor Suppressor	SuperArray Oncogene
Total number of sequences	676	335	26
signal transduction	81 [81]	54 [54]	6 [6]
oncogenesis	55 [54]	73 [71]	8 [8]
cell cycle control	43 [43]	19 [19]	1 [1]
cell proliferation	38 [38]	24 [24]	8 [8]
developmental processes	21 [21]	13 [13]	3 [3]
protein phosphorylation	21 [21]	19 [19]	5 [5]
negative control of cell proliferation	19 [19]	16 [15]	0 [0]
cell-cell signalling	17 [16]	13 [13]	0 [0]
signal transduction	10 [10]	6 [6]	0 [0]

TABLE 15.3:   A comparison of the ClonTech set of arrays (Atlas Select Human Oncogene 7831-1 and Human Oncogene/Tumor Suppressor 7745-1) with the Perkin-Elmer and SuperArray (Human Cancer/Oncogene) arrays. When the two ClonTech arrays are used together, there is a good representation of general signal transduction, protein phosphorylation and negative control of cell proliferation. However, oncogenesis is still better represented on the PE array. The numbers represent sequences present on the arrays; the numbers in brackets represent distinct UniGene clusters. Reprinted from Drăghici, S. et al., Microarrays and cancer: research and applications, Biotechniques, p. 55–61, 2003. With permission from Eaton Publishing.

driven research often concentrates on a few biological mechanisms and pathways. However, even a single biological mechanism may still involve hundreds of genes which may make the microarray approach the preferred tactic. If microarrays are to be involved in any follow-up, focused, hypothesis driven research, we argue that one should use the array(s) that best represent the corresponding pathways. Functional analysis can suggest the best array or set of arrays to be used to test a given hypothesis. This can be accomplished by analyzing the list of genes on all existing arrays and providing information about the pathways, biological mechanisms and molecular functions represented by the genes on each array. Onto-Compare is a tool that allows such comparisons. This tool is available at: http://vortex.cs.wayne.edu/Projects.html.

Chapter 16

Commercial applications

16.1 Introduction

This book was intended, designed and written as an unbiased technical text, independent of any particular software package or vendor. The methods discussed are very general and they can be implemented in many ways. The main goal of the book was to describe problems, algorithms, techniques and approaches as opposed to software packages. However, most readers would not want to re-invent the wheel and will probably use a commercial package. It is important to make a connection between the abstract algorithms and techniques discussed and their implementation in the mainstream software currently available. The aim of this chapter is to showcase some illustrative commercial packages.

The commercial packages can be roughly divided into several categories as follows:

- **Enterprise level packages**. These are typically very expensive software suites that are designed to cover many if not all the processes of microarray design, production, hybridization, data analysis and data storage. Examples include Rosetta Resolver (Rosetta Biosoftware), Gene Director (BioDiscovery), Gene Traffic (Iobion), Silicon Genetics Enterprise Solution (Silicon Genetics), Array Scout (Lion Bioscience AG), etc. These packages are usually very large, very often include a relational database, and have to accomplish a large number of very heterogeneous tasks. As such, they go well beyond the scope of this book which is focused on data analysis.

- **"Point and click" analysis tools.** These tend to be rather sophisticated products, highly tailored and customized for the specific use in the analysis of microarray data. In most cases, such packages will include several tools designed to help with the analysis of the biological aspects of the data (e.g. chromosome and pathways viewers, analysis of biological function, etc.). Examples might include GeneSight (BioDiscovery), GeneSpring (Silicon Genetics), Array Analyzer (Insightful), dChip (Harvard Univ.), ArrayStat (Imaging Research), AMIADA (Hong Kong University), Expression Profiler (EBI), TIGR Array Viewer and Multiple Experiment Viewer (TIGR), XCluster (Stanford Univ.), etc. In many cases, such packages are included as the data analysis engine in the enterprise solutions offered by their respective companies (e.g.

393

Silicon Genetics' GeneSpring is part of their GeNet enterprise suite, BioDiscovery's GeneSight is a module in their enterprise solution GeneDirector, etc.).

- **Analysis/visualization environments.** Usually, they are general purpose statistical or numerical processing packages that allow the user to write scripts describing the processing to be done in a language resembling a programming language. This language usually includes operators for higher level analysis functions and thus allows a compact description of rather sophisticated analysis sequences. Examples include Excel, S-PLUS, SAS, MATLAB, etc. Depending on the environment, third party packages or analysis routines may be available. For instance, SAM [268] is implemented as an Excel add on (Tibshirani, Stanford) and various ANOVA methods [180, 182, 183] are available as a collection of MATLAB routines (Churchill, Jackson Laboratories). In many cases, the company that developed the original, general purpose package also made available a more customized add-on package for microarray data analysis. A typical example would be Spotfire Inc. whose first offering (Spotfire) was a general visualization package targeted at a very wide market. More recently, the company added a number of customized packages (e.g. Decision Site) that also include some analysis tools useful for the analysis of microarray data.

The advantage of a "point and click" package is that the user is not expected to have a lot of mathematical expertise. Note that a basic understanding of the various tools, their characteristics and limitations is still crucial to a successful analysis (hence this book). However, this knowledge is only needed to guide the user through the analysis, without the need to control the minute details of every step from every procedure. The price paid for this is a fixed functionality and a relative rigidity. The software will only do the types of processing envisaged by its designers. In most cases, it is difficult, if not impossible, to customize the analysis by using a completely novel normalization technique, for instance.

At the other extreme, the analysis environments offer extreme flexibility and limitless functionality. Any procedure or analysis technique can be implemented and changed at any time. The downside is, of course, that expertise in numerical and statistical methods, as well as good programming skills, are needed in order to take advantage of this flexibility and power.

An excellent compromise that has started to become available more recently is to have a "point and click" tool running on top of an open analysis environment. This combines the advantages of the two levels: the life scientist can perform the analysis using high level tools without worrying about the details, while the statistician or computer scientist can dive in, program away, and customize to their heart's desire. It is interesting to note that companies tend to converge to this compromise from both directions. Formerly closed software packages are currently enriched with application programming interfaces (APIs) that effectively give the user the possibility to implement custom routines. At the same time, analysis environments are enriched with modules and GUIs that group together the most often used analysis tools, thus adding some point and click capabilities.

This chapter will showcase a few packages, from each of the above categories. In essence, we contacted several companies offering widely used software for the analysis of microarrays and asked them to provide a brief description that would showcase their solution. We suggested taking a public data set and using it to illustrate the main features of their packages. However, we also accepted a shorter description simply discussing the most prominent features in the respective packages. The descriptions received from the companies were included here with a minimum of editing aimed mainly at obtaining a formatting consistent with the rest of the book. As such, *the opinions expressed in these sections are those of the respective authors.*

GeneSight 5.0 (BioDiscovery Inc., Los Angeles, CA) is a very comprehensive and sophisticated "point and click" tool. S-PLUS (Insightful Inc., Seattle, WA) is a statistical analysis environment. Array Analyzer is the higher level package running on top of the S-PLUS environment. SAS has been the brand name in statistical analyis packages for a number of years. Recently, the company has released a number of tools for the data analysis of microarray data. Finally, Spotfire, a brand name in data visualization, has recently released its DecisionSite platform for functional genomics. The following sections are aimed at showcasing the features of these products and, at the same time, illustrate the flow of analysis on some real world data sets.

The GeneSight and ArrayAnalyzer packages are included with this book. The reader is invited to use them and actually perform the analyzes described in the examples that follow, as well as any of the analyzes described in earlier chapters. Both packages include full functionality and a number of other real world data sets.

16.2   Significance testing among groups using GeneSight

Dr. Bruce Hoff, Director of Analytical Sciences, BioDiscovery Inc.

16.2.1   Problem description

A common application of microarray technology is comparing gene expression patterns among groups of individuals with a particular phenotype or genotype. The goals of these experiments are typically either to 1) identify patterns of expression suitable for discriminating different phenotypes/genotypes; 2) identify potential molecular targets for therapies when the phenotypes are diseases; or 3) elucidate the underlying mechanistic connections between genotype and phenotype. An example of this application is the work of Hedenfalk and coworkers [135] who surveyed gene expression in breast tumors obtained from individuals with the BRCA1 or BRCA2 mutations along with tumors from individuals who were wild-type at those loci. With these tumors, histopathological and molecular evidence relating to estrogen and progesterone receptor status suggested that tumors originating in individuals with BRCA1 mutations are molecularly distinct from those in individuals with BRCA2 mutations. The goal of the work was to test this hypothesis.

In this application, GeneSight is used to analyze the data of Hedenfalk et al. with the goal of developing a list of genes whose expression differ significantly among the three types of tumors analyzed. This list is then used to cluster the groups and visually compare that result to a clustering based on the list of genes identified by the authors.

16.2.2 Experiment design

Hedenfalk et al. analyzed expression levels in a total of 22 tumors obtained from seven individuals with BRCA1 mutations (one tumor from each individual), seven individuals with BRCA2 mutations (two tumors from one individual and one tumor each from the remaining six individuals), and seven individuals without mutated alleles at either locus. The authors used spotted cDNA arrays containing 6512 sequences representing 5361 unique genes. Tumor mRNA was isolated and reversed transcribed into cDNA labelled with Cy5. A reference mRNA was obtained from MCF-10 cells and reverse transcribed and labelled with Cy3. Slides were hybridized with the Cy5-labelled tumor samples and the Cy3-labelled reference sample and fluorescence intensity measurements for the two fluorochromes obtained with a scanning confocal microscope. The data file that the authors provide on their web site contains the expression ratios for 3226 genes (spots). This set of 3226 genes was selected based on an average intensity of greater than 2500 pixels among all samples, an average spot area of greater than 40 pixels, and no more than one spot having an area of zero pixels. The majority of the clone IDs in the data set are unique; there are 160 clone IDs with two sets of measurements and five clone IDs with triplicate measurements. Included with the ratios are textual annotations for essentially all of the clone IDs. The tab-delimited text file contains 25 columns (22 columns of data and 3 columns containing the IMAGE Clone ID number, plate position, and annotations). This file was downloaded from: http:// research.nhgri.nih.gov/ microarray/ NEJM_Supplement/ Images/ nejm_brca_release.txt. The first two rows were deleted to leave the experiment ID row to label the columns. The modified data file was named nejm_brca_release_mod.txt. The set of genes the authors believe are most significantly altered among the three classes of tumors is also available on their web site as a PDF file. That file was downloaded and used to construct a text file suitable for defining those genes as a separate group in GeneSight.

16.2.3 Data analysis

16.2.3.1 Creating the data set

We imported the data file described above into the GeneSight data analysis software. The file is not in a standard format (i.e. that created by one of the common microarray image analysis systems) but GeneSight allows the user to read from any custom file so

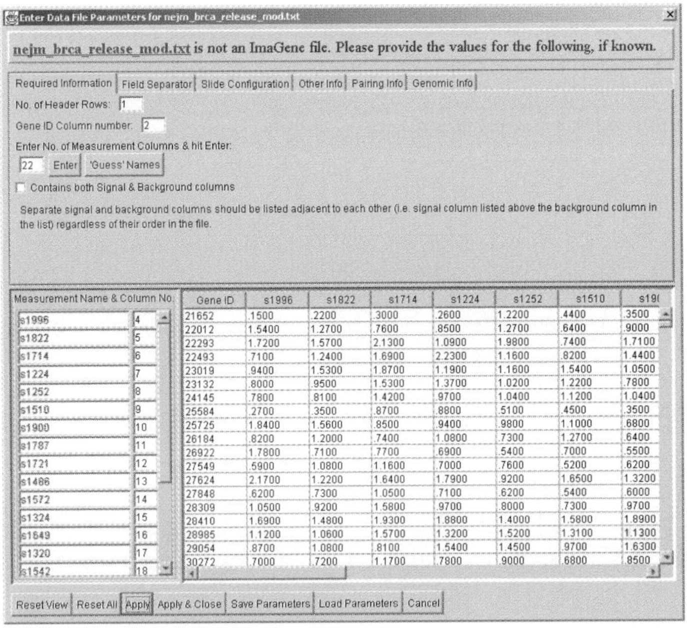

FIGURE 16.1: Data set builder window.

this was not a problem. We specify which column contains the gene IDs (the IMAGE Clone ID column), that contains annotation information (column 3) which contain the data (columns 4 through 25). Fig. 16.1 shows the Data set builder window in which this information is specified.

16.2.3.2 Data preprocessing and normalization

GeneSight allows a number of data preprocessing steps to be applied to raw expression data including background correction, merging of replicates, normalization across channels and arrays and a number of other operations. This is done by graphically building a pipeline of numerical operations, and observing the results in a spreadsheet style table of data values. The graphical pipeline offers a good compromise between flexibility and power. Each icon represents a normalization procedure whose parameters can be specified when the icon is placed in the pipeline. The user is allowed to combine the icons in an arbitrary order which translates into an "on-the-fly" creation of user specified custom pre-processing and normalization procedures.

In this case, the data have already been normalized, and the Cy5/Cy3 ratios calculated, so we only need to apply a log transform to the ratio values. The result is a unimodal, two-tailed distribution of differential expression values. The data preparation window is shown in Fig. 16.2.

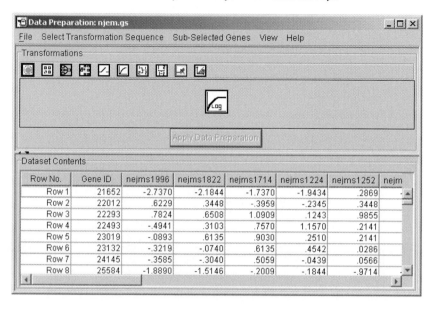

FIGURE 16.2:   Data preparation window.

16.2.3.3   Creating groups of conditions

GeneSight allows the user to define categories of genes and/or experimental conditions. There is a construct called a partition, which is a list of groups (categories) of elements. Within a partition each group has a unique name and color code. The group colors can be used to color-code the graphic elements in various plots. We create a "Mutation" partition having three groups, called "BRCA1", "BRCA2" and "Sporadic", and classify the experiments accordingly. For example, the data file indicates that experiments s1224, s1252, s1510, s1714, s1822, s1905, and s1996 are of type "BRCA1," thus we add those experiments to the "BRCA1" group. Fig. 16.3 shows the Partition Editor windows with this partition and its groups.

16.2.3.4   Creating Hedenfalk's gene list

We exploit GeneSight's ability to create gene partitions as well as condition partitions, and import the list of genes identified by Hedenfalk et al. which differentiate tumor types (see Fig. 16.4). This list will be used at the end of our analysis to compare the list of genes found by us with the list of genes found by the authors of the original paper.

Having imported and pre-processed the expression data, and defined the relevant groups of experiments and genes, we may proceed with the expression analysis.

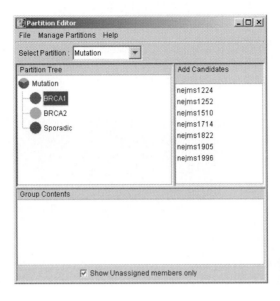

FIGURE 16.3: Mutation condition partition.

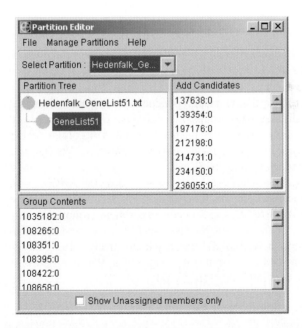

FIGURE 16.4: Hedenfalk gene list.

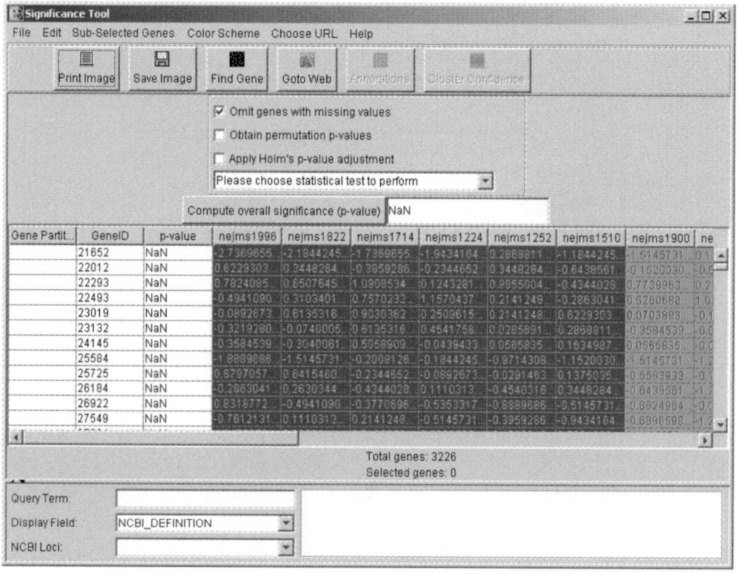

FIGURE 16.5:   The Significance Analysis Tool.

16.2.3.5   Finding differentially regulated genes using the Significance Analyzer

GeneSight's Significance Analyzer implements a number of statistical tests including t-test (for 2 categories), Mann-Whitney test (for 2 categories, short samples), ANOVA (more than 2 categories) and Kruskal-Wallis (more than 2 categories, short samples). The results of each test will include a p-value for each gene indicating the significance with which it differentiates groups of conditions. The appearance is that of a spread sheet, with a column for gene IDs and p-values, and with columns of expression values color-coded according to the chosen condition partition. We choose the "Mutation" partition defined above (Fig. 16.5).

Since in this problem there are 3 groups, we choose the ANOVA test. The ANOVA uses the f-statistic to compute, for each gene, the significance with which it differentiates the BRCA1, BRCA2 and Sporadic experiment groups. Clicking on the header of the "p-value" column sorts the rows such that the most significant row (lowest p-value) comes first (Fig. 16.6). We may interactively select genes with $p < 0.001$ which are the top 48 genes in this list (Fig. 16.7). We save the list of 48 genes in a new partition called "pLT.001" ($p < 0.001$).

16.2.3.6   Exporting the gene list

GeneSight has a report generation tool, which creates a customizable text file describing the incorporated data, the data preparation protocol, the generated partitions, and the output of the analytical tools used. Fig. 16.8 shows an excerpt of the

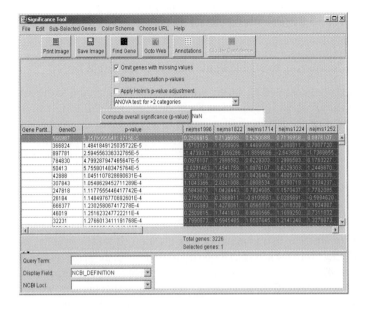

FIGURE 16.6:   Significance analysis results.

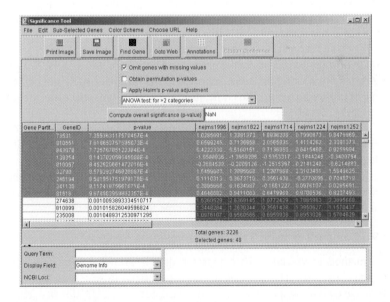

FIGURE 16.7:   Selection of significant genes.

report generated in this session, including the list of significant genes (indicated by the IMAGE clone ID from the input data file), their annotations, and their calculated p-values.

16.2.3.7 Comparing the GeneSight list to the Hedenfalk list

The list of genes generated by the significance analysis is similar to the list generated by the Hedenfalk et al. [135]. GeneSight provides the ability to compare lists of genes, computing the intersection or union of any number of groups. In the figure below we select the Hedenfalk list and the "pLT.001" list and press "Intersection", to see that 43 of the 48 genes in the "pLT.001" list are in common with the Hedenfalk list Fig. 16.9 shows the main GUI of the application including the two lists of genes (bottom left) and the result of their intersection (bottom right).

16.2.3.8 Comparison using hierarchical clustering

The utility of the gene lists is in tumor classification. If done correctly, the small selection of genes from the original 3000 will provide a "fingerprint" or "profile" for the type of cancer each patient has. Below we verify that such a genetic profile has been created, by using two different clustering tools to show that the selected genes cleanly identify and group the breast cancer types. First, we perform unsupervised hierarchical clustering of the 23 tissue samples using all 3000 genes. The result is shown in Fig. 16.10 in a standard graphic form in which each matrix row is one gene, each column is one sample, a color-coded heat map shows relative differential expression for each gene in each sample, and a dendrogram shows the clustering. Further, we color coded the leaves of the dendrogram according to the cancer type (blue=BRCA1, green=BRCA2, red=Sporadic). Clearly there is no clean differentiation of sample types in the clustering result obtained using all genes.

GeneSight allows us to sub-select the genes used by its analytical tools. We perform such a sub-selection using the genes reported in the original paper, and perform the same type of clustering again. This time, the three sample types are cleanly separated as shown in Fig. 16.11.

Finally, we sub-select the list of 48 genes from the significance analysis, and compute the hierarchical clustering. The resulted dendrogram shown in Fig. 16.12 exhibits the same clean separation of the three sample types.

If we performed a microarray analysis on a tissue sample from a new patient, the gene expression signature generated by the selected genes would cluster it along with one of the three categories, predicting the cancer type for the new patient.

16.2.3.9 Comparison using Principal Component Analysis

Another method to visualize the expression data is principal component analysis (PCA) (see Chap. 10). Here, a scatter plot is generated in which each point corresponds to one patient sample. Conceptually, the data form a 3000 dimensional scatter plot in which each point is represented by a vector of 3000 expression values. The goal is to project the points into a two-dimensional plane which can be displayed in

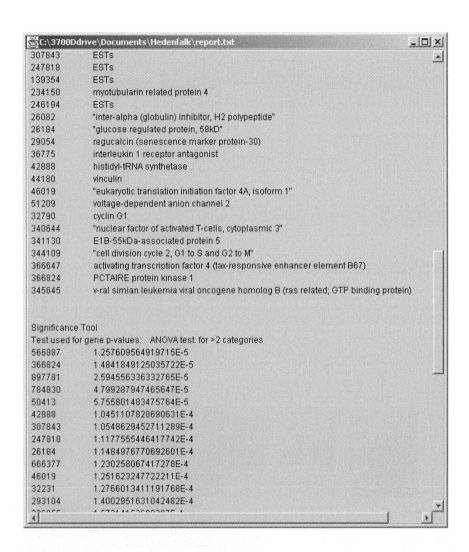

FIGURE 16.8:   GeneSight report.

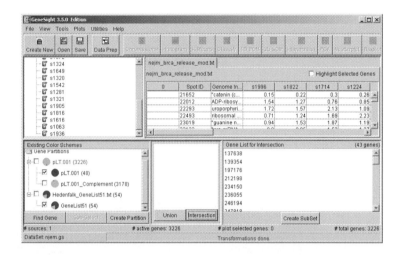

FIGURE 16.9: GeneSight main window, showing comparison of gene lists.

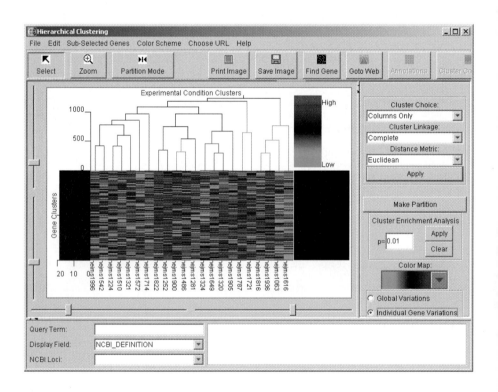

FIGURE 16.10: Hierarchical clustering result using all genes.

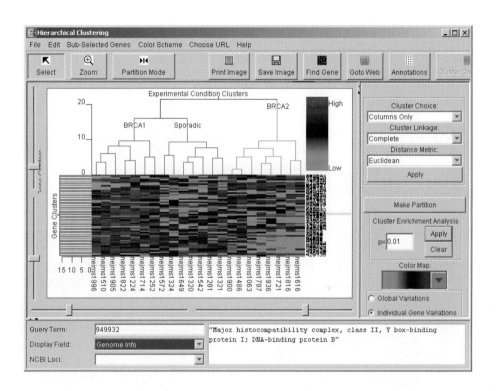

FIGURE 16.11: Hierarchical clustering result using the genes reported in the original paper.

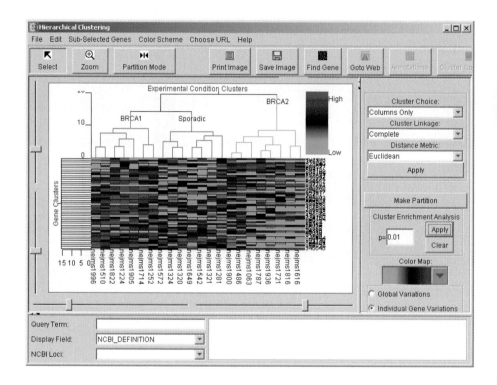

FIGURE 16.12: Hierarchical clustering result using gene list from significance analysis.

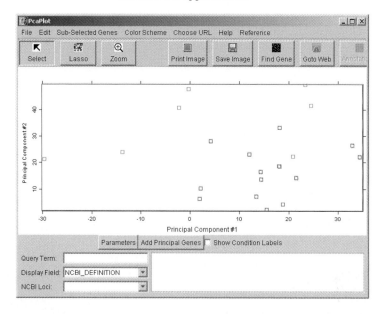

FIGURE 16.13: Principal component analysis using all genes.

such a way that the loss of information due to the dimensionality reduction is minimal. As discussed in Chap. 10, PCA can be used to achieve this optimal projection. The result of PCA using all genes is shown in Fig. 16.13. As before, the points are color-coded according to the cancer category. Clearly the three types of cancer are intermingled when all genes are used.

Fig. 16.14 shows the results after sub-selecting just the gene list from the original paper [135]. Clearly the categories are now cleanly separated. Finally, PCA is performed after sub-selecting the list of 48 genes from the significance analysis. Fig. 16.15 shows the samples projected in the space of the first two principal components. Once more, the delineation of the cancer types is clear. The two-dimensional scatter plot is quite similar to that generated using the list of genes from the original paper.

Again, if microarray expression measurements for an unclassified patient were introduced, adding the additional point to the PCA scatter plot would probably allow us to classify that patient into one of the existing categories.

16.2.4   Conclusion

Significance analysis is a powerful tool for selecting genetic markers for cancer types. The clustering and PCA algorithms and visualization techniques shown here are powerful tools allowing the researcher to quickly and intuitively examine the results of said marker identification. The comparison between GeneSight's signif-

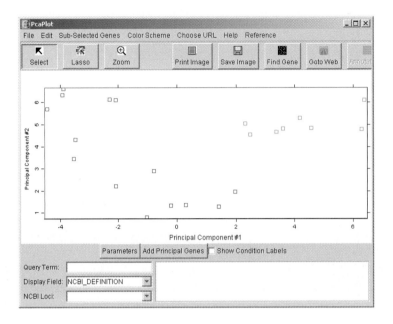

FIGURE 16.14: Principal component analysis using Hedenfalk's gene list.

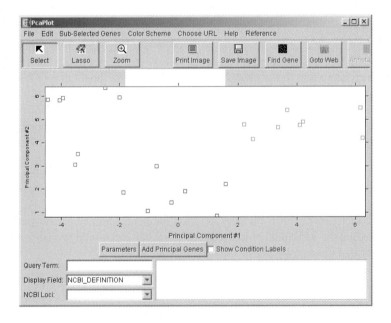

FIGURE 16.15: Principal component analysis using gene list from significance analysis.

icance analysis and the Hedenfalk analysis informally validates the computational components of the Significance Analyzer. The exercise highlights the ease-of-use of GeneSight: From start to finish, the analysis presented here took less than 60 minutes to perform.

16.3 Statistical analysis of microarray data using S-PLUS and Insightful ArrayAnalyzer

Dr. Michael O'Connell, Director, BioPharm Solutions Insightful Corporation

S-PLUS® includes out-of-the-box functionality to address most of the many data management and analysis issues that arise in the analysis of microarray data. Some of these statistical issues include: experimental design, pre-processing e.g. normalization, differential expression testing, clustering and prediction, and annotation. In addition, Insightful, the makers of S-PLUS, offers the S+ArrayAnalyzer™ solution for microarray data analysis. S+ArrayAnalyzer is available as a desktop analysis workbench and an enterprise client-server and web-based application. S+ArrayAnalyzer provides an end-to-end solution for microarray data analysis, as well as a toolkit and development environment for extending and customizing microarray analysis implementations. S+ArrayAnalyzer includes S-PLUS libraries and scripts, including libraries based on packages from the Bioconductor project, www.bioconductor.org; web scripts; data connections and S-PLUS Graphlets™ for interactive annotation.

S-PLUS is the premier software solution for exploratory data analysis and statistical modeling. S-PLUS offers point-and-click capabilities to import various data formats, select statistical functions and display results. S-PLUS includes more than 4,200 data analysis functions, including the most comprehensive set of robust and modern methods. In addition to out-of-the-box functionality, users may modify existing methods or develop new ones using the S programming language [61].

Graphics are a strong point of S-PLUS. Graphics are highly interactive and the user has control over every detail of all graphs. Trellis™ graphics, unique to S-PLUS, can display many variables informatively. Further, all S-PLUS graphs can be displayed as Graphlets, lightweight applets that may be simply created with java and XML-based graphics classes using the java.graph graphics device that is new to S-PLUS 6. In addition to providing interactive graphs in a web browser, Graphlets enable connection to external databases such as NCBI Genbank[32, 33], Onto-Express [90, 185] and Onto-Compare [91] databases. Such connectivity facilitates incorporation of annotation information into graphical and tabular summaries, captured as Graphlets, via database querying on the URL.

...ent design

...ital design is crucial for a good microarray experiment. Issues are
... 250, 219]. S-PLUS code for sample size calculations and power
... [219] is available from:
http://www.biostat.umn.edu/ weip/paper/mmmSampSize.s.

16.3.2   Data preparation and exploratory data analysis

Exploratory data analysis (EDA), quality control and normalization are achieved
quite simply in S-PLUS. The S-PLUS functions hist(), boxplot(), plot() and im-
age() are particularly useful for EDA and quality control. There have been many
normalization methods proposed and many in current use. Chap. 12 and [292, 40]
discuss various approaches and methods. Simple normalization of replicate chips to
the same interquartile range and median is achieved as follows for chip data in an
S-PLUS dataframe x with rows as genes and columns as expression intensities:

```
iqrfn <- function(xx)
quantile(xx,0.75,na.rm=T)-quantile(xx,0.25,na.rm=T)
medianIQR.norm <- function(tmp){
    #Adjust IQ ranges to be the same as max of IQRs
    divisor <- matrix(rep(apply(tmp,2,iqrfn)/max(apply(tmp,2,iqrfn)),
    dim(tmp)[1]), nrow=dim(tmp)[1],byrow=T)
    tmp.adj <- tmp/divisor
    #Adjust medians to be the same as max of medians
    adjustment <- matrix(rep(max(apply(tmp.adj,2,median,na.rm=T))
        -apply(tmp.adj,2,median,na.rm=T),
    dim(tmp.adj)[1]),nrow=dim(tmp.adj)[1],byrow=T)
    tmp.adj2 <- tmp.adj+adjustment
    return(tmp.adj2)}
x.norm <- medianIQR.norm(x)
```

Signal extraction is a little more involved, but there are segmentation algorithms
(e.g. seeded region growing) that are available in S-PLUS. Probe-level analysis of
oligonucleotide chip data, for example as described by [165, 194], is also available
in S+ArrayAnalyzer.

16.3.3   Differential expression analysis

Many methods for differential expression testing are available in S-PLUS. Two-
group comparisons (within-gene) using t-tests and permutation methods and corrections for multiple comparisons are readily available using the S-PLUS functions
t.test() and multicomp(). Chap. 13 and [97] provide a good review of this area, and
much of the functionality described in this paper is implemented in S+ArrayAnalyzer.
e.g. two-sample t-test, Welch t-test, Wilcoxon test, F-test, paired t-test, blocked F
tests.

Multi-group comparisons such as those described in [287], are achieved using the
S-PLUS functions varcomp(), lme() and nlme(). This mixed model analysis of de-

signed microarray experiments is particularly appropriate for well replicated factorial experiments and time course experiments.

Direct calculation of posterior probabilities of differential expression using a hierarchical Bayes approach is described by [215] and S-PLUS code for this approach is available from: ftp://ftp.biostat.wisc.edu/pub/newton/Arrays/code/.

Whatever the significance test chosen for assessing differential expression, adjustment of the p-values obtained, to take account of the multiple (many genes) testing procedures, is needed (see Chap. 9). Procedures in S+ArrayAnalyzer for such adjustment include the Bonferroni, [149, 145] and Sidak procedures for strong control of the family-wise Type I error rate (FWER), and the [30, 31] procedures for (strong) control of the false discovery rate (FDR).

16.3.4   Clustering and prediction

There are many clustering and prediction methods available in S-PLUS. The hierarchical clustering methods include agglomerative: hclust(), agnes(); and divisive: diana(). Partitioning clustering methods include pam() and kmeans(). The dendrograms produced by the hierarchical methods may be visualized using plclust() and layered over heatmaps of the expression intensity data produced using image() to produce the now familiar visualization of microarray experimental data [104]. Model based clustering methods, using mixture models, are available using the function mclust() [113]. A repository of related S-PLUS code is available for download from: http://www.stat.washington.edu/fraley/mclust/.

There are many other supervised and unsupervised learning approaches to microarray data analysis. S-PLUS code for some of these methods is available from the GeneClust site: http://odin.mdacc.tmc.edu/ kim/geneclust/. This includes S-PLUS code for gene shaving [132, 85].

16.3.5   Analysis summaries, visualization and annotation of results

Typical summaries of microarray data analyses include short-lists of genes in functional genomics studies, longer lists of genes for unsupervised learning studies and targeted sets of genes that discriminate between levels of experimental conditions subject covariables. These results are presented in S+ArrayAnalyzer as html/xml tables of gene lists and interactive Graphlets. In addition to providing interactive graphs in a web browser, Graphlets enable connection to external databases such as NCBI Genbank[32, 33], Onto-Express [90, 185] and Onto-Compare [91] databases. Such connectivity facilitates incorporation of annotation information into graphical and tabular summaries, captured as Graphlets, via database querying on the URL.

In addition to hotlinked tabular and graphical summaries, S+ArrayAnalyzer includes specific functions for searching Genbank and other locations, e.g. Pubmed, for annotation and publication information.

16.3.6   S+ArrayAnalyzer example: Swirl Zebrafish experiment

The S+ArrayAnalyzer is available as a desktop analysis workbench and an enterprise client-server and web-based application. In this section we show the S+ArrayAnalyzer desktop workflow using gene expression data from an experiment involving swirl zebrafish. The analysis and experiment description is included directly from that presented by [293]. The data were originally provided by Katrin Wuennenberg-Stapleton from the Ngai Lab at UC Berkeley.

As described by [293], "this experiment was carried out using zebrafish as a model organism to study early development in vertebrates. Swirl is a point mutant in the BMP2 gene that affects the dorsal/ventral body axis. A goal of the Swirl experiment is to identify genes with altered expression in the swirl mutant compared to wild-type zebrafish. Two sets of dye-swap experiments were performed, for a total of four replicate hybridizations. For each of these hybridizations, target cDNA from the swirl mutant was labeled using one of the Cy3 or Cy5 dyes and the target cDNA wild-type mutant was labeled using the other dye. Target cDNA was hybridized to microarrays containing 8,448 cDNA probes, including 768 controls spots (e.g. negative, positive, and normalization controls spots). Microarrays were printed using 4 x 4 print-tips and are thus partitioned into a 4 x 4 grid matrix. Each grid consists of a 22 x 24 spot matrix that was printed with a single print-tip. Each row of spots corresponds to probe sequences from the same 384 well-plate. Each of the four hybridizations produced a pair of 16-bit images, which were processed using the image analysis software package Spot [294]. The data set includes four output files swirl.1.spot, swirl.2.spot, swirl.3.spot, and swirl.4.spot from the Spot package. The file fish.gal is a gal file generated by the GenePix program; it contains information on individual probe sequences, such as gene names, spot ID, spot coordinates."

To load the swirl data set we use the Import cDNA Data from the S+ArrayAnalyzer menu in S-PLUS (see Fig. 16.16). Once the data have been imported, MIAME data captured and the variables for analyis specified, the data may be normalized. There are many available options for normalization in S-PLUS including median alignment, loess adjustment on MvA plots, loess adjustment by print tips and median absolute deviation (MAD) scale adjustment. Following Yang and Dudoit (2002), we normalize using the printTipLoess method, for within-print-tip-group intensity dependent location normalization using the loess function.

Once the data have been imported, MIAME data have been captured and the variables for analysis have been specified, the data may be normalized. There are many available options for normalization in S-PLUS including median alignment, loess adjustment on MvA plots, loess adjustment by print tips and median absolute deviation (MAD) scale adjustment (see Fig. 16.17). Following [293], we normalize using the printTipLoess method, for within-print-tip-group intensity dependent location normalization using the loess function. Results of the normalization are shown by default using a set of boxplots. Note that the boxplots of the differences $M = \log(R/G)$ in the post-normalization are all centered nicely at zero indicating a successful location adjustment.

Once the data have been normalized they are ready for differential expression testing.

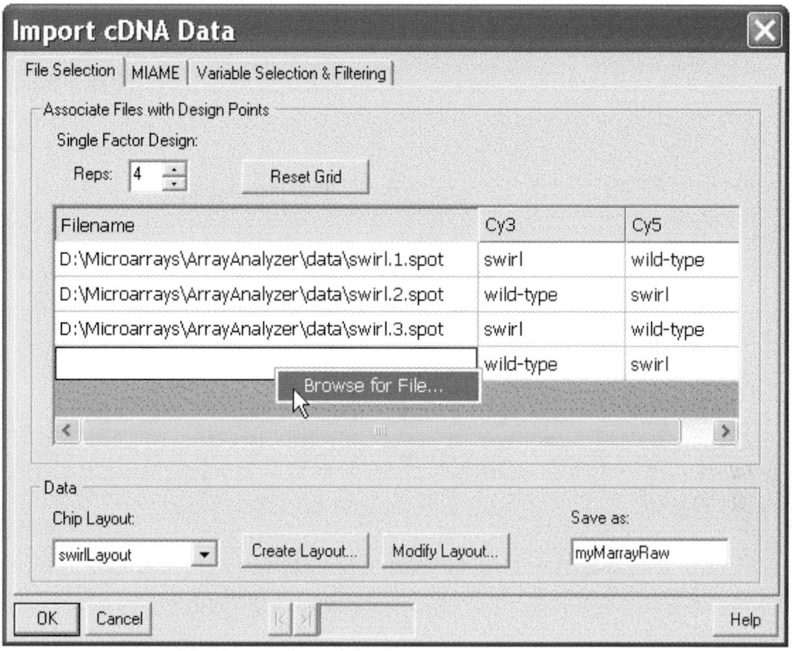

FIGURE 16.16: The import cDNA window in S+ArrayAnalyzer

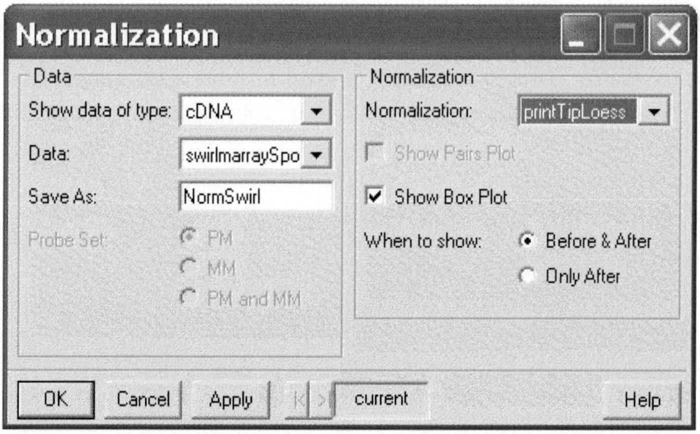

FIGURE 16.17: Choosing the normalization parameters in S+ArrayAnalyzer

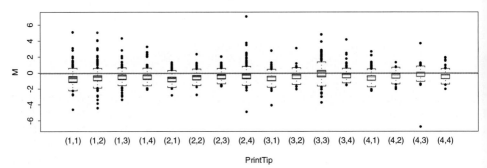

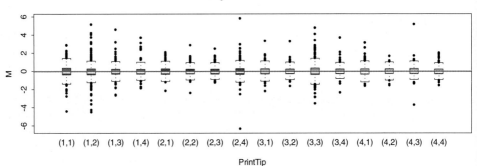

FIGURE 16.18: Box plots of the differences $M = \log(R/G)$ before and after normalization.

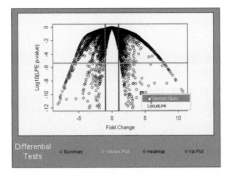

FIGURE 16.19: A volcano plot is a plot of the p-value (adjusted or unadjusted) versus the fold change.

There are a number of S-PLUS functions available for differential expression testing. These include two-sample t-test, Welch t-test, Wilcoxon test, F-test, paired t-test, blocked F tests and local pooled error tests. These 2-color cDNA data are naturally blocked via the chips (blocks of size 2) so that the paired t-test and blocked F test are well suited to the design. Note however, that there are only 4 replicates, so methods that leverage groups of genes such as the local pooled error approach [192] may be useful in this case. Whatever the significance test chosen for assessing differential expression, adjustment of the p-values obtained to take account of the multiple (many genes) testing procedures is needed. Procedures in S+ArrayAnalyzer for such adjustment include the Bonferroni, [145, 149] and Sidak procedures for strong control of the family-wise Type I error rate (FWER), and procedures for (strong) control of the false discovery rate (FDR) [30, 31].

One useful plot offered in S+ArrayAnalyzer for summarizing differentially expressed genes is the volcano plot. This is a plot of the p-value (adjusted or unadjusted) versus the fold change. Typically a horizontal line at a p-value corresponding to control of the FWER is included on the plot as well as vertical lines indicating 2x or 3x fold-change. Most graphical summaries in S+ArrayAnalyzer are presented as Graphlets. In addition to providing interactive graphs in a web browser, Graphlets enable connection to external databases such as NCBI Genbank, Onto-Express [90, 185] and Onto-Compare [91] databases. In the volcano plot shown in Fig. 16.19, the user has clicked on a particular gene and chosen to obtain annotation information for this gene from the Unigene database on Genbank.

16.3.7 Summary

Microarray technology is complex, and experiments using microarrays are resource-intensive. As such, there is an urgent need for rigorous, statistical design and analysis of microarray experiments. Some of the important statistical issues in microarray experiments include:

1. Experimental design

2. Pre-processing e.g. normalization

3. Differential expression testing

4. Clustering and prediction

5. Annotation

S-PLUS® includes out-of-the-box functionality to address most of these many data management and analysis issues that arise in the analysis of microarray data. In addition, the S+ArrayAnalyzer™ solution for microarray data analysis provides an end-to-end solution for microarray data analysis, as well as a toolkit and development environment for extending and customizing microarray analysis implementations. S+ArrayAnalyzer includes S-PLUS libraries and scripts, including libraries based on packages from the Bioconductor project, www.bioconductor.org; web scripts; data connections and S-PLUS Graphlets™ for interactive annotation. S+ArrayAnalyzer is available as a desktop analysis workbench and an enterprise client-server and web-based application.

S-PLUS code snippets for some of the methods described above are available from the Insightful web site at www.insightful.com/ArrayAnalyzer. This site also includes information regarding S+ArrayAnalyzer solutions and updates regarding analysis of microarray data using S-PLUS and S+ArrayAnalyzer.

16.4   SAS software for genomics

Dr. Russ Wolfinger, SAS Institute Inc., Cary, NC

SAS has long been recognized as a premiere provider of data management and statistical analysis software. Dozens of various generic SAS software packages have bearing in the life sciences, and SAS has also recently begun pursuing genomics-specific offerings. For sake of space, we will just overview these new components here.

16.4.1   SAS research data management

SAS RDM (see Fig. 16.20) is the data management core of the SAS Scientific Discovery Solutions product line. It is a Java client-server application that utilizes SAS warehousing technologies to provide a centralized repository for your organization's discovery research data, and other ancillary information.

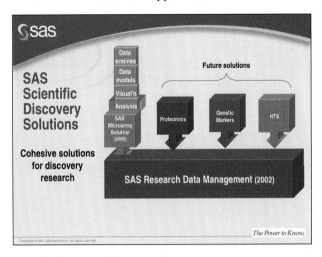

FIGURE 16.20: RDM as a foundation for SAS SDS

16.4.1.1 Datamarts

The concept behind datamarts involves extracting selected data and reorganizing that data in a form well suited to specific analysis and reporting task. As with larger data warehouses, a key aspect of datamarts is metadata, or data about the primary data. Examples of metadata include a data file name, location, storage format, and structure. A datamart can capture metadata throughout the analysis process: where the data came from, what transformations were performed, and information about the context of the data. Managing the metadata and controlling the transformation processes are the key benefits of datamarts.

16.4.1.2 Pooled metadata repository

SAS RDM provides a platform for centralizing access and managing discovery research data. At its core is the Pooled Metadata Repository (PMR), which provides a means of consolidating metadata from any number of datamarts to create a single searchable (Figure 2) repository. The PMR does not move the physical data, but rather gathers distributed metadata into one location.

16.4.1.3 Security model

Among the regulatory compliance features is a security model that controls access to the system by requiring a user name and password. RDM controls access to warehouse objects as well as access to certain areas of the system at the user level or at a user-group level. Permissions range from no access to read-only access to full edit access. The security model also includes an audit trail of actions performed in the system and versioning of all data loaded in the warehouse. The audit trail and version control simplify the ability to trace back the source of data and/or modifications

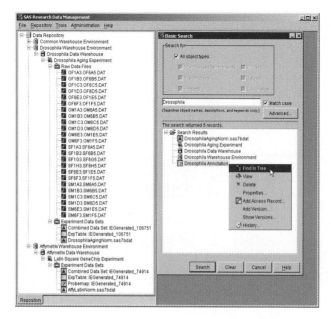

FIGURE 16.21:   Perform a search on the Pooled Metadata Repository.

made along the way.

16.4.1.4   Data models

Built using add-ins to SAS/Warehouse Administrator, RDM enables you to implement virtually an unlimited number of data schemas. For microarray data such as those from our Drosophila example, the recently OMG-sanctioned MAGE-ML standard (http://www.mged.org) provides a useful foundation. You can use the SAS XML engine to import data from MAGE-ML.

16.4.2   SAS microarray solution

SAS MAS is the first vertical deployment of the RDM platform, and integrates seamlessly with it to produce a whole product solution for managing, analyzing, and visualizing microarray data. The additional functionality of MAS includes input engines and analytical processes.

16.4.2.1   Input engines

Input engines are software routines that pull data from instrumentation systems output and load the data into a SAS MAS warehouse. Input engines are specific to input data structures, and you can customize them to suit the vendors you select to perform your microarray experiments. Input engines enable you to directly import an entire collection of raw numerical array files with a few mouse clicks. In addition to the raw data files, you must provide an experimental design file that shows how to map the experimental factors to the arrays. This kind of approach allows you to input designs of arbitrary complexity. A different input engine enables you to input annotation data for the genes on a particular chip, and you can use the resulting SAS data set in any experiment for which it is appropriate.

16.4.2.2 Analytical processes

Analytical processes are special SAS macro programs that perform data manipulations and statistical calculations on the experimental data you have loaded into RDM. These processes employ the power of the SAS System to generate analysis data sets, listings, statistical results and graphs. Analytical processes are reusable and flexible. They can range in functionality from very simple data displays to complex statistical modeling. Scientists and statisticians alike can run analytical processes against any experiment data set by simply providing appropriate values for input parameters. After you select an analytical process to run, a parameter input window requests information required for successful execution. The parameters values are specific to the data set at hand, but you do not need to edit the code itself once you have written and loaded the analytical process into SAS MAS.

In its initial release, SAS Microarray Solution provides four analytical processes for use with your experiment data:

- **DataContents** displays the contents of a SAS data set in HTML format.

- **ArrayGroupCorrelation** takes all of the array data from an experiment, divides it into groups selected by the user, and then performs a multivariate correlation analysis on each group.

- **MixedModelNormalization** normalizes microarray data by fitting a mixed linear model across all of the arrays in an experiment.

- **MixedModelAnalysis** provides a comprehensive look at results from fitting mixed models on a gene-by-gene basis. References for this kind of approach include [67], [78], [122], [151], [167], [287].

Upon submission, an Analytical Process sends its macro code via SAS/Integration Technologies to a preactivated SAS server. The code can execute any SAS routine you wish, and typically includes calls to SAS data steps and procedures. You can also use it to create a JMP Scripting Language (JSL) file. SAS MAS then executes this file, producing predefined graphical and analytical displays in JMP (Fig. 16.22). JMP is a dynamic interactive module that offers loads of features for scientific discovery. The preceding examples represent a small fraction of the capabilities SAS offers you for analyzing microarray and related genomics data. RDM and MAS provide a framework that enables you to tap into the full power of the SAS System, including the following capabilities:

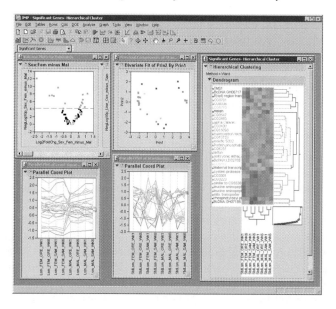

FIGURE 16.22:   Exploring SAS analytical process results in JMP

- The SAS Data Step, SAS Macro Language, and JMP Scripting Language are very flexible and extensive environments for manipulating and processing genomics data.

- Proc Optex, Proc Factex, Proc Plan, the ADX Menu System, and the DOE features in JMP help you to create optimal designs.

- Over half of the 50+ procedures in SAS/Stat are applicable to array data. These include procedures for analysis of variance, clustering, density estimation, discriminant analysis, multidimensional scaling, multiple comparisons, nonparametric statistics, partial least squares, power and sample size, principal components and smoothing. New residual diagnostics will be available in Release 9.1 of Proc Mixed, enabling better quality control of your array data. Procedures from SAS/QC, SAS/OR, SAS/ETS, and SAS/IML (including IML Workshop) can also be useful. Scores of examples from other scientific disciplines are in the SAS Sample Library and at www.sas.com/techsup/download/stat.

- You can use RDM and MAS to preprocess and create analysis-ready data sets for SAS Enterprise Miner. Its incredibly powerful collection of data mining methodologies and intuitive process-flow interface enable you to efficiently generate a wide range of cross-validated predictions. Furthermore, you can use SAS Text Miner to process collections of scientific abstracts you create from important gene lists.

- SAS/MP-Connect enables parallelization of suitable methods across multiple CPUs.

- SAS/Genetics, introduced in 2002, is a stand-alone product containing procedures for the statistical processing of DNA marker data.

Details about these and other SAS offerings are available at www.sas.com/sds.

16.5 Spotfire's DecisionSite

Matthew Anstett, M.Sc., Application Specialist, Spotfire Inc.

16.5.1 Introduction

Spotfire's DecisionSite for Functional Genomics is a platform for gene expression analysis encompassing each of the categories of the commercial applications previously mentioned in this chapter, including enterprise level package (extensive database access and development platform), "Point and click" analysis (guided workflows), and Analysis/Visualization environment (dynamic data analysis and visualizations). This platform combines state-of-the-art data access, gene expression analysis methods, guided workflows, dynamic visualizations, and extensive development tools. Tools and Guides are components of this platform. Tools are the analytic access components added to the interface, while Guides connect Tools together to initiate suggested analysis paths, allowing the user to deviate as the results determine.

With rapidly emerging microarray analysis methods, scientists are faced with the challenge of incorporating new information into their analysis quickly and easily. Spotfire DecisionSite Developer offers the ability to add new microarray data sources, normalization methods, and algorithms within a guided workflow that can be rapidly deployed to all users. The platform can be updated and extended to incorporate the expertise of the user and the rapid advances in analysis methodology, and is therefore appropriate and useful for a broad range of expertise from novice to expert.

16.5.2 Experiment description

In this overview of Spotfire's DecisionSite for Functional Genomics, we will explore cardiac development, aging and maturation microarray data sets from the National Center for Biotechnology Information (NCBI), Gene Expression Omnibus (GEO) repository (GEO Accession # GSE75). This data set contains benchmark gene expression data from normal wild type mouse ventricular tissue harvested at six time points from multiple individuals. Samples were prepared from three individuals at each time point for a total of 18 samples. A subset of the study group contains samples from three female and male individuals at 3 to 5 months and 1 year. We will determine which genes are up- and down-regulated over the course of the experiments and may be implicated in aging and cardiac disease. Furthermore, it is suspected that there are gender-specific differences in cardiac gene expression that contribute to lower risk of cardiac disease in females; we will use treatment comparison tools to identify these gender-specific differences.

16.5.3   Microarray data access

Microarray data exist in a variety of formats which often depend on the particular array technology and detection instruments used. This data can be easily loaded into DecisionSite by copy/paste from spreadsheets, direct loading of text or comma separated values (.csv) files, via pre-configured or *ad hoc* queries of relational databases, and from proprietary databases and export file formats from microarray manufacturers such as Affymetrix.

During the course of analysis, annotations can be added at any time by several methods. Web links allow the user to select gene clusters within a visualization and to retrieve data from external or internal web sites for genes that are members of the selected cluster. Web links can be easily configured by the end-user for their particular search engine of choice. In addition, internal annotations can be added from files or from a relational database to incorporate proprietary information and expertise for the target genes or pathways of interest.

Support of standard microarray data platforms, such as Affymetrix and Rosetta, is integrated within DecisionSite for Functional Genomics. An Affymetrix QA/QC query tool is part of DecisionSite and facilitates inspection of QA/QC parameters of Affymetrix GeneChips, based on control probes. Microarrays that pass inspection are selected prior to retrieval of the complete data set from the Affymetrix AADM database. The Rosetta Resolver integration provides the means to export data directly to DecisionSite from the Resolver client, identify subsets of genes by merging additional columns of data from other sources, and send results back to the Resolver system.

When data is loaded into DecisionSite, columns of data are parsed and interpreted on-the-fly. The user can plot, filter or perform calculations on all of the data or subsets of the data. Available visualizations include 2-D and 3-D scatter plots, profile charts, heat maps, pie charts, histograms, and bar charts. Visualizations are linked and can be dynamically filtered using query devices that are automatically created for each column of data. Visualizations enable the exploration of the data in multiple dimensions including x, y and z-axis and marker properties such as color, size and shape. Each of these dimensions can be easily assigned to any field in the data set. Query devices are assigned to each field of data so that, for example, using a statistic to filter out significant genes results in a display of the gene profiles over experimental conditions and an updated histogram of the major biological processes linked to these genes.

In our example data set, gene expression data from 1-year time points in the cardiac development study is opened directly into DecisionSite from Affymetrix Microarray Suite export files (see Fig. 16.23). We begin by loading data from the 1-year time

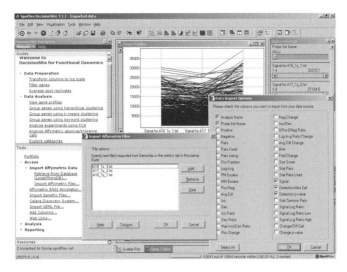

FIGURE 16.23:   Import of Affymetrix Microarray Suite export files. From the tool menu, data access options allow users to select specific columns within Affymetrix MAS export files. Additionally, database links to Affymetrix AADM walk a user through a sequential query to retrieve their experiments, and include built-in QA/QC tools that plot control probe parameters.

point, and then add the previous time points to the analysis using the Add Columns from File Tool (Fig. 16.24). Data can also be added from Spotfire files, the clipboard, or database links, provided a common identifier exists between the data sets.

16.5.4   Data transformation

Data transformations are often required to effectively analyze microarray data (see Chap. 12). Methods that are available include: transpose and pivot data, base 2 and base 10 logarithms, as well as multiple normalization and summarization methods for replicate genes or experiments. The gene expression data that we have retrieved is plotted in two profile charts. One chart displays recent 1-year time point experiments and the second displays all samples and time points (Fig. 16.25). Multiple samples for each time point are summarized by taking the average value of the samples. From this summarization, new columns are added to the data set that can then be plotted as summary profiles. Additional summarization methods that are more resistant to extreme outliers are also available, such as median. The genes for our summarized time points are normalized using a Z-score calculation to center the mean at 0, which standardizes expression levels to facilitate comparison (Fig. 16.26) [261].

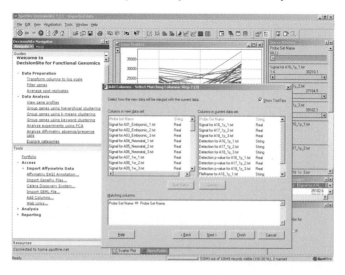

FIGURE 16.24:   Merging Additional Data. The Add Column function allows users to select a matching ID column and merge data from previous experiments that exist as text or .csv files, Spotfire files, or clipboard. Additional data, such as annotations, can be added during analysis from a file or database provided there is a matching ID column.

16.5.5   Filtering and visualizing gene expression data

Adding information and automatically generating the appropriate view of the results can be performed through guided analysis, such as the analysis of Affymetrix detection calls (Fig. 16.27). Query devices are assigned to every field of data and allow the user to perform filtering with multiple selection criteria, resulting in updates of all visualizations to display the results of this cumulative filtering. DecisionSite for Functional Genomics Guides connect analytic and data access Tools in order to perform repetitive tasks quickly or to initiate a series of specific steps in the analysis. Throughout the analysis, filtering using any data field query device can be used to subset data and limit the genes that are included in further calculations and visualizations. Significant genes can also be filtered by standard deviation, maximum fold change, and modulation (frequency crossing a threshold). For example, filtering genes on modulation by setting a 2-fold change threshold (base 10 logarithm = 0.3) will split genes out by the number of times they fall above 2-fold change over the selected experiments.

In our example data set, a Guide was used to analyze Affymetrix detection calls. This Guide summarizes all of the detection columns so that each gene has a new column of information that displays the frequency of Absent, Present or Marginal calls from the Affymetrix MAS export files (see Chap. 12). These new columns in the data set are assigned query devices that we then use to select only the genes that

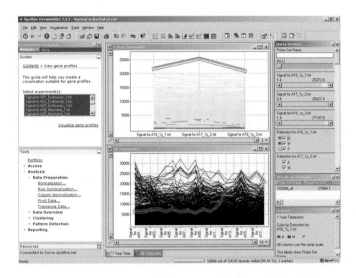

FIGURE 16.25: Profile chart visualizations. A Guide, shown in the upper left window of DecisionSite walks novice users through creating profile charts. In the upper visualization, the three new samples collected at 1 year are shown on the x-axis and the profiles are colored by Affymetrix detection call for the first sample as indicated in the legend for the active view. The lower visualization displays the triplicate samples at each of the five time points in the study. Linking of visualizations is demonstrated by a single profile colored red and linked to the corresponding gene profile in both charts by an outline.

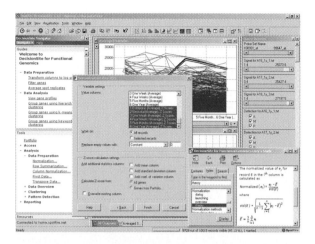

FIGURE 16.26: Summarization and normalization. Average values for triplicate samples at each time point are displayed in a new profile chart. Z-score statistic is used to standardize the gene expression levels and facilitate their comparison over the cardiac development time points. Context sensitive help is available within all computation dialogs that display details of the method including the specific equation selected.

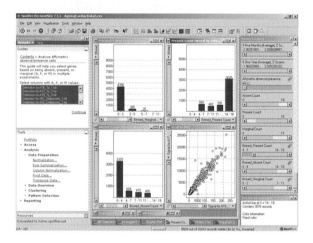

FIGURE 16.27: Summarizing Affymetrix A/P calls. Analyzing Affymetrix absence/presence calls can be performed by stepping through a Guide which calculates the frequency of each call for the selected experiments. Frequency histograms for each call can be easily created from the Guide and new query devices can be used to select only those genes with present calls in every experiment to use in further analysis.

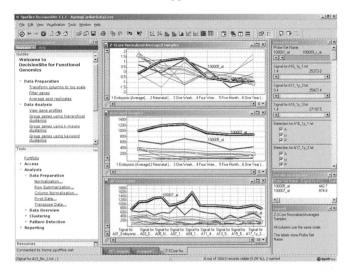

FIGURE 16.28: Comparing raw and transformed data. Individual views of raw, summarized, and normalized data can be displayed together.

were determined to be present in every sample. The raw, summarized, and normalized data are compared by using multiple visualizations (Fig. 16.28). Three profile charts were created using the original cardiac samples, averaged sample data, and Z-score normalized data at the five development time points. For the selected genes we observe the overall profile trend for summarized samples is reflected in averaged profiles which may have been influenced by variations within replicates. The Z-score normalization has standardized the profiles for more accurate comparisons.

16.5.6 Finding gene expression patterns

Profile searching can be used to find similar genes or an expected pattern within large data sets. A Guide can be used that prompts the user to first identify the profile or pattern. For example, a particular gene can be selected by entering its gene name or other identifier in a full text search to find genes that behave similarly in the experiment. Available similarity measures include Euclidean distance, Pearson's correlation, Cosine Correlation and City block/Manhattan distance (see Chap. 11).

In the cardiac development data set, we are interested in identifying genes that increase or decrease over the study. To find these genes, we create a gene expression profile in the profile editor that increases over the development time points and compare it with all gene profiles in our data set (Fig. 16.29)

The Pearson's correlation similarity measure was used in profile searching of the cardiac data set. A similarity value was calculated for every gene in our data set and was then used to filter genes that are up-regulated or down-regulated over the development time points (Fig. 16.30). Correlation similarity measure has a range from

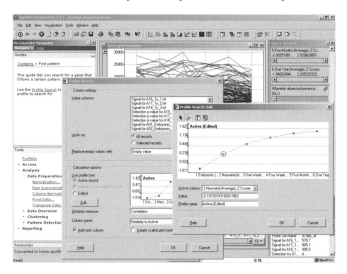

FIGURE 16.29: Profile searching. A profile template is edited to create a master profile used in similarity searching. For every gene in the data set, a similarity measure is calculated and added to the data set based on similarity to the master profile selected. In the cardiac development study, we are looking for genes that increase in samples over five time points.

-1 to +1. By taking the absolute value of the correlation similarity column, a new column and the corresponding query device can be filtered in one direction to select genes that are both most similar (up-regulated) and most dissimilar (down-regulated) to our edited profile. The profiles of the raw, averaged and Z-score normalized data are all updated to reflect our query device setting. Up- and down-regulated genes are easily identified in the Z-score normalized data set (Fig. 16.30).

16.5.7   Using clustering and data reduction techniques to isolate group of genes

K-means clustering and hierarchical clustering are methods for grouping genes by their expression patterns over the experimental variables. Both of these methods are available in DecisionSite for Functional Genomics. K-means clustering uses an iterative method for comparing gene profiles, creating a new cluster centroid at each iteration until a steady state is achieved. K-means clustering parameters include maximum number of clusters, cluster initialization, and similarity measure. Once the calculation is completed, gene profiles are then displayed in a trellis plot based on cluster. This reduces the data set to manageable groups that can be explored with annotations and additional information. This implementation of the hierarchical clustering uses a hierarchical agglomerative method (see Chap. 11). The results are displayed within a heat map visualization with integrated dendrogram tree. The

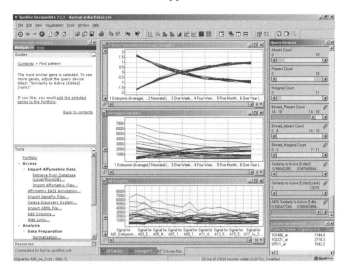

FIGURE 16.30:   Finding gene expression patterns. Within a Guide, text prompts provide instruction on filtering similar genes and adding them to a Portfolio gene list.

branches of the tree represent the clusters and can be explored within the visualization. Initialization parameters include clustering methods and similarity measures. 2-D hierarchical clustering can be used to identify clusters within samples as well as genes. This is useful when investigating the relationship between samples such as tumor types or cell lines. Clustering can also be performed on keywords using a guided work flow to aggregate text information, perform clustering, and split out genes by common annotations.

Both clustering methods, as with all computation dialogs within DecisionSite, allow the user to specify the samples to include in analysis and to work only on selected records (those that have not been filtered out using query devices). Thus, data that are determined to be out of detection range, do not satisfy certain statistical criteria, or are not the focus of a specific question can be removed from the analysis.

Principal Components Analysis (PCA) can be used when a large number of genes are analyzed to reduce dimensionality of the data while preserving variability. This allows the user to confirm categorization of genes across diverse samples, such as tumor types, in fewer dimensions. Results of PCA are typically plotted in a 3-D scatter plot where a "cloud" of genes can be observed in the new PCA space and categories of genes or experiments can be revealed as tightly clustered markers (see Chap. 10 for a full discussion of PCA).

In our example data set, a Guide was used to group genes using hierarchical clustering. In the Guide, visualizations are automatically generated, annotations added, and number of clusters selected and added to the data set (Fig. 16.31). Genes were grouped using all the sample values, Pearson's correlation similarity measure and complete linkage clustering method (see Chap. 11 for a full discussion of various

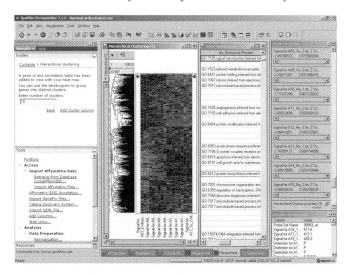

FIGURE 16.31: Hierarchical clustering. A heat map and dendrogram display the results of hierarchical clustering and corresponding table provides annotation details for clusters selected from the heat map. Coloring from green to red reflect low to high expression, respectively. A Guide was used to walk a user through the analysis which includes the ability to add annotations from flat text files.

distances and linkages as well as criteria to select them). The Guide was used to add annotations from text files downloaded from Affymetrix NetAffx website. These text files contain annotations and matching Probe Set ID information for the corresponding Affymetrix Murine GeneChip U74v2 used in the study. The results of hierarchical clustering display roughly three large blocks of up-regulated genes (shown in red) specific for the early embryonic stage, neonatal through 1 week, and 4 weeks through 1 year in the heat map visualization in Figure 16.31. These groups are also members of the first three nodes of the dendrogram and can be categorically grouped by pruning three clusters from the dendrogram using the Guide or a level selector within the dendrogram window. The dendrogram branches can also be explored by double-clicking on nodes to zoom in on clusters of leaves of interest in the dendrogram. The Gene Ontology (GO)biological processes displayed in the corresponding table view associated with down-regulated genes reflect processes that may decrease from embryonic stages through adulthood such as intracellular signaling cascade, fatty acid synthesis, regulation of cell growth, and cell growth.

The Portfolio Tool in DecisionSite for Functional Genomics stores gene lists. Gene lists were created from the cardiac development data set by selecting 50 genes that were observed to be increasing and decreasing in the Z-score normalized average sample profile chart. These Portfolio lists were then used to make selections in the visualizations, confirm groupings and identify possible subsets within the hierarchical clustering results (Fig. 16.32). Using the similarity rank query device and

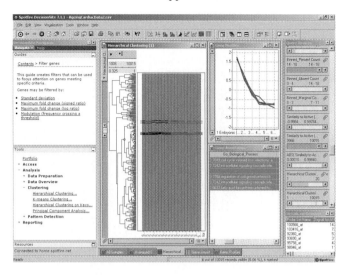

FIGURE 16.32: Creating portfolio lists. the list of top down-regulated genes is created by using the similarity query device to select genes in visualizations and add them to a Portfolio list. While querying on similarity measure the scientist can explore genes and annotations in the table view, confirm the expression pattern in the Gene Profiles chart, and confirm whether these these genes were grouped using the hierarchical clustering method selected.

observing the status bar for the number of genes selected, 50 of the most similar and dissimilar genes were selected and added to the Portfolio list. Portfolio lists are persistent and can be exported, imported and shared. These lists can be added to the data set as categories and can be applied to analysis by filtering or plotting axes to quickly isolate the top 50 up- or down-regulated cardiac development genes in new data sets and future experiments (Fig. 16.33).

16.5.8 Comparing sample groups

In our example data set, cardiac ventricular tissue RNA was isolated from male and female mice at 3 to 5 months and 1 year and used in microarray experiments. Using treatment comparison tools in DecisionSite for Functional Genomics, gender-specific genes at different development stages will be isolated and explored. The Distinction Calculation is used to prioritize genes based on their difference from an idealized profile that is low in one group and high in another [127]. These values are then used to select the top genes that have high expression in each gender at a particular time point. These gene lists were stored in the Portfolio and used to order heat map visualizations. A column was also added to the heat map colored categorically based on the added GO annotation Biological Process to look for groupings within the gender specific genes.

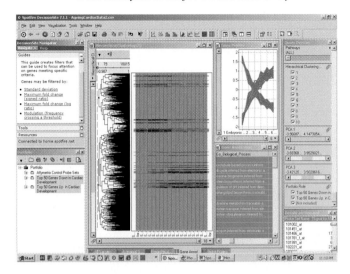

FIGURE 16.33:   Using lists to update visualizations. Portfolio lists of the top 50 up- and down-regulated genes are selected and used to filter all of the tiled visualizations. The scientist can explore genes and annotations in the table view, confirm the expression pattern in the profile chart and can quickly confirm expected patterns or evaluate project gene lists in other experiments.

16.5.9   Using Portfolio Lists to isolate significant genes

Portfolio lists are persistent and interactive lists that are used to store genes categorized by sample type, experimental variable and significance.

In our example data set, gene lists were compared to identify genes in males or females at both time points measured that were found to be significant. Any number of lists can be compared using Boolean logic to create new lists of genes. These lists were used to select corresponding genes in visualizations and add categorical data to the analysis as a Portfolio Role query device. This new data was used to order the heat map to display only the genes found to be significantly up-regulated in females at both time points (Fig. 16.35).

Genes found to be gender-specific by distinction calculation at both time points are selected in a profile chart using combined lists (Fig. 16.36). Annotations added from Affymetrix NetAffx annotation files were used to identify gene title for a gene expressed at higher levels in females called "matrix gamma-carboxyglutamate (gla protein)." By selecting this gene in the visualization and launching Weblinks, the Affymetrix Probe Set ID for this gene is automatically used to query the NetAffx website. Data is retrieved for this Probe Set and its target including links to additional data such as PubMed. From PubMed links we find an article describing gla protein mutants having increased vascular calcification in mice. The higher level of the mRNA for this protein in females may indicate a protective function in lowering

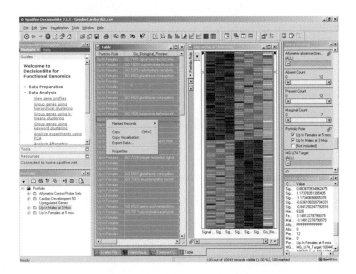

FIGURE 16.34: Display of gender-specific genes. Heat map ordered by significant gene lists that result from filtering on gender comparison statistic. Genes up-regulated in females and males at 3 and 5 months are displayed. A categorical column of GO Biological Process annotation was added to the heat map to look for clusters in either gender pool that may have a specific biological process associated with their gender-specific expression pattern. The table view containing the selected genes, gene list category, and any other desired data fields, can be exported as a text file report directly from the table view (pop-up menu is shown).

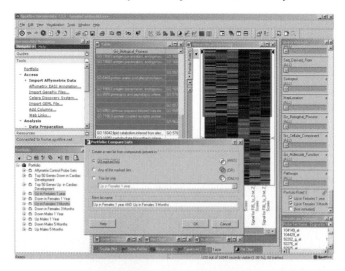

FIGURE 16.35:   Comparing lists. List logic is used to create summary lists. Any number of lists can be compared using Boolean logic to generate new subsets of genes. Genes found to be up-regulated in females at both 3 months and 1 year are isolated and displayed on the y-axis, respectively.

cardiac disease rates in females [203, 227].

Genes of interest in cardiac development and gender-specific samples are identified in the analysis and saved as lists in the Portfolio. Table visualizations can be created and ordered using any of the data from the analysis, including added statistics, cluster categories and annotations. From this table visualization a report can be exported as a text file containing the genes found to be significant in our analysis. Using reporting tools, visualizations can be sent directly to MS Word, PowerPoint, or a webpage.

16.5.10   Summary

Spotfire's DecisionSite for Functional Genomics is a powerful platform for gene expression analysis encompassing each of the categories of the commercial applications mentioned in this chapter. This is demonstrated in the analysis of a cardiac data set, from data access and use of guided workflows deployed from a server, to isolation of genes that are involved in cardiac development and gender-specific resistance to cardiac disease using dynamic visualizations and analysis tools.

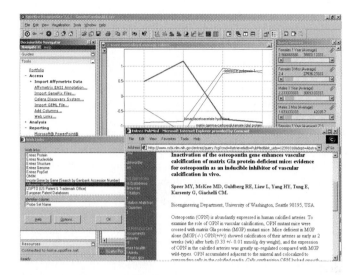

FIGURE 16.36: Using Web Links to explore significant genes. A gene found to be up-regulated at both 3 months and 1 year in females is isolated using list logic and explored via web links to Affymetrix NetAffx and NCBI PubMed.

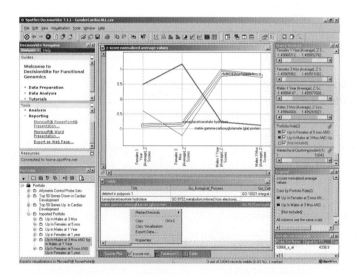

FIGURE 16.37: Identifying gender-specific genes. Genes are isolated using lists of significant genes created using distinction calculation on male and female groups. Reporting tools are used to export text file reports from the table view and visualizations are sent directly to MS Word, PowerPoint or displayed as a web page.

16.6 Summary

This chapter discussed very briefly the characteristics of the software tools currently available for the data analysis of microarrays. The classification introduced distinguished between enterprise level, "point and click," analysis environments and hybrid approaches.

The chapter also included a number of sections that showcased some of the software packages currently available. BioDiscovery's GeneSight was showcased on a breast tumor data set [135]. The goal of the analysis was to identify those genes that are differentially expressed between tumors originating in individuals with BRCA1 mutations and those from individuals with BRCA2 mutations. The S-Plus Array Analyzer was showcased on a swirl zebrafish experiment aimed at identifying genes with altered expression in the swirl mutant compared to wild-type zebra fish. A short section presented some of the software available from SAS and discussed its capabilities. Finally, the Spotfire DecisionSite was showcased on a cardiac development, aging and maturation microarray data sets from the National Center for Biotechnology Information (NCBI). This data set contains benchmark gene expression data from normal wild type mouse ventricular tissue harvested at six time points from multiple individuals. The goal of this analysis example is to determine which genes are up- and down-regulated over the course of the experiments and may be implicated in aging and cardiac disease.

Chapter 17

The road ahead

It is difficult to say what is impossible, for the dream of yesterday is the hope of today and the reality of tomorrow.

—Robert Goddard

17.1   What next?

This book barely scratched the surface of the things that can be achieved using microarrays. This is a very new field even by the standards of the 21st century. Five short years ago, there were a handful of laboratories using this technology. Today, most life scientists are familiar with microarrays and thousands of laboratories are using them routinely. Because of the breathtaking speed with which this technology has been adopted, and the large number of researchers just entering this field, this book has focused on the most basic questions that can be answered using high throughput gene expression techniques. However, there are several important issues that seem to be the next logical steps in this field. Some results in these areas have already been obtained while others seem just around the corner. This chapter will briefly sketch some of the most promising directions of current and future research.

17.2   Molecular diagnosis

Arguably, the one most important direction of future research related to microarrays is molecular diagnosis. The main idea behind this direction of research is that the gene expression profile constructed using high throughput techniques such as microarrays can be used to identify sets of genes that are intimately related to the development and onset of various ailments. A natural target for this approach is the range of illnesses that are both life threatening and difficult to diagnose. Many cancers, for instance, fall in this category. A stronger motivation is related to the fact that most cancers are easier to treat if they are discovered early. Since cancer development is a complex process that usually involves several disruptions in the normal biological functioning of the organism, gene and/or protein microarrays can be potentially used to identify markers for early cancer detection. Furthermore, genomics and proteomics screening can also be used to differentiate between different types of cancer or for treatment selection and/or prognostic purposes.

The ultimate goal of this approach is to use a subset of genes that have patterns of expression highly correlated with the disease to construct a molecular classifier. Recently, a number of papers addressed the issue of using gene expression to distinguish between classes of conditions such as leukemia [127], colon cancer [10], malignant melanoma [38] and B-cell lymphoma [9]. In a machine learning framework, this *modus operandi* is known as **supervised learning**.

Cluster analysis is currently by far the most frequently used multivariate technique to analyze gene sequence expression data. Clustering is appropriate when there is no *a priori* knowledge about the data. In such circumstances, the only possible approach is to study the similarity between different samples or experiments. In a machine learning framework, such analysis process is known as **unsupervised learning** since there is no known desired answer for any particular gene or experiment. In contrast, in the case of supervised learning, the classifier is constructed based on a set of expression profiles for which the desired output is known (e.g. cancer or healthy). In many cases, e.g. neural networks, the construction of the classifier is performed in an iterative process called **training**. This involves repeated presentations of the molecular profiles together with their desired classification. The desired classification is considered the target output. The actual output is compared with the target output and the structure and parameters of the classifier are changed in such a way that the actual output is brought closer to the desired output.

A discussion about supervised learning techniques in the context of gene expression data should include the "curse of dimensionality" [27]. In short, the curse of dimensionality refers to the fact that the difficulty of building such a classifier increases exponentially with the dimensionality of the problem, i.e. the number of genes involved. Another important issue is the relationship between the number of parameters of the model used and the number of data points available. Essentially, the amount of data (i.e. the number of samples) has to be much larger than the number of parameters of the model. This will allow the model to generalize well. Most algorithms and models will be able to produce the desired output for those patterns used to construct the model. However, this will only reflect the **memorization abilities** of the constructed classifier. The **generalization abilities** of a model refer to its abilities to predict correctly the output for inputs that have not been used during the construction of the model (training).

When exploring such supervised learning techniques, a crucial issue is the **performance assessment**. Related to this issue, there are two contradictory requirements. The first requirement is to use as much data as possible. The motivation here is very intuitive: the more information is available at the time the model is constructed, the better the model will perform. The second requirement is to always test the model on data that were not used during the model construction. The motivation here is that one is interested in the generalization abilities of the model and testing it on data used during the training will only provide information about its memorization abilities. Since in most cases the amount of data (e.g. patient samples) is limited, the two requirements conflict. **Cross-validation** techniques must be used in order to maximize the amount of data available during the construction of the model while still obtaining a realistic estimate of its performance.

Reviews of various supervised and unsupervised techniques are presented in [49, 51, 60]. Other results using supervised learning have been recently presented in the literature [55, 56, 131, 141]. Other classical machine learning techniques such as nearest neighbor [95], boosting [28, 114], Bayesian classifiers [175, 282], support vector machines [56, 72, 116, 257], singular value decomposition [11, 282] have shown to be successful in genomics and proteomics.

The rich existing literature in machine learning and neural networks should be the starting point for the researcher interested to explore this direction [52, 130, 134, 188, 286].

17.3 Gene regulatory networks

Another important direction of future research is the inference of gene regulatory networks. Pioneering work has already been done in this area. One of the first studies focused on the exploration of the metabolic and genetic control of gene expression on a genomic scale in yeast [66, 80].

The basic thinking is that one could make inferences about the way genes interact with each other by analyzing large amounts of data in which many genes are screened over a number of time points or different conditions. Fundamentally, a gene regulatory network is a graph in which genes are represented by nodes and edges between nodes represent regulatory interactions between genes. Mathematically, this can be expressed as a weight matrix in which the weight w_{ij} between genes i and j is a quantitative representation of the stimulatory (when $w_{ij} > 0$) or inhibitory ($w_{ij} < 0$) effect of gene i on gene j [278].

Several approaches have been proposed in order to construct such weight matrices. Butte et al. proposed to use the correlation coefficient as the weight between any two genes [57, 58]. Such networks built using high correlation coefficient are called relevance network and have been shown to be able to retrieve known regulatory links between individual genes. A variation of this method is to transform the data into time frequency domain by using the Fourier transform, then calculate the correlation between genes is calculated in frequency domain. This method can eliminate the effect of time delay existing in time series data.

Ando et al. proposed genetic algorithms to find the optimum weight matrix for the model above [12]. The idea is that a large number of random networks are considered as individuals of an initial population. Operations similar to cross-over, selection and mutations are performed on the member of this population in order to generate the next generation of individuals. Each such individual is assessed using a fitness criterion that involves the goodness of fit with the given expression data as well as the minimum description length (MDL) principle [228].

Yeung et al. proposed to use the singular value decomposition (SVD) to find a family of solutions [297]. Essentially, SVD uses an idea very similar to that of the PCA. The weight matrix above is decomposed in such a way that some principal components can be constructed for the space of gene interactions. By using a limited number of these new directions, one implements a dimensionality reduction. Based on the empirical observation that the interaction matrix is usually sparse, Yeung choose the sparsest one from all the possible solutions.

A two-way clustering approach has also been used in order to identify subsets of genes and samples that yield stable and significant clustering partitions [120]. This approach has been shown to be able to identify conditionally correlated genes on some colon cancer and leukemia data.

An approach inspired from metabolic control analysis [76] has also been proposed in order to infer quantitative gene networks from microarray data. This method is particularly elegant inasmuch as it calculates co-control coefficients by using directly relative expression values as typically provided by two-channel microarrays [46, 75, 74].

The gene regulatory mechanism can also be seen as a dynamical system. Hence, it can be modelled by a set of differential equations [63]. Two methods have been proposed to use this approach and construct a model from experimental data: i) minimum weight solutions to linear equations (MWSLE) and ii) Fourier transform for stable systems (FTSS). The MWSLE approach determines gene regulation by solving under-determined linear equations. The FTSS approach adds cell cycle constraints. Other models including the RNA model, the protein model and a time delay model have also been studied using this approach.

It is important to note that some research seems to indicate that the information available at the mRNA level might be insufficient to construct complete gene regulatory networks [133]. Non-linear stability analysis has been used to show that the gene expression information at the mRNA level must be complemented by information at protein level in order to capture the entire complexity of the gene regulation phenomena.

Finally, clustering has been combined with a mutual information minimization to construct gene regulatory networks from expression data [302]. A probabilistic Boolean network is constructed as follows. Firstly, the number of possible parent gene sets and the input sets of gene variables corresponding to each gene is determined using a mutual information minimization clustering. Simulated annealing is then used to solve the optimization. Finally, each function is modelled by a perceptron consisting of a linear term and a nonlinear term. A reversible-jump Markov-chain-Monte-Carlo (MCMC) technique is used to calculate the model order and the parameters. Finally, the coefficient of determination (CoD) is employed to compute the probability of selecting different predictors for each gene.

Good but somehow dated reviews of the approaches proposed to infer gene networks can be found in the literature [82] and [251].

17.4   Conclusions

The wealth of software and resources available is a clear indication of the amount of effort currently directed at developing new tools for gene and protein expression data analysis. Microarrays have already proven invaluable tools able to shed new light on subtle and complex phenomena that happen at the genetic level. However, in order to realize its full potential, the laboratory work using microarrays needs to be complemented by a careful and detailed analysis of the results. The early years of this technology were characterized by a wealth of spectacular results obtained with very simple data analysis tools. Those were some of the low-hanging fruits of the field. A careful analysis performed with suitable tools may reveal that those low-hanging fruits were only a very small fraction of the potential crop. Furthermore, it is likely that the most spectacular results, deeply affecting the way we currently understand the genetic mechanism, are further up in the knowledge tree and their gathering will require a close symbiosis between biologists, computer scientists and statisticians.

I not only use all the brains I have, but all that I can borrow.

—Woodrow Wilson

References

[1] ***. Yeast sporulation expression data set. Technical report, Harvard-Lipper Center for Computational Genetics, 2002. http://arep.med.harvard.edu/cgi-bin/ExpressDByeast/EXDDisplayEDS?EDSNo=0.

[2] J. Aach, W. Rindone, and G. M. Church. Systematic management and analysis of yeast gene expression data. *Genome Research*, 10:431–445, 2000. http://arep.med.harvard.edu/ExpressDB.

[3] M. D. Adams, S. E. Celniker, R. A. Holt, C. A. Evans, J. D. Gocayne, P. G. Amanatides, S. E. Scherer, et al. The genome sequence of Drosophila melanogaster. *Science*, 287:2185–2195, 2000.

[4] Affymetrix. Genechip analysis suite. User guide, version 3.3, Affymetrix, 1999.

[5] Affymetrix. Expression analysis technical manual. Technical report, Affymetrix, 2000.

[6] Affymetrix. Genechip expression analysis. Technical manual, Affymetrix, 2000.

[7] Affymetrix. Statistical algorithms description document. Technical report, Affymetrix Inc., 2002.

[8] A. Aharoni, L. C. P. Keizer, H. J. Bouwneester, Z. Sun, et al. Identification of the SAAT gene involved in strawberry flavor biogenesis by use of DNA microarrays. *The Plant Cell*, 12:647–661, May 1975.

[9] A. A. Alizadeh et al. Distinct type of diffuse large B-cell lymphoma identified by gene expression profiling. *Nature*, 403:503–510, 2000.

[10] U. Alon, N. Barkai, D. A. Notterman, K. Gish, S. Ybarra, D. Mack, and A. J. Levine. Broad patterns of gene expression revealed by clustering of tumor and normal colon tissues probed by nucleotide arrays. *Proc. Natl. Acad. Sci.*, 96:6745–6750, 1999.

[11] O. Alter, P. Brown, and D. Botstein. Singular value decomposition for genome-wide expression data processing and modeling. *Proc. Natl. Acad. Sci. USA*, 97(18):10101–10106, 2000.

[12] H. Ando, S. Iba. Inference of gene regulatory model by genetic algorithms. In *Proc. of the 2001 Congress on Evolutionary Computation*, volume 1, pages 712–719, 2001.

[13] R. Aqeilan, R. Kedar, A. Ben-Yehudah, and H. Lorberboum-Galski. Mechanism of action of il2-bax; an apoptosis-inducing chimeric protein targeted against interleukin-2 receptor expressing cells. *Biochem J.*, Oct. 2002. e-pub ahead of print.

[14] Arabidopsis. Normalization method comparison. Technical report, Stanford University, 2001. http://afgc.stanford.edu/ finkel/talk.htm.

[15] S. Armstrong, J. Staunton, L. Silverman, R. Pieters, M. den Boer, M. Minden, S. Sallan, E. Lander, T. Golub, and S. Korsmeyer. MLL translocations specify a distinct gene expression profile that distinguishes a unique leukemia. *Nature Genetics*, 30(1):41–47, Jan. 2002.

[16] M. Ashburner, C. A. Ball, J. A. Blake, D. Botstein, et al. Gene ontology: tool for the unification of biology. *Nature Genetics*, 25:25–29, 2000.

[17] M. Ashburner, C. A. Ball, J. A. Blake, D. Botstein, et al. Creating the gene ontology resource: Design and implementation. *Genome Research*, 11(8):1425–1433, August 2001.

[18] S. Audic and J.-M. Claverie. The significance of digital gene expression profiles. *Genome Research*, 7(10):986–995, Oct. 1997.

[19] S. Audic and J.-M. Claverie. Visualizing the competitive recognition of TATA-boxes in vertebrate promoters. *Trends in Genetics*, 14:10–11, 1998.

[20] A. Bairoch and R. Apweiler. The SWISS-PROT protein sequence database and its supplement TrEMBL in 2000. *Nucleic Acids Research*, 28(1):45–48, Jan. 2000.

[21] P. Baldi and A. D. Long. A Bayesian framework for the analysis of microarray expression data: regularized t-test and statistical inferences of gene changes. *Bioinformatics*, 17(6):509–519, 2001. Accompanying web page at 128.200.5.223/CyberT/.

[22] L. R. Baugh, A. A. Hill, E. L. Brown, and H. C. P. Quantitative analysis of mRNA amplification by in vitro transcription. *Nucleic Acids Research*, 29(5):E29, March 2001.

[23] T. Bayes. An essay towards solving a problem in the doctrine of chances. *Philosophical Transactions of the Royal Society*, 53:370–418, 1763.

[24] T. Bayes. An essay towards solving a problem in the doctrine of chances. *Philosophical Transactions of the Royal Society*, 54:298–310, 1764. Reprinted in [25].

[25] T. Bayes. An essay towards solving a problem in the doctrine of chances – reprint of the 1763 article. *Biometrica*, 45:293–298, 1958.

[26] A. J. Bell and T. H. Sejnowski. An information-maximisation approach to blind separation and blind deconvolution. *Neural Computation*, 7(6):1004–1034, 1995.

[27] R. Bellman. *Adaptive Control Processes: A Guided Tour.* Princeton University Press, 1961.

[28] A. Ben-Dor, L. Bruhn, N. Friedman, I. Nachman, M. Schummer, and Z. Yakhini. Tissue classification with gene expression profiles. In *Proc. Fourth Annual Inter. Conf. on Computational Molecular Biology (RECOMB 2000)*, pages 54–64, Tokyo, Japan, April 2000.

[29] A. Ben-Dor, R. Shamir, and Z. Yakhini. Clustering gene expression patterns. *J. of Computational Biology*, 6(3/4):281–297, 1999.

[30] Y. Benjamini and Y. Hochberg. Controlling the false discovery rate: A practical and powerful approach to multiple testing. *Journal of The Royal Statistical Society B*, 57(1):289–300, 1995.

[31] Y. Benjamini and D. Yekutieli. The control of the false discovery rate in multiple testing under dependency. *Annals of Statistics*, 29(4):1165–1188, August 2001.

[32] D. A. Benson, I. Karsch-Mizrachi, D. J. Lipman, J. Ostell, B. A. Rapp, and D. L. Wheeler. Genbank. *Nucleic Acids Research*, 28(1):15–18, Jan. 2000.

[33] D. A. Benson, I. Karsch-Mizrachi, D. J. Lipman, J. Ostell, B. A. Rapp, and D. L. Wheeler. Genbank. *Nucleic Acids Research*, 30(1):17–20, 2002.

[34] C. Bernard. *An introduction to the Study of Experimental Medicine.* Dover Publications, New York, 1957.

[35] A. Bhattacharjee, W. Richards, J. Staunton, C. Li, S. Monti, P. Vasa, C. Ladd, J. Beheshti, R. Bueno, M. Gillette, M. Loda, G. Weber, E. Mark, E. Lander, W. Wong, B. Johnson, T. Golub, D. Sugarbaker, and M. M. Classification of human lung carcinomas by mrna expression profiling reveals distinct adenocarcinoma subclasses. *Proc. Natl. Acad. Sci. USA*, 98(24):13790–5, Nov. 2001.

[36] BioDiscovery. Imagene – User's manual. Technical report, BioDiscovery Inc., 2001.

[37] B. Biosciences. BD PowerBlot Western Array Screening Service. Technical report, BD Biosciences, 2002. http://www.bdbiosciences.com.

[38] M. Bittner, P. Meltzer, Y. Chen, Y. Jiang, E. Seftor, M. Hendrix, M. Radmacher, R. Simon, Z. Yakhini, A. Ben-Dor, N. Sampas, E. Dougherty, E. Wang, F. Marincola, C. Gooden, J. Lueders, A. Glatfelter, P. Pollock, J. Carpten, E. Gillanders, D. Leja, K. Dietrich, C. Beaudry, M. Berens, D. Alberts, V. Sondak, N. Hayward, and J. Trent. Molecular classification of cutaneous malignant melanoma by gene expression profiling. *Nature*, 406(6795):536–540, 2000.

[39] T. Blundell and S. Fortier. Data mining in crystallography – report on the 29th Crystallographic Course and Summer School. Technical report, E. Majorana Center, 1999. http://www.geomin.unibo.it/orgv/erice/DataMini.htm.

[40] B. M. Bolstad, R. A. Irizarry, M. Astrand, and T. P. Speed. A comparison of normalization methods for high density oligonucleotide array data based on variance and bias. *Bioinformatics*, 19(2):185–193, Jan. 2003.

[41] C. E. Bonferroni. *Il calcolo delle assicurazioni su gruppi di teste*, chapter "Studi in Onore del Professore Salvatore Ortu Carboni", pages 13–60. Rome, 1935.

[42] C. E. Bonferroni. Teoria statistica delle classi e calcolo delle probabilità. *Pubblicazioni del Istituto Superiore di Scienze Economiche e Commerciali di Firenze*, 8:3–62, 1936.

[43] C. Bouton, G. W. Henry, and J. Pevsner. Database referencing of array genes online – DRAGON. Technical report, Kennedy Krieger Institute, 2001. http://pevsnerlab.kennedykrieger.org/dragon.htm.

[44] C. M. Bouton and J. Pevsner. DRAGON View: information visualization for annotated microarray data. *Bioinformatics*, 18:323–324, 2002.

[45] J. Boyd. Adenovirus E1B 19 kDa and Bcl-2 proteins interact with a common set of cellular proteins. *Cell*, 79(2):341–351, Dec. 1994.

[46] P. Brazhnik, A. de la Fuente, and P. Mendes. Gene networks: how to put the function in genomics. *Trends Biotechnol.*, 20(11):467–472, Nov. 2002.

[47] A. Brazma. On the importance of standardisation in life sciences. *Bioinformatics*, 17(2):113–114, 2001.

[48] A. Brazma, P. Hingamp, J. Quackenbush, G. Sherlock, et al. Minimum information about a microarray experiment (MIAME)-toward standards for microarray data. *Nature Genetics*, 29(4):365–371, Dec. 2001.

[49] A. Brazma, I. Jonassen, I. Eidhammer, and D. Gilbert. Approaches to the automatic discovery of patterns in biosequences. *Journal of Computational Biology*, 5(2):279–305, 1998.

[50] A. Brazma, I. Jonassen, J. Vilo, and E. Ukkonen. Predicting gene regulatory elements in silico on a genomic scale. *Genome Research*, 8(11):1202–1215, 1998.

[51] A. Brazma and J. Vilo. Gene expression data analysis. *Federation of European Biochemical Societies Letters*, 480(23893):17–24, 2000.

[52] L. Breiman, J. H. Friedman, R. A. Olsen, and C. J. Stone. *Classification and regression trees*. Wadsworth and Brooks, New York, 1984.

[53] H. Brightman and H. Schneider. *Statistics for Business Problem Solving*. Southwestern Publishing Co., 1994.

[54] C. C. Brown and T. R. Fears. Exact significance levels for multiple binomial testing with application to carcinogenicity screens. *Biometrics*, 37:763 –774, 1981.

[55] M. P. S. Brown, W. B. Grundy, D. Lin, N. Christianini, C. W. Sugnet, M. Ares, and D. Haussler. Support vector machine classification of microarray gene expression data. Technical report, UCSC, 1999.

[56] M. P. S. Brown, W. B. Grundy, D. Lin, N. Cristianini, C. W. Sugnet, T. S. Furgey, M. A. Manuel, and D. Haussler. Knowledge-based analysis of microarray gene expression data by using support vector machines. *Proc. Natl. Acad. Sci. USA*, 97(1):262–267, 2000.

[57] A. Butte, P. Tamayo, D. Slonim, T. Golub, and I. Kohane. Discovering functional relationships between RNA expression and chemotherapeutic susceptibility using relevance networks. *Proc. Natl. Acad. Sci.*, 97(22):12182–12186, 2000.

[58] A. J. Butte, P. Tamayo, D. Slonim, T. R. Golub, and I. S. Kohane. Discovering functional relationships between RNA expression and chemotherapeutic susceptibility using relevance networks. *PNAS*, 97(22):12182–12186, 2000.

[59] G. Casella. *Statistical Inference*. Duxbury, Pacific Grove, 2002.

[60] J. E. Celis, M. Kruhoffer, I. Gromova, C. Frederiksen, M. Ostergaard, T. Thykjaer, et al. Gene expression profiling: monitoring transcription and translation products using DNA microarrays and proteomics. *Federation of European Biochemical Societies Letters*, (23892):1–15, 2000.

[61] J. M. Chambers. *Programming with Data: A Guide to the S Language*. Springer, 1998.

[62] F. Chen, R. Chang, M. Trivedi, Y. Capetanaki, and V. Cryns. Caspase proteolysis of desmin produces a dominant negative inhibitor of intermediate filaments and promotes apoptosis. *J. Biol. Chem.*, Dec. 2002. e-pub ahead of print.

[63] T. Chen, H. L. He, and G. M. Church. Modeling gene expression with differential equations. In *Pacific Symposium on Biocomputing*, volume 4, pages 29–40, 1999. Available online at http://psb.stanford.edu.

[64] Y. Chen, E. R. Dougherty, and M. L. Bittner. Ratio-based decisions and the quantitative analysis of cDNA microarray images. *Journal of Biomedical Optics*, 2(4):364–374, 1997.

[65] R. Cho, M. Huang, M. Campbell, H. Dong, et al. Transcriptional regulation and function during the human cell cycle. *Nature Genetics*, 27:48–54, 2001.

[66] S. Chu, J. DeRisi, M. Eisen, J. Mulholland, D. Botstein, P. Brown, and I. Herskowitz. The transcriptional program of sporulation in budding yeast. *Science*, 282(5393):699–705, 1998.

[67] T.-M. Chu, B. Weir, and R. Wolfinger. A systematic statistical linear modeling approach to oligonucleotide array experiments. *Mathematical Biosciences*, 176(1):35–51, 2002.

[68] J.-M. Claverie. Computational methods for the identification of differential and coordinated gene expression. *Human Molecular Genetics*, 8(10):1821–1832, 1999. Available online at http://biosun01.biostat.jhsph.edu/ gparmigi/688/claverie1999.pdf.

[69] W. Cleveland. Robust locally weighted regression and smoothing scatterplots. *Journal of the American Statistical Association*, 74:829–836, 1979.

[70] W. Cleveland and S. Devlin. Locally weighted regression: An approach to regression analysis by local fitting. *Journal of the American Statistical Association*, 83:596–610, 1983.

[71] CNRS. Microarray databases. Technical report, Centre National de la Recherche Scietifique, 2001. http://www.biologie.ens.fr/en/genetiqu/puces/bddeng.html.

[72] C. Cortes and V. Vapnik. Support-vector networks. *Machine Learning*, 20(3):273–297, 1995.

[73] K. Dahlquist, N. Salomonis, K. Vranizan, S. Lawlor, and B. Conklin. GenMAPP, a new tool for viewing and analyzing microarray data on biological pathways. *Nature Genetics*, 31(1):19–20, May 2002.

[74] A. de la Fuente, P. Brazhnik, and P. Mendes. Linking the genes: inferring quantitative gene networks from microarray data. *Trends in Genetics*, 18(8):395–398, Aug. 2002.

[75] A. de la Fuente and P. Mendes. Quantifying gene networks with regulatory strengths. *Mol Biol Rep.*, 29(1-2):73–77, 2002.

[76] A. de la Fuente, J. Snoep, H. Westerhoff, and P. Mendes. Metabolic control in integrated biochemical systems. *Eur J Biochem.*, 269(18):4399–4408, Sept. 2002.

[77] J. Delehanty and F. S. Ligler. Method for printing functional protein microarrays. *Biotechniques*, 34(2):380–385, Feb 2003.

[78] S. Deng, T.-M. Chu, and R. D. Wolfinger. Transcriptome variability in the normal mouse.

[79] S. Der, B. Williams, and R. Silverman. Identification of genes differentially regulated by interferon alpha, beta, or gamma using oligonucleotide arrays. *Proc. Natl. Acad. of Sci.*, 26(95):15623–15628, 1998.

[80] J. L. DeRisi, V. R. Iyer, and P. O. Brown. Exploring the metabolic and genetic control of gene expression on a genomic scale. *Science*, 278:680–686, 1997.

[81] J. L. DeRisi, L. Penland, P. O. Brown, M. L. Bittner, et al. Use of a cDNA microarray to analyse gene expression patterns in human cancer. *Nature Genetics*, 14(4):457–460, 1996.

[82] P. D'haeseleer, S. Liang, and R. Somogyi. Genetic network inference: From co-expression clustering to reverse engineering. *Bioinformatics*, 16(8):707–726, 2000.

[83] P. D'haeseller. *Genetic Network Inference: From Co-Expression Clustering to Reverse Engineering*. PhD Thesis, University of New Mexico, 2000.

[84] P. D'haeseller, S. Liang, and R. Somogyi. Genetic network inference: From co-expression clustering to reverse engineering. *Bioinformatics*, 8(16):707–726, 2000.

[85] K. Do and W. Broom. *The Analysis of Gene Expression Data: Methods and Software*, chapter GeneClus. Springer, New York, 2003.

[86] K. Drlica. *Understanding DNA and Gene Cloning*. John Wiley and Sons, 1996.

[87] S. Drăghici. Piecewise linearization method for the normalization of cDNA and protein microarrays in multi-channel experiments. Technical report, Biodiscovery Inc., 2001. Patent Application.

[88] S. Drăghici. Statistical intelligence: effective analysis of high-density microarray data. *Drug Discovery Today*, 7(11):S55–S63, 2002.

[89] S. Drăghici and P. Khatri. Onto-Express web site. Technical report, Wayne State University, 2002. http://vortex.cs.wayne.edu.

[90] S. Drăghici, P. Khatri, R. P. Martins, G. C. Ostermeier, and S. A. Krawetz. Global functional profiling of gene expression. *Genomics*, 81(2):98–104, Feb. 2003.

[91] S. Drăghici, P. Khatri, A. Shah, and M. Tainsky. Assessing the functional bias of commercial microarrays using the Onto-Compare database. *BioTechniques*, Microarrays and Cancer: Research and Applications, March 2003.

[92] S. Drăghici, A. Kuklin, B. Hoff, and S. Shams. Experimental design, analysis of variance and slide quality assessment in gene expression arrays. *Current Opinion in Drug Discovery and Development*, 4(3):332–337, 2001.

[93] S. Drăghici, O. Kulaeva, A. Petrov, B. Hoff, A. Kuklin, S. Shams, and M. Tainsky. Noise sampling method: an ANOVA approach allowing robust selection of differentially regulated genes measured by dna microarrays. *to appear in Bioinformatics*, (accepted), 2003.

[94] S. Drăghici and B. Potter. Data mining HIV drug resistance with neural networks. *Bioinformatics*, 19(1):98–107, Jan 2003.

[95] R. Duda and P. Hart. *Pattern Classification and Scene Analysis*. John Wiley and Sons, New York, 1973.

[96] R. Duda, P. Hart, and D. Stork. *Pattern classification*. John Wiley and Sons, New York, 2000.

[97]   S. Dudoit and J. Fridlyand. A prediction-based resampling method to estimate the number of clusters in a dataset. *Genome Biology*, 3(7):0036.1–0036.21, 2002.

[98]   S. Dudoit, Y. H. Yang, M. Callow, and T. Speed. Statistical models for identifying differentially expressed genes in replicated cDNA microarray experiments. Technical Report 578, University of California, Berkeley, 2000. Available at www.stat.berkeley.edu/tech-reports/index.html.

[99]   D. Duggan, M. Bittner, Y. Chen, , P. Meltzer, and J. Trent. Expression profiling using cdna microarrays. *Nat Genet.*, 21(1 Suppl):10–14, 1999.

[100] D. Durand. A dictionary for statimagicians. *The American Statistician*, 24(3):21, 1970.

[101] EBI. ArrayExpress. Technical report, European Bioinformatics Institute, 2001. http://www.ebi.ac.uk/arrayexpress/index.html.

[102] EBI. Microarray gene expression database group. Technical report, European Bioinformatics Institute, 2001. http://www.mged.org/.

[103] B. Efron and R. J. Tibshirani. *An Introduction to Bootstrap*. Chapman and Hall, London, UK, 1993.

[104] M. Eisen, P. Spellman, P. O. Brown, and D. Botstein. Cluster analysis and display of genome-wid expression patterns. In *Proc. Natl. Acad. Sci. USA*, volume 95, pages 14863–14868, December 8 1998. Available online at http://biosun01.biostat.jhsph.edu/ gparmigi/688/eisen1998.pdf.

[105] D. Eisenberg, E. M. Marcotte, loannis Xenarios, and T. O. Yeates. Protein function in the post-genomic era. *Nature*, 405:823–826, 2000.

[106] R. M. Ewing, A. B. Kahla, O. Poirot, F. Lopez, S. Audic, and J.-M. Claverie. Large-scale statistical analyses of rice ESTs reveal correlated patterns of gene expression. *Genome Research*, 9:950–959, 1999.

[107] J.-B. Fan, X. Chen, M. Halushka, A. Berno, X. Huang, T. Ryder, R. Lipshutz, D. Lockhart, and A. Chakravarti. Parallel genotyping of human SNPs using generic high-density oligonucleotide tag arrays. *Genome Research*, (10):853–860, 2000.

[108] J. Felsenstein. Confidence limits on phylogenies: An approach using the bootstrap. *Evolution*, 39:783–791, 1985.

[109] D. B. Finkelstein, R. Ewing, J. Gollub, F. Sterky, S. Somerville, and J. M. Cherry. Iterative linear regression by sector. In S. M. Lin and K. F. Johnson, editors, *Methods of Microarray Data Analysis*, pages 57–68. Kluwer Academic, 2002.

[110] L. D. Fisher and G. van Belle. *Biostatistics: A Methodology for Health Sciences*. John Wiley and Sons, New York, 1993.

[111] R. A. Fisher. *The Design of Experiments*. Oliver and Boyd, 1942.

[112] W. Fleischmann, S. Moller, A. Gateau, and R. Apweiler. A novel method for automatic functional annotation of proteins. *Bioinformatics*, 15(3):228–233, March 1999.

[113] C. Fraley and A. E. Raftery. Mclust: Software for model-based clustering, density estimation and discriminant analysis. Technical Report 415, Department of Statistics, University of Washington, Oct. 2002.

[114] Y. Freung. A decision-theoretic generalization of on-line learning and an application to boosting. *J. Computer and System Science*, 55:119–139, 1997.

[115] S. H. Friend. How DNA microarrays and expression profiling will affect clinical practice. *British Medical Journal*, 319:1–2, 1999.

[116] T. Furey, N. Cristianini, N. Duffy, D. Bednarski, M. Schummer, and D. Haussler. Support vector machine classification and validation of cancer tissue samples using microarray expression data. *Bioinformatics*, 16(10):906–914, 2000.

[117] C. C. Gaither and A. E. Cavazos-Gaither. *Statistically Speaking*. Institute of Physics, 1996.

[118] T. Galitski, A. J. Saldanha, C. A. Styles, E. S. Lander, and G. R. Fink. Ploidy regulation of gene expression. *Science*, 285:251–254, 1999.

[119] A. Gavin, M. Bosche, K. R., P. Grandi, et al. Functional organization of the yeast proteome by systematic analysis of protein complexes. *Nature*, 415(6868):141–147, 2002.

[120] G. Getz, E. Levine, and E. Domany. Coupled two-way clustering analysis of gene microarray data. *Proc. Natl. Acad. Sci.*, 97(22):12079–12084, 2000.

[121] D. Ghosh and A. M. Chinnaiyan. Mixture modelling of gene expression data from microarray experiments. *Bioinformatics*, 18:275–286, 2002.

[122] G. Gibson. Mmanmada tutorial. *http://statgen.ncsu.edu/ggibson/Pubs.htm*, 2002.

[123] G. Gigerenzer and U. Hoffrage. How to improve bayesian reasoning without instruction: Frequency formats. *Psychological Review*, 102(4):684–704, 1995.

[124] A. Giodini, M. Kallio, N. Wall, G. Gorbsky, S. Tognin, P. Marchisio, M. Symons, and D. Altieri. Regulation of microtubule stability and mitotic progression by survivin. *Cancer Research*, 62(9):2462–2467, May 2002.

[125] T. Glover and K. Mitchell. *An Introduction to Biostatistics*. McGraw-Hill, New York, 2002.

[126] GO. Gene ontology. Technical report, Gene Ontology Consortium, 2001. http://www.geneontology.org/.

[127] T. R. Golub, D. K. Slonim, P. Tamayo, C. Huard, M. Gaasenbeek, J. P. Mesirov, H. Coller, H. Lo, J. R. Downing, and e. a. M. A. Caligiuri. Molecular classification of cancer: Class discovery and class predication by gene expression monitoring. *Science*, 286:531–537, 1999.

[128] W. Gosset. The probable error of a mean. *Biometrika*, 6:1–25, 1908.

[129] D. Gusfield. Bioinformatics FTE proposal. Technical report, University of California, Davis, 1999. http://genomics.ucdavis.edu/~gusfield/bioinfomaster.

[130] M. T. Hagan, H. B. Demuth, and M. H. Beale. *Neural Network Design*. Brooks Cole, Boston, 1995.

[131] T. Hastie, R. Tibshirani, D. Botstein, and P. Brown. Supervised harvesting of expression trees. *Genome Biology*, 2(1):3.1–3.12, 2001.

[132] T. Hastie, R. Tibshirani, M. B. Eisen, A. Alizadeh, R. Levy, L. Staudt, W. Chan, D. Botstein, and P. Brown. 'Gene shaving' as a method for indentifying distinct sets of genes with similar expression patterns. *Genome Biology*, 1(2):1–21, 2000.

[133] V. Hatzimanikatis and K. H. Lee. Dynamical analysis of gene networks requires both mRNA and protein expression information. *Metabolic Engineering*, 1(4):275–281, Oct. 1999.

[134] S. Haykin. *Neural Networks – A comprehensive foundation*. Prentice Hall, 1999.

[135] I. Hedenfalk, D. Duggan, Y. Chen, M. Radmacher, M. Bittner, R. Simon, P. Meltzer, B. Gusterson, M. Esteller, M. Raffeld, Z. Yakhini, A. Ben-Dor, E. Dougherty, J. Kononen, L. Bubendorf, W. Fehrle, S. Pittaluga, S. Gruvberger, N. Loman, O. Johannsson, H. Olsson, B. Wilfond, G. Sauter, O. Kallioniemi, A. Borg, and J. Trent. Gene expression profiles in hereditary breast cancer. *New England Journal of Medicine*, 244(8):539–548, 2001.

[136] P. Hedge, R. Qi, K. Abernathy, C. Gay, S. Dharap, R. Gaspard, J. Earle-Hughes, E. Snasrud, N. Lee, and J. Quackenbush. A concise guide to cDNA microarray analysis. *Biotechniques*, 29(3):548–562, 2000.

[137] J. Herrero, A. Valencia, and J. Dopazo. A hierarchical unsupervised growing neural network for clustering gene expression patterns. *Bioinformatics*, 17(2):126–136, Feb 2001.

[138] J. Hertz, A. Krogh, and R. G. Palmer. *Introduction to the Theory of Neural Computation*. Perseus Books, 1991.

[139] R. Herwig, A. Poustka, C. Muller, C. Bull, H. Lehrach, and J. O'Brien. Large-scale clustering of cDNA-fingerprinting data. *Genome Research*, 9(11):1093–1105, 1999.

[140] C. Hetz, M. Hunn, P. Rojas, V. Torres, L. Leyton, and A. Quest. Caspase-dependent initiation of apoptosis and necrosis by the fas receptor in lymphoid cells: onset of necrosis is associated with delayed ceramide increase. *J. Cell. Sci.*, 115:4671–4683, Dec. 2002.

[141] L. J. Heyer, S. Kruglyak, and S. Yooseph. Exploring expression data: Identification and analysis of coexpressed genes. *Genome Research*, 9:1106–1115, 1999.

[142] J. Heyse and D. Rom. Adjusting for multiplicity of statistical tests in the analysis of carcinogenicity studies. *Biometrical Journal*, 30:883–896, 1988.

[143] A. A. Hill, C. P. Hunter, B. T. Tsung, G. Tucker-Kellogg, and E. L. Brown. Genomic analysis of gene expression in *C. elegans*. *Science*, 290:809–812, 2000.

[144] S. Hilsenbeck, W. Friedrichs, R. Schiff, P. O'Connell, R. Hansen, C. Osborne, and S. W. Fuqua. Statistical analysis of array expression data as applied to the problem of Tamoxifen resistance. *Journal of the National Cancer Institute*, 91(5):453–459, 1999.

[145] Y. Hochberg. A sharper Bonferroni procedure for multiple tests of significance. *Biometrika*, 75:800–802, 1988.

[146] Y. Hochberg and A. C. Tamhane. *Multiple Comparison Procedures*. John Wiley and Sons, Inc., New York, 1987.

[147] S. Hodge, G. Hodge, R. Flower, P. Reynolds, R. Scicchitano, and M. Holmes. Up-regulation of production of tgf-beta and il-4 and down-regulation of il-6 by apoptotic human bronchial epithelial cells. *Immunol. Cell. Biol.*, 80(6):537–543, Dec. 2002.

[148] B. Holland and M. D. Copenhaver. An improved sequentially rejective Bonferroni test procedure. *Biometrica*, 43:417–423, 1987.

[149] S. Holm. A simple sequentially rejective multiple test procedure. *Scandinavian Journal of Statistics*, 6:65–70, 1979.

[150] T. M. Houts. Improved 2-color Exponential normalization for microarray analyses employing cyanine dyes. In S. Lin, editor, *Procedings of CAMDA 2000, "Critical Assesment of Techniques for Microarray Data Mining"*, December 18-19, Durham, NC, 2000. Duke University Medical Center.

[151] W.-P. Hsieh, T.-M. Chu, and R. D. Wolfinger. Who are those strangers in the latin square? *in CAMDA proceedings, Duke University*, 2003.

[152] http://ep.ebi.ac.uk/EP. Expression profiler – analysis and clustering of gene expression and sequence data.

[153] http://ep.ebi.ac.uk/EP/GO. EP:GO - Browser and analysis for Gene Ontology.

[154] http://www.ebi.ac.uk/ego. GO at EBI - QuickGO.

[155] http://www.ebi.ac.uk/interpro. Ebi interpro database.

[156] http://www.geneontology.org. Gene Ontology Consortium.

[157] http://www.geneontology.org/doc/GO.annotation.html.

[158] http://www.geneontology.org/doc/GO.defs. GO Term Definitions.

[159] http://www.geneontology.org/doc/GO.doc.html. GO General Documentation.

[160] http://www.geneontology.org/doc/GO.evidence.html. GO Evidence Codes.

[161] http://www.geneontology.org/doc/GO.xrf_abbs. GO Collaborative Database Abbreviations.

[162] http://www.godatabase.org/dev. Gene Ontology Tools.

[163] http://www.informatics.jax.org/searches/GO_form.shtml. GO Browser at Mouse Genome Informatics.

[164] A. Inc. *GeneChip Expression Analysis Algorithm Tutorial*. Affymetrix Inc., Santa Clara, CA, 1999.

[165] R. A. Irizarry, B. Hobbs, F. Collin, Y. D. Beazer-Barclay, K. J. Antonellis, U. Scherf, and T. P. Speed. Exploration, normalization, and summaries of high density oligonucleotide array probe level data. *Biostatistics*, to appear, 2003.

[166] C. H. Jiang, J. Tsien, P. Schultz, and Y. Hu. The effects of aging on gene expression in the hypothalamus and cortex of mice. *PNAS*, 98(4):1930–1934, 2001.

[167] W. Jin, R. Riley, R. D. Wolfinger, K. P. White, G. Passador-Gurgel, and G. Gibson. Contributions of sex, genotype and age to transcriptional variance in drosophila melanogaster. *Nature Genetics*, 29:389–395, 2001.

[168] R. A. Johnson and D. W. Wichern. *Applied Multivariate Statistical Analysis*. Prentice-Hall, 1998.

[169] G. Kamberova and S. Shah, editors. *DNA Array Image Analysis: Nuts and Bolts*. DNA Press, Eagleville, PA, 2002.

[170] M. Kanehisa and S. Goto. KEGG: Kyoto encyclopedia of genes and genomes. *Nucleic Acids Research*, 28(1):27–30, January 2000.

[171] M. Kanehisa, S. Goto, S. Kawashima, and A. Nakaya. The KEGG databases at GenomeNet. *Nucleic Acids Research*, 30(1):42–46, January 2002.

[172] L. Kari. DNA computing: arrival of biological mathematics. *The Mathematical Intelligencer*, 19(2):9–22, 1997.

[173] L. Kari. Computing with DNA. *Methods in Molecular Biology*, 132:413–430, 2000.

[174] L. Kari. DNA computing in vitro and in vivo. *FUTURE GENER COMP SY*, 17(7):823–834, May 2001.

[175] A. Keller, M. Shummer, L. Hood, and W. Ruzzo. Bayesian classification of DNA array expression data. Technical report, University of Washington, 2000. UW-CSE-2000-08-01.

[176] J. W. Kennedy, G. W. Kaiser, L. D. Fisher, J. K. Fritz, W. Myers, J. Mudd, and T. Ryan. Clinical and angiographic predictors of operative mortality from the collaborative study in coronary artery surgery (CASS). *Circulation*, 63(4):793–802, 1981.

[177] T. Kepler, L. Crosby, and K. Morgan. Normalization and analysis of DNA microarray data by self-consistency and local regression. *Submitted to Nucleic Acids Research*, 2001.

[178] K. Kerr, E. Leiter, and G. Churchill. Analysis of a designed microarray experiment. In *Proceedings of the IEEE-Eurasip Nonlinear Signal and Image Processing Workshop*, June 3-6 2001.

[179] M. K. Kerr, C. A. Afshari, L. Bennett, P. Bushel, J. Martinez, N. J. Walker, and G. A. Churchill. Statistical analysis of a gene expression microarray experiment with replication. *Statistica Sinica*, 12(1):203–218. www.jax.org/research/churchill/pubs/index.html.

[180] M. K. Kerr and G. A. Churchill. Bootstrapping cluster analysis: Assessing the reliability of conclusions from microarray experiments. *Proceedings of the National Academy of Science USA*, 98(16):8961–8965, July 2001. www.jax.org/research/churchill/pubs/index.html.

[181] M. K. Kerr and G. A. Churchill. Bootstrapping cluster analysis: Assessing the reliability of conclusions from microarray experiments. *Proceedings of the National Academy of Science USA*, 98(16):8961–8965, July 2001. www.jax.org/research/churchill/pubs/index.html.

[182] M. K. Kerr and G. A. Churchill. Statistical design and the analysis of gene expression microarray data. *Genetical Research*, 77(2):123–128, Apr 2001. www.jax.org/research/churchill/pubs/index.html.

[183] M. K. Kerr, M. Martin, and G. A. Churchill. Analysis of variance for gene expression microarray data. *Journal of Computational Biology*, 7(6):819–837, 2000.

[184] J. Khan, L. H. Saal, M. L. Bittner, Y. Chen, J. M. Trent, and P. S. Meltzer. Expression profiling in cancer using cDNA microarrays. *Electrophoresis*, 20(2), 1999.

[185] P. Khatri, S. Drăghici, G. C. Ostermeier, and S. A. Krawetz. Profiling gene expression using Onto-Express. *Genomics*, 79(2):266–270, Feb. 2002.

[186] M. Kimura, S. Kotani, T. Hattori, N. Sumi, T. Yoshioka, K. Todokoro, and Y. Okano. Cell cycle-dependent expression and spindle pole localization of a novel human protein kinase. *Journal of Biological Chemistry*, 272(21):13766–13771, May 1997.

[187] T. Kohonen. Learning vector quantization. *Neural Networks*, 1(suppl. 1):303, 1988.

[188] T. Kohonen. *Self-Organizing Maps*. Springer, Berlin, 1995.

[189] E. Kretschmann, W. Fleischmann, and A. R. Automatic rule generation for protein annotation with the C4.5 data mining algorithm applied on SWISS-PROT. *Bioinformatics*, 17(10):920–926, Oct. 2001.

[190] W. Kuo, T. Jenssen, A. Butte, L. Ohno-Machado, and I. Kohane. Analysis of matched mRNA measurements from two different microarray technologies. *Bioinformatics*, 18(3):405–412, 2002.

[191] A. E. Lash, C. M. Tolstoshev, L. Wagner, G. D. Shuler, R. L. Strausberg, G. J. Riggins, and S. F. Altschul. SAGEmap: A public gene expression resource. *Genome Research*, 10:1051–1060, 2000.

[192] J. K. Lee and M. O'Connell. *The analysis of gene expression data: methods and software*, chapter An S-PLUS Library for the Analysis and Visualization of Differential Expression. Springer, N.Y., 2003.

[193] M.-L. T. Lee, F. C. Kuo, G. A. Whitmore, and J. Sklar. Importance of replication in microarray gene expression studies: Statistical methods and evidence from repetitive cDNA hybridizations. *Proc. Natl. Acad. Sci.*, 97(18):9834–9839, 2000.

[194] C. Li and W. H. Wong. Model-based analysis of oligonucleotide arrays: Expression index computation and outlier detection. *Proc. Natl. Acad. Sci.*, 98(1):31–36, Jan 2001. Software available at http://www.dchip.org.

[195] C. Li and W. H. Wong. Model-based analysis of oligonucleotide arrays: Model validation, design issues and standard error application. *Genome Biology*, 2(8):1–11, 2001. Software available at http://www.dchip.org.

[196] F. Li, G. Ambrosini, E. Chu, J. Plescia, S. Tognin, P. Marchisio, and D. Altieri. Control of apoptosis and mitotic spindle checkpoint by survivin. *Nature*, 396(6711):580–584, Dec. 1998.

[197] J. Li, M. Pankratz, and J. Johnson. Differential gene expression patterns revealed by oligonucleotide versus long cDNA arrays. *Toxicol Sci*, 69(2):383–390, Oct. 2002.

[198] W. Liebermeister. Independent component analysis of gene expression data. In *Proc. of German Conference on Bioinformatics GCB'01*, 2001. http://www.bioinfo.de/isb/gcb01/poster/index.html.

[199] R. Lipshutz, S. Fodor, T. Gingeras, and D. Lockhart. High density synthetic oligonucleotide arrays. *Nature Genetics*, 21(1):20–24, Jan. 1999.

[200] D. J. Lockhart, H. Dong, M. Byrne, M. Folletie, M. V. Gallo, M. S. Chee, M. Mittmann, C. Want, M. Kobayashi, H. Horton, and E. L. Brown. DNA expression monitoring by hybridization of high density oligonucleotide arrays. *Nature Biotechnology*, 14(13):1675–1680, 1996.

[201] D. J. Lockhart and E. A. Winzeler. Genomics, gene expression and DNA arrays. *Nature*, 405:827–836, 2000.

[202] A. Long, H. Mangalam, B. Chan, L. Tolleri, G. W. Hatfield, and P. Baldi. Improved statistical inference from DNA microarray data using analysis of variance and a Bayesian statistical framework. *J. Biol. Chem.*, 276(23):19937–19944, 2001.

[203] G. Luo, P. Ducy, M. D. McKee, G. J. Pinero, E. Loyer, R. R. Behringer, and G. Karsenty. Spontaneous calcification of arteries and cartilage in mice lacking matrix gla protein. *Nature*, 6(386, 6620):78–81, 1997.

[204] A. M. On the representation of gene function in genetic databases. In *ISMB, Montreal*, 1998.

[205] M. Magrane and R. Apweiler. Organisation and standardisation of information in SWISS-PROT and TrEMBL. *Data Science Journal*, 1(1):13–18, 2002.

[206] M. Z. Man, Z. Wang, and Y. Wang. POWER_SAGE: Comparing statistical tests for SAGE experiments. *Bioinformatics*, 16(11):953–959, 2000.

[207] E. Manduchi, G. R. Grant, S. E. McKenzie, G. C. Overton, S. Surrey, and C. J. Stoeckert. Generation of patterns from gene expression data by assigninig confidence to differentially expressed genes. *Bioinformatics*, 16(8):685–698, 2000.

[208] H. Mangalam, J. Stewart, J. Zhou, K. Schlauch, M. Waugh, G. Chen, A. D. Farmer, G. Colello, and J. W. Weller. GeneX: An open source gene expression database and integrated tool set. *IBM Systems Journal*, 40(2):552–569, 2001. http://www.ncgr.org/genex/.

[209] M. J. Marton, J. L. DeRisi, H. A. Bennett, V. R. Iyer, M. R. Meyer, C. J. Roberts, R. Stoughton, J. Buchard, D. Slade, H. Dia, D. Bassett, Jr., L. H. Hartwell, P. O. Brown, and S. H. Friend. Drug target validation and identification of secondary drug effects using DNA microarrays. *Nature Medicine*, 4:1293–1302, 1998.

[210] Merriam-Webster, editor. *Merriam-Webster's Collegiate Dictionary*. Merriam-Webster, Inc., 1998.

[211] M. Meyerson, G. H. Enders, C. L. Wu, L. K. Su, C. Gorka, C. Nelson, E. Harlow, and L. H. Tsai. A family of human cdc2-related protein kinases. *Embo Journal*, 11(8):2909–2917, August 1992.

458   *References*

[212] G. Michaels, D. B. Carr, X. Wen, S. Fuhrman, M. Askenazi, and R. Somogyi. Cluster analysis and data visualization of large-scale gene expression data. In *Pac. Symp. Biocomp.*, volume 3, pages 20–29, 1998.

[213] R. Michelmore, K. Burtis, and D. Gusfield. The UC Davis genomics initiative. Technical report, University of California, Davis, 2000. http://genomics.ucdavis.edu/what.html.

[214] D. C. Montgomery. *Design and analysis of experiments*. John Wiley and Sons, New York, 2001.

[215] M. Newton, C. Kendziorski, C. Richmond, F. R. Blattner, and K. Tsui. On differential variability of expression ratios: Improving statistical inference about gene expresison changes from microarray data. *Journal of Computational Biology*, 8(1):37–52, 2001.

[216] NHGRI. ArrayDB. Technical report, National Human Genome Research Institute, 2001. http://genome.nhgri.nih.gov/arraydb/schema.html.

[217] H. Ogata, S. Goto, K. Sato, W. Fujibuchi, H. Bono, and M. Kanehisa. KEGG: Kyoto encyclopedia of genes and genomes. *Nucleic Acids Research*, 27(1):29–34, 1999.

[218] G. Ostermeier, D. Dix, D. Miller, P. Khatri, and S. Krawetz. Spermatozoal rna profiles of normal fertile men. *The Lancet*, 360(9335):773–777, Sept. 2002.

[219] W. Pan, J. Lin, and C. Le. How many replicates of arrays are required to detect gene expression changes in microarray experiments? a mixture model approach. *Genome Biology*, 3(5):research0022.1– research0022.10, 2002.

[220] C. Perou, T. Sørlie, M. Eisen, M. van de Rijn, S. Jeffrey, C. Rees, J. Pollack, D. Ross, H. Johnsen, et al. Molecular portraits of human breast tumours. *Nature*, 406(6797):747–752, 2000.

[221] G. Pietu, R. Mariage-Samson, N.-A. Fayein, C. Matingou, E. Eveno, et al. The genexpress IMAGE knowledge base of the human brain transcriptome: A prototype integrated resource for functional and computational genomics. *Genome Research*, 9:195–209, 1999.

[222] Proteome. Proteome BioKnowledge Library. Technical report, Incyte Genomics, 2002. http://www.incyte.com/sequence/proteome.

[223] K. D. Pruitt, K. S. Katz, H. Sicotte, and D. R. Maglott. Introducing refseq and locuslink: curated human genome resources at the ncbi. *Trends in Genetics*, 16(1):44–47, January 2000.

[224] S. Ramaswamy, P. Tamayo, R. Rifkin, S. Mukherjee, C. Yeang, M. Angelo, C. Ladd, M. Reich, E. Latulippe, J. Mesirov, T. Poggio, W. Gerald, M. Loda, E. Lander, and T. Golub. Multiclass cancer diagnosis using tumor gene expression signatures. *Proc. Natl. Acad. Sci. USA*, 98(26):15149–15154, Dec. 2001.

[225] S. Raychaudhuri, J. M. Stuart, and R. Altman. Principal components analysis to summarize microarray experiments: Application to sporulation time series. In *Proceedings of the Pacific Symposium on Biocomputing*, volume 5, pages 452–463, 2000. Available online at http://psb.stanford.edu.

[226] C. S. Richmond, J. D. Glasner, R. Mau, H. Jin, and F. R. Blattner. Genome-wide expression profiling in *Escherichia coli* K-12. *Nucleic Acids Research*, 27(19):3821–3835, 1999.

[227] R. D. Rifkin, A. F. Parisi, and E. Folland. Coronary calcification in the diagnosis of coronary artery disease. *American Journal of Cardiology*, 44:141–147, 1979.

[228] J. Rissanen. *Stochastic complexity in statistical inquiry*. Word Scientific, Singapore, 1989.

[229] C. J. Roberts, B. Nelson, M. J. Marton, R. Stoughton, M. R. Meyer, H. A. Bennett, Y. D. He, H. Dia, W. L. Walker, T. R. Hughes, M. Tyers, C. Boone, and S. H. Friend. Signaling and circuitry of multiple MAPK pathways revealed by a matrix of global gene expression profiles. *Science*, 287(5454):873–880, Feb. 2000.

[230] A. D. Roses. Pharmacogenetics and the practice of medicine. *Nature*, 405:857–865, 2000.

[231] D. Ross, M. Eisen, C. Perou, C. Rees, P. Spellman, V. Iyer, S. Jeffrey, M. V. de Rijn, M. Waltham, A. Pergamenschikov, J. Lee, D. Lashkari, D. Shalon, T. Myers, J. Weinstein, D. Botstein, and P. Brown. Systematic variation in gene expression patterns in human cancer cell lines. *Nature Genetics*, 24(3):227–235, 2000.

[232] P. E. Ross. The making of a 24 billion gene machine. *Forbes*, February 21:98–104, 2000.

[233] M. Sapir and G. A. Churchill. Estimating the posterior probability of differential gene expression from microarray data. Technical Report http://www.jax.org/research/churchill/pubs/, Jackson Labs, Bar Harbor, ME, 2000.

[234] R. Sasik, T. Hwa, N. Iranfar, and W. F. Loomis. Percolation clustering: A noval algorithm applied to the clustering of gene expression patterns in dictyostelium development. In *Pacific Symposium on Biocomputing*, volume 6, pages 335–347, 2001. Available online at http://psb.stanford.edu.

[235] E. E. Schadt, L. Cheng, C. Su, and W. H. Wong. Analyzing high-density oligonucleotide gene expression array data. *Journal of Cellular Biochemistry*, 80(2):192–202, Oct. 2000.

[236] M. Schena. *DNA Microarrays: A Practical Approach*. Practical Approach Series, 205. Oxford University Press, Oxford, UK, 1999.

[237] M. Schena. *Microarray Biochip Technology*. Eaton Publishing, Sunnyvale, CA, 2000.

[238] M. Schena, D. Shalon, R. Davis, and P. Brown. Quantitative monitoring of gene expression patterns with a complementary DNA microarray. *Science*, 270:467–470, 1995.

[239] M. Schena, D. Shalon, R. Heller, A. Chai, P. Brown, and R. Davis. Parallel human genome analysis: microarray-based expression monitoring of 1000 genes. *Proc. Natl. Acad. of Sci. USA*, 93:10614–10519, 1996.

[240] M. Scherf, A. Klingenhoff, and T. Werner. Highly specific localization of promoter regions in large genomic sequences by PromoterInspector: A novel context analysis appraoch. *Journal of Molecular Biology*, 297(3):599–606, March 2000.

[241] J. Schuchhardt, D. Beule, E. Wolski, and H. Eickhoff. Normalization strategies for cDNA microarrays. *Nucleic Acids Research*, 28(10):e47i–e47v, May 2000.

[242] S. Schuhknecht, S. Duensing, I. Dallmann, J. Grosse, M. Reitz, and J. Atzpodien. Interleukin-12 inhibits apoptosis in chronic lymphatic leukemia (cll) b cells. *Cancer Biother. Radiopharm.*, 17(5):495–499, Oct. 2002.

[243] G. D. Schuler. Pieces of puzzle: Expressed sequence tags and the catalog of human genes. *Journal of Molecular Medicine*, 75(10):694–698, Oct. 1997.

[244] J. P. Shaffer. Modified sequentially rejective multiple test procedures. *Journal of American Statistical Association*, 81:826–831, 1986.

[245] J. P. Shaffer. Multiple hypothesis testing. *Annual Reviews in Psychology*, 46:561–584, 1995.

[246] D. Shalon, S. J. Smith, and P. O. Brown. A DNA microarray system for analyzing complex DNA samples using two-color fluorescent probe hybridization. *Genome Research*, 6:639–645, 1996.

[247] G. Sherlock, T. Hernandez-Boussard, A. Kasarskis, G. Binkley, et al. The Stanford Microarray Database. *Nucleic Acid Research*, 29(1):152–155, January 2001.

[248] L. Shi. DNA microarray - monitoring the genome on a chip. Technical report, 2001. http://www.gene-chips.com/.

[249] M. Shipp, K. Ross, P. Tamayo, A. Weng, J. Kutok, R. Aguiar, M. Gaasenbeek, M. Angelo, M. Reich, G. Pinkus, T. Ray, M. Koval, K. Last, A. Norton, T. Lister, J. Mesirov, D. Neuberg, E. Lander, J. Aster, and T. Golub. Diffuse large B-cell lymphoma outcome prediction by gene-expression profiling and supervised machine learning. *Nature Medicine*, 8(1):68–74, Jan. 2002.

[250] R. Simon, M. D. Radmacher, and K. Dobbin. Design of studies using DNA microarrays. *Genetic Epidemiology*, 23(1):21–36, 2002.

[251] P. Smolen, D. A. Baxter, and J. H. Byrne. Modeling transcriptional control in gene networks–methods, recent results, and future directions. *Bull Math Biol*, 62(2):247–292, March 2000.

[252] D. L. Souvaine and J. M. Steele. Efficient time and space algorithms for least median of squares regression. *J. Amer. Stat. Assn.*, 82:794–801, 1987.

[253] T. P. Speed. Hints and prejudices – always log spot intensities and ratios. Technical report, University of California, Berkeley, 2000. http://www.stat.berkeley.edu/users/terry/zarray/Html/log.html.

[254] P. T. Spellman, G. Sherlock, M. Q. Zhang, W. Iyer, K. Anders, M. B. Eisen, P. O. Brown, D. Bostein, and B. Futcher. Comprehensive identification of cell cycle-regulated genes of the yeast Saccharomyces cerevisiae by microarray hybridization. *Mol. Biol. Cell*, 9(12):3273–3297, 1998.

[255] Stanford. SMD – Stanford Microarray Database. Technical report, Stanford University, 2001. http://genome-www4.Stanford.EDU/MicroArray/SMD/.

[256] J. Staunton, D. Slonim, H. Coller, P. Tamayo, M. Angelo, J. Park, U. Scherf, J. Lee, W. Reinhold, J. Weinstein, J. Mesirov, E. Lander, and T. Golub. Chemosensitivity prediction by transcriptional profiling. *Proc. Natl. Acad. Sci. USA*, 98(19):10787–92, Sept. 2001.

[257] M. O. Stitson, J. A. E. Weston, A. Gammerman, V. Vovk, and V. Vapnik. Theory of support vector machines. Technical Report CSD-TR-96-17, Royal Holloway U. London, Dec. 1996.

[258] M. E. Stokes, C. S. Davis, and G. G. Koch. *Categorical Data Analysis Using the SAS System*. SAS Institute, Cary, NC, 1995.

[259] T. Strachan and A. P. Read. *Human Molecular Genetics*. Wiley-LISS, New York, NY, 2nd edition, 1999.

[260] P. Sudarsanam, V. R. Iyer, P. O. Brown, and F. Winston. Whole-genome expression analysis of snf/swi mutants of *Saccharomyces cerevisiae*. *Proc. Natl. Acad. Sci.*, 97(7):3364–3369, 2000.

[261] A. M. Taalat, S. T. Howard, W. Hale, R. Lyons, H. Ganner, and S. A. Johnson. Genomic dna standard for gene expression profiling in mycobacterium tuberculosis. *Nucleic Acids Research*, 30(20):E104, 2002.

[262] P. Tamayo, D. Slonim, J. Mesirov, Q. Zhu, S. Kitareewan, E. Dmitrovsky, E. S. Lander, and T. R. Golub. Interpreting patterns of gene expression with self-organizing maps: Methods and application to hematopoietic differentiation. *Proc. Natl. Acad. Sci. USA*, 96(6):2907–2912, 1999.

[263] H. Tao, C. Bausch, C. Richmond, F. R. Blattner, and T. Conway. Functional genomics: Expression analysis of *Escherichia coli* growing on minimal and rich media. *Journal of Bacteriology*, 181(20):6425–6440, 1999.

[264] S. Tavazoie, J. D. Hughes, M. J. Campbell, R. J. Cho, and G. M. Church. Systematic determination of genetic network architecture. *Nature Genetics*, 22:281–285, 1999.

[265] J. J. M. ter Linde, H. Liang, R. W. Davis, H. Y. Steensma, J. P. V. Dijken, and J. T. Pronk. Genome-wide transcriptional analysis of aerobic and anaerobic chemostat cultures of *Saccharomyces cerevisiae*. *Journal of Bacteriology*, 181(24):7409–7413, 1999.

[266] R. Tibshirani, G. Walther, D. Botstein, and P. Brown. Cluster validation by prediction strength. Technical report, Statistics Department, Stanford University, 2000. Manuscript available at http://www-stat.stanford.edu/ tibs/research.html.

[267] S. Tsoka and C. A. Ouzounis. Recent developments and future directions in computational genomics. *Federation of European Biochemical Societies Letters*, (23897):1–7, 2000.

[268] V. G. Tusher, R. Tibshirani, and G. Chu. Significance analysis of microarrays applied to the ionizing radiation response. *Proc. Natl. Acad. Sci.*, 98(9):5116–5121, Apr 2001. Software available for download at http://www-stat.standford.edu/ tibs/SAM/index.html.

[269] J. van Helden, A. F. Rios, and J. Collado-Vides. Discovering regulatory elements in non-coding sequences by analysis of spaced dyads. *Nucleic Acids Research*, 28(8):1808–1818, 2000.

[270] L. J. van 't Veer, H. Dai, M. J. van de Vijver, Y. D. He, et al. Gene expression profiling predicts clinical outcome of breast cancer. *Nature*, 415:530–536, Jan. 2002.

[271] A. R. Venkitaraman. Cancer susceptibility and the functions of BRCA1 and BRCA2. *Cell*, 108(2):171–182, January 2002.

[272] Z. Šidák. Rectangular confidence regions for the means of multivariate normal distributions. *Journal of the American Statistical Association*, 62:626–633, 1967.

[273] O. G. Vukmirovic and S. M. Tilghman. Exploring genome space. *Nature*, 405:820–822, 2000.

[274] D. G. Wang, J. B. Fan, C. J. Siao, A. Berno, et al. Large-scale identification, mapping, and genotyping of single-nucleotide polymorphisms in the human genome. *Science*, 280(5366):1077–1082, 1998.

[275] M. L. Wang, S. Belmonte, U. Kim, M. Dolan, J. W. Morris, and H. M. Goodman. A cluster of ABA-regulated genes on *Arabidopsis Thaliana* BAC T07M07. *Genome Research*, 9:325–333, 1999.

[276] W. Wang, J. Lu, R. Lee, Z. Gu, and R. Clarke. Iterative normalization of cDNA microarray data. *IEEE Transactions on Information Technology in Biomedicine*, 6(1):29–37, March 2002.

[277] M. E. Waugh, D. L. Bulmore, A. D. Farmer, P. A. Steadman, et al. PathDB: A metabolic database with sophisticated search and visualization tools. In *Proc. of Plant and Animal Genome VIII Conference*, San Diego, CA, January 9-12 2000.

[278] D. Weaver, C. Workman, and G. Stormo. Modeling regulatory networks with weight matrices. *Pacific Symp. Biocomp. 99*, 4:112–123, 1999.

[279] P. L. Welcsh, M. K. Lee, R. M. Gonzalez-Hernandez, D. J. Black, M. Mahade-vappa, E. M. Swisher, J. A. Warrington, and M.-C. King. BRCA1 transcriptionally regulates genes involved in breast tumorigenesis. *Proc. Natl. Acad. Sci. USA*, 99(11):7560–7565, 2002.

[280] S. Welle, A. I. Brooks, and C. A. Thornton. Computational method for reducing variance with Affymetrix microarrays. *BMC Bioinformatics*, 3:23, 2002.

[281] A. Wellmann, C. Thieblemont, S. Pittaluga, A. Sakai, et al. Detection of differentially expressed genes in lymphomas using cDNA arrays: identification of *clusterin* as a new diagnostic marker for anaplastic large-cell lymphomas. *Blood*, 96(2):398–404, 2000.

[282] M. West, J. Nevins, J. Marks, R. Spang, C. Blanchette, and H. Zuzan. Bayesian regression analysis in the "large p, small n" paradigm with application in DNA microarray studies. Technical report, Duke University, 2000.

[283] P. H. Westfall and S. S. Young. *Resampling-based multiple testing: Examples and Methods for p-value adjustment*. Wiley, New York, 1993.

[284] K. P. White, S. A. Rifkin, P. Hurban, and D. S. Hogness. Microarray analysis of Drosophila development during metamorphosis. *Science*, 286:2179–2184, 1999.

[285] S. E. Wildsmith, G. E. Archer, A. J. Winkley, P. W. Lane, and P. J. Bugel-ski. Maximizing of signal derived from cDNA microarrays. *BioTechniques*, 30:202–208, 2000.

[286] I. H. Witten and E. Frank. *Data Mining*. Morgan Kaufmann, 1999.

[287] R. D. Wolfinger, G. Gibson, E. D. Wolfinger, L. Bennett, H. Hamadeh, P. Bushel, C. Afshari, and R. S. Paules. Assessing gene significance from cdna microarray data via mixed models. *Journal of Computational Biology*, 8:625–637, 2001.

[288] L. Wu, T. Hughes, A. Davierwala, M. Robinson, R. Stoughton, and S. Altschuler. Large-scale prediction of saccharomyces cerevisiae gene function using overlapping transcriptional clusters. *Nature Genetics*, 31(3):255–265, July 2002.

[289] E. P. Xing and R. M. Karp. Clustering of high-dimensional microarray data via iterative feature filtering using normalized cuts. *Bioinformatics*, 17(Suppl. 1):S306–S315, 2001.

[290] Y. Yang, M. J. Buckley, S. Dudoit, and T.P.Speed. Comparison of methods for image analysis on cDNA. Technical report, University of California, Berkeley, 2000. http://www.stat.berkeley.edu/users/terry/zarray/Html/log.html.

[291] Y. Yang, S. Dudoit, P. Luu, and T.P.Speed. Normalization for cDNA microarray data. In *Proc. of SPIE BiOS*, volume 4266, page 31, San Jose, CA, 2001.

[292] Y. H. Yang, M. J. Buckley, S. Dudoit, and T. P. Speed. Comparison of methods for image analysis on cDNA microarray data. *Journal of Computational and Graphic Statistics*, 11:108–136, 2002.

[293] Y. H. Yang and S. Dudoit. Bioconductor multtest package. www.bioconductor.org, 2002.

[294] Y. H. Yang, S. Dudoit, P. Luu, D. M. Lin, V. Peng, J. Ngai, and T. P. Speed. Normalization for cDNA microarray data: a robust composite method addressing single and multiple slide systematic variation. *Nucleic Acids Research*, 30(4):e15, 2002. Available online at http://linkage.rockefeller.edu/wli/microarray/yang02.pdf.

[295] K. Y. Yeung, C. Fraley, A. Murua, A. E. Raftery, and W. L. Ruzzo. Model-based clustering and data transformation for gene expression data. *Bioinformatics*, 17(10):977–987, Oct. 2001.

[296] K. Y. Yeung, D. R. Haynor, and W. L. Ruzzo. Validating clustering for gene expression data. *Bioinformatics*, 17(4):309–318, April 2001.

[297] M. K. S. Yeung, J. Tegner, and J. J. Collins. Reverse engineering gene networks using singular value decomposition and robust regression. *PNAS*, 99(9):6163–6168, 2002.

[298] T. Yoshida, N. Koide, T. Sugiyama, I. Mori, and T. Yokochi. A novel caspase dependent pathway is involved in apoptosis of human endothelial cells by shiga toxins. *Microbiol. Immunol.*, 46(10):697–700, 2002.

[299] H. Yue, P. Eastman, B. Wang, J. Minor, M. Doctolero, R. L. Nuttall, R. Stack, J. W. Becker, J. R. Montgomery, M. Vainer, and R. Johnston. An evaluation of the performance of cDNA microarrays for detecting changes in global mRNA expression. *Nucleic Acids Research*, 29(8):e41, 2001.

[300] T. Yuen, E. Wurmbach, R. Pfeffer, B. Ebersole, and S. Sealfon. Accuracy and calibration of commercial oligonucleotide and custom cDNA microarrays. *Nucleic Acids Research*, 30(10):e48, May 2002.

[301] M. Q. Zhang. Large-scaled gene expression data analysis: A new challenge to computational biologists. *Genome Research*, 9:681–688, 1999.

[302] Y. Zhou and R. Abagyan. Match-only integral distribution (MOID) algorithm for high-density oligonucleotide array analysis. *BMC Bioinformatics*, 3:3, 2002. Available at http://linkage.rockefeller.edu/wli/microarray/zhou02.pdf.

[303] H. Zhu, J. Cong, G. Mamtora, T. Gingeras, and T. Shenk. Cellular gene expression altered by human cytomegalovirus: global monitoring with oligonucleotide arrays. *Proc. Natl. Acad. Sci.*, 24(95):14470–14475, 1998.

[304] J. Zhu and M. Zhang. Cluster, function and promoter: Analysis of yeast expression array. In *Pacific Symposium on Biocomputing*, volume 5, pages 476–487, 2000. Available online at http://psb.stanford.edu.

Index